AF581850

Complexity Science in Human Change: Research, Models, Clinical Applications

Complexity Science in Human Change: Research, Models, Clinical Applications

Editors

Franco Orsucci
Wolfgang Tschacher

MDPI • Basel • Beijing • Wuhan • Barcelona • Belgrade • Manchester • Tokyo • Cluj • Tianjin

Editors
Franco Orsucci
University College London
UK

Wolfgang Tschacher
University of Bern
Switzerland

Editorial Office
MDPI
St. Alban-Anlage 66
4052 Basel, Switzerland

This is a reprint of articles from the Special Issue published online in the open access journal *Entropy* (ISSN 1099-4300) (available at: https://www.mdpi.com/journal/entropy/special_issues/Clinical_Applications).

For citation purposes, cite each article independently as indicated on the article page online and as indicated below:

LastName, A.A.; LastName, B.B.; LastName, C.C. Article Title. *Journal Name* **Year**, *Volume Number*, Page Range.

ISBN 978-3-0365-6219-3 (Hbk)
ISBN 978-3-0365-6220-9 (PDF)

Cover image courtesy of Wolfgang Tschacher

Contents

About the Editors

Franco Orsucci

Received his medical degree in 1980 and psychiatry post-graduate degree in 1984 at the Sapienza University of Rome (both 1st Class Hons). He became a psychoanalyst of the International Psychoanalytical Association in 1992. He has been professor at the universities Catholic of Rome, Siena, Chieti-Pescara, Cusano (London). Visiting professor at the University of Kent and, currently, at University College London. He is Clinical Director at the Norfolk and Suffolk NHS Foundation Trust. He is Fellow of the Royal Society of Medicine and the European Academy of Arts and Sciences. He has been founding president of the Italian Society for Chaos and Complexity Science and, currently, of the Mind Force Society. He has organized international conferences and workshops with these societies and the Society for Psychotherapy Research (SPR), where he is chairing the Complexity Science Interest Section.

Wolfgang Tschacher

Received his Ph.D. in psychology 1990. Psychotherapy training in systemic therapy at the Institute of Family Therapy, Munich. Habilitation in psychology and Venia legendi 1996 at University of Bern, Switzerland, professorship in 2002, head of department of psychotherapy at the University Hospital of Psychiatry and Psychotherapy, now professor emeritus. His main interests are in psychotherapy process research and empirical aesthetics, with an emphasis on complexity science, embodied cognition, and self-organization. Organizer of the series of 'Herbstakademie' conferences on systems theory in psychology and philosophy of mind. He is member of the board of directors of the Society for Mind-Matter Research. From 2007 to 2010 he was president of the European Chapter of the Society for Psychotherapy Research (SPR).

Editorial

Complexity Science in Human Change: Research, Models, Clinical Applications

Franco Orsucci [1,2,*] and Wolfgang Tschacher [3]

[1] Department of Psychology, University College London, London WC1E 6BT, UK
[2] Norfolk and Suffolk NHS Foundation Trust, Drayton High Road, Norwich NR6 5BE, UK
[3] Department of Experimental Psychology, University Hospital of Psychiatry and Psychotherapy, CH-3060 Bern, Switzerland
* Correspondence: f.orsucci@ucl.ac.uk

Citation: Orsucci, F.; Tschacher, W. Complexity Science in Human Change: Research, Models, Clinical Applications. *Entropy* **2022**, *24*, 1670. https://doi.org/10.3390/e24111670

Received: 8 November 2022
Accepted: 8 November 2022
Published: 17 November 2022

Publisher's Note: MDPI stays neutral with regard to jurisdictional claims in published maps and institutional affiliations.

Complexity and entropy prevail in human behavior and social interaction because the systems underlying behavior and interaction are, without a doubt, highly complex. The human brain, body, language, society, and culture consist of vast numbers of components, and the degrees of freedom in behavior, cognition, and experience are just as immense. So why do we usually experience the world around us as structured and well-organized instead of disorganized and random? How do the patterns emerge? We are witnessing self-organization and pattern-formation processes, which organize and modulate complexity.

Increasingly, such processes are acknowledged as essential for human affairs and are gradually coming to the fore in psychology and social science research. Research informed by dynamical systems theory, synergetics, and complexity theory has introduced concepts such as attractor, synchrony, and coupling to psychology. In psychotherapy research, empirical findings show that regular patterns of interaction arise in all therapeutic relationships. The therapist–patient alliance is a paradigmatic case to highlight further how interactions evolve and can be changed and how humans can change. Attractors describe the stable states of a process, e.g., the stability or instability of personalities and disorders. They can be detected and described based on empirical time-series.

In the first paper, Orsucci [1] examines certain theoretical implications of empirical studies developed over recent years by his research groups. These experiments have explored the biosemiotic nature of communication streams from emotional neuroscience and embodied mind perspectives. Information combinatorics analysis enabled a deeper understanding of the coupling and decoupling dynamics of biosemiotics streams. They investigated intraindividual and interpersonal relations as the coevolution dynamics of hybrid couplings, synchronizations, and desynchronizations. Cluster analysis and Markov chains produced evidence of chimera states and phase transitions. A probabilistic and nondeterministic approach clarified the properties of these hybrid dynamics. As a result, multidimensional theoretical models can better represent the hybrid nature of human interactions.

In the second contribution, a study by Tomashin et al. [2], the authors consider how fractal properties in time series of human behavior and physiology are ubiquitous, and several methods to capture such properties have been proposed in the past decades. The paper takes this suggestion as a point of departure to propose and test several approaches to quantifying fractal fluctuations in synthetic and empirical time-series data using recurrence-based analysis. They show that such measures can be extracted based on recurrence plots and contrast the different approaches in terms of their accuracy and range of applicability.

In the third paper, Altmann et al. [3] compare eight algorithms that quantify synchronization in time series. The authors use a benchmark dataset that describes body movement in 30 dyadic interviews on somatic complaints conducted by medical students with 15 depressed and 15 healthy interviewees. Twenty-one different synchrony measures are tested, derived from four classes of algorithms: (windowed) cross-correlations, local regressions with or without peak-picking, mutual information, and cross-recurrence quantification. The intercorrelations of the results show that synchrony estimations are highly divergent, and no convergent validity is manifest. However, measures from the same class tend to be correlated, and cross-correlation-based measures form a factor. In contrast, the mutual information and the peak-picking measures load on a different factor. Most measures do not support the assumption underlying predictive validity that depression should have lowered synchrony measures. The authors conclude that more analyses and sensitivity studies are needed to clarify the psychological meaning of synchrony.

In the fourth contribution, research by Stamovlasis et al. [4], the authors investigate and propose a nonlinear model that might explain empirical data better than ordinary linear ones and elucidate the role of depression in a financial capacity. Financial incapacity is one of the cognitive deficits observed in amnestic mild cognitive impairment and dementia, while the combined interference of depression remains unexplored. Cusp catastrophe analysis was applied to the data, which suggested that the nonlinear model was superior to the linear and logistic alternatives, demonstrating that depression contributes to a bifurcation effect. Depressive symptomatology induces nonlinear effects, and a sudden decline in financial capacity is observed beyond a threshold value. Implications for theory and practice are discussed.

In the fifth contribution, Zubek et al. [5] reflected on the pandemic that forced our daily interactions to move into the virtual world. People had to adapt to new communication media that afford different ways of interaction. Remote communication decreases the availability and salience of some cues but also may enable and highlight others. Importantly, basic movement dynamics, which are crucial for any interaction as they are responsible for the informational and affective coupling, are affected. It is therefore essential to discover exactly how these dynamics change. In this exploratory study of six interacting dyads, they used tradi-tional variability measures and cross recurrence quantification analysis to compare the movement coordination dynamics in quasi-natural dialogues in four situations: (1) re-mote video-mediated conversations with a self-view mirror image present, (2) remote video-mediated conversations without a self-view, (3) face-to-face conversations with a self-view, and (4) face-to-face conversations without a self-view. They discovered that in remote interactions movements pertaining to communicative gestures were exagger-ated, while the stability of interpersonal coordination was greatly decreased. The presence of the self-view image made the gestures less exaggerated but did not affect the coordination. The dynamical analyses clarified the interaction processes and may be useful in explaining phenomena connected with video-mediated communication, such as "Zoom fatigue".

In the sixth contribution, Laudańska et al. [6] clarify how infants' limb movements evolve from disorganized to more selectively coordinated during the first year of life as they learn to navigate and interact with an ever-changing environment more efficiently. However, how these coordination patterns change during the first year of life and across different contexts is unknown. Here, they used wearable motion trackers to study the developmental changes in the complexity of limb movements (arms and legs) at 4, 6, 9, and 12 months of age in two different tasks: rhythmic rattle-shaking and free play. They applied multidimensional recurrence quantification analysis (MdRQA) to capture the nonlinear changes in infants' limb complexity. They show that the MdRQA parameters (entropy, recurrence rate, and mean line) are task-dependent only at 9 and 12 months of age, with higher values in rattle-shaking than free play. Infants' motor system becomes more stable and flexible with age, allowing for the flexible adaptation of behaviors to task demands.

The seventh contribution by Ganesh and Gabora [7] take a human dynamical systems approach to modeling therapeutic change, using reflexively autocatalytic food set-derived (RAF) networks. RAFs have been used to model the self-organization of adaptive networks associated with the origin and early evolution of both biological life and the development of the kind of cognitive structure necessary for cultural evolution. The RAF approach is applicable in these seemingly disparate cases because it provides a theoretical framework for formally describing under what conditions systems com-posed of elements that interact and "catalyze" the formation of new elements collec-tively become integrated wholes. This contribution develops in line with the growing recognition of the role of embodiment, affect, and implicit processes in psychotherapy, and several recent studies have examined the role of physiological synchrony in the process and outcome of psychotherapy. This study aims to introduce partial directed coherence (PDC) as a novel approach to calculating psychophysiological synchrony and examines its potential to contribute to our understanding of the therapy process. The study adopts a single-case, mixed-method design and examines physiological syn-chrony in one-couple therapy in relation to the therapeutic alliance and a narrative analysis of meaning construction in the sessions. The findings of this study point to the complex interplay between explicit and implicit levels of interaction and the potential contribution of including physiological synchrony in the study of interactional pro-cesses in psychotherapy.

The paper by Avdi et al. [8] addresses physiological synchrony in one family therapy of fifteen sessions and two physiological measurement sessions in which cardiac measures were recorded. The sessions concern a couple and two female psychotherapists. Physiological data are transformed into an index of sympathetic activity, and synchrony is computed using partial directed coherence. This method detects the direction of influence ("pacing/leading") in each pair of participants. In addition to the quantitative findings on synchrony, rating scales depict therapeutic alliance, and qualitative coding separates the measurement sessions into topical episodes that are semantically similar. Finally, the therapy process is described by the percentage of time windows synchronized concerning the couple's sympathetic activity, which is found to be reduced in the second measurement session, where the patterns of pacing and leading have changed towards a more balanced embodied relatedness. The authors conclude that a mixed-methods approach allows the linking of the quantitative synchrony findings to the qualitative clinical process in this successful couple therapy.

In the ninth paper, a study by Nkomidio et al. [9], the authors investigate the response characteristics of a two-dimensional neuron model exposed to an externally applied extremely low frequency (ELF) sinusoidal electric field and the synchronization of neurons weakly coupled with gap junction. They find, by numerical simulations, that neurons can exhibit different spiking patterns, which are well observed in the structure of the recurrence plot (R.P.). Then they further study the synchronization between weakly coupled neurons in chaotic regimes under the influence of a weak ELF electric field. In general, detecting the phases of chaotic spiky signals is not easy when using standard methods. Recurrence analysis provides a reliable tool for defining phases, even for noncoherent regimes or spiky signals. Recurrence-based synchronization analysis reveals that, even in the range of weak coupling, the phase synchronization of the coupled neurons occurs. By adding an ELF electric field, this synchronization increases depending on the amplitude of the externally applied ELF electric field. Authors further suggest a novel measure for RP-based phase synchronization analysis, which better considers the probabilities of recurrences.

In the tenth contribution, Webber [10] clarifies how the recurrence analyses of dynamical systems can only process the short sections of signals that may be infinitely long. By necessity, the recurrence plot and its quantifications are constrained within a truncated triangle that clips the signals at its borders. Recurrence variables defined within these confining borders can be influenced by truncation effects depending on the system under evaluation. In this study, the question being asked is, if the boundary borders were tilted, what would be the effect on all recurrence variables? This question is examined by comparing recurrence variables computed with the triangular recurrence area versus the boxed recurrence area. Examples include the logistic equation (mathematical series), the Dow Jones Industrial Average over a decade (real-world data), and a square wave pulse (toy series). Good agreement among the variables in terms of timing and amplitude was found for most, but not all, variables. These significant results are discussed.

In the eleventh paper, Ciompi and Tschacher [11] develop an account of modeling schizophrenia based on four different but related complexity theories: affect-logic, 4E-cognition/embodiment, synergetics, and the free-energy principle. All theories have in common that they are built on loop dynamics, so-called circular causality. In affect-logic, the loop is given by circular interactions between emotion ('affect') and cognition ('logic'), where emotion is the energy source for cognitive dynamics. Such interactions occur at the individual level, the level of micro-social interaction, and the societal level, which are structurally coupled. In synergetics, emotions act as control parameters that drive the system toward pattern formation. The embodiment and the free-energy principles are likewise built on circular dynamics between mind and body, respectively, between a generative model and sensory evidence. The article uses these commonalities for insights into the dynamics of schizophrenia spectrum symptoms: overly strong emotional tension then forces the cognitive system into dysfunctional patterns, which, however, are functional insofar as free energy is reduced. Ideas for therapeutic guidelines, as in the Soteria model, are also derived.

In the twelfth paper, Prinz et al. [12] compute the synchrony of electrodermal activity in psychotherapy interventions, focusing on the technique of imagery rescripting. This therapeutic technique was developed to modulate traumatic memories in a positive and desired direction. The activation of such memories commonly also affects the therapist involved in the session. Therefore, client-therapist synchrony based on cross-correlations is explored in 50 clients. Client-led synchrony is differentiated from therapist-led synchrony by the sign of the lags of cross-correlations. It is found that therapist-led synchrony is significantly associated with clients' emotional experiences of greater contentment, lower anxiety, and lower depression. In contrast, client-led synchrony is linked to the clients' more significant anxiety. The authors interpret their findings as supportive of the therapists' role in regulating mood.

In their contribution, Gennaro et al. [13] introduce a novel lexical method called the affective saturation index (ASI) to assess affectivity based on interview transcripts. Affect is semiotically defined as a sign that makes sense of the world in terms of patterns of bodily activation. Affect saturation in the ASI is then defined based on a "phase space of meaning". The ASI was correlated with several measures of semantic complexity, students' emotion regulation, and heart-rate variability in a sample of 40 students who participated in semi-structured interviews on neutral issues. The study shows that affective saturation is significantly and inversely linked to the semantic entropy index and heart-rate variability, consistent with the expectation that the ASI can detect the lexical-syntactic complexity of the interview text as well as physiological signs of affective arousal. It is concluded that ASI thus has potential applicability in clinical and community interventions, social communication, marketing, and media monitoring. Most approaches to computing interpersonal coupling are dyadic in that they focus on bivariate synchrony, such as that between client and therapist.

Meier and Tschacher [14] developed an algorithm for multivariate surrogate synchrony (mv-SUSY), which is based on the eigen-decomposition of the correlation matrix of multiple time series, like principal component analysis (mv-SUSY variant lambdamax). A further variant labeled omega is derived from the determinant of the variance-covariance matrix as a measure of actual entropy and standardized by potential entropy, the product of all variances. Computation is carried out in time-series segments, and segment-shuffled surrogate datasets are used as a control condition. The authors apply mv-SUSY to the simulated multivariate time series that realize the various types of regularities (random data, autocorrelations, trends, oscillatory behavior, intercorrelated random data) and to empirical multivariate time series (motion capture data from persons dancing and from a group discussion). It was found that mv-SUSY correctly identifies whether regular patterns exist in the datasets. The results of the multivariate algorithms are additionally validated by conventional dyadic synchrony methods.

We believe that change is generally studied in phase transitions when the dynamics move between different attractors, as evident in behaviors, mental states, and neurobiology. Theoretical models can represent dynamical change maps in mathematical equations and topological structures. Mapping theory to empirical research and vice versa is challenging but heuristic. Nevertheless, it paves the road to a future discipline of a general complexity theory of human change.

One feature of complexity and self-organization is the presence of scaling and fractal dynamics with the emergence of higher-order organizations. Moreover, heterogeneous human networks present specific kinds of self-similarity in the embodied mind in individual and social dynamics. Finally, translational processes and procedures from research to applications and vice versa are particularly relevant as they frequently include interdisciplinary collaborations.

Based on these thoughts, the Special Issue "Complexity Science in Human Change" has addressed an interdisciplinary community of scientists and practitioners interested in dynamical systems theory, especially approaches considering complex systems and applications to psychology and psychotherapy. Most of the contributions in this Issue analyze empirical data, predominantly time series. Some contributions contain theoretical models or methodological topics of complexity science.

This Special Issue closely represents the current work of complexity researchers in human behavior and change. Psychotherapy and communication systems are the backgrounds of six articles, mental health and psychopathology of four articles, and one paper concerns developmental psychology. Six papers put forward methods for the computation of interpersonal synchrony and innovations of existing synchrony algorithms. Regarding methodology, recurrence quantification analysis is frequently applied in articles on this topic. Four put forward correlation-based analyses of synchrony. Finally, network modeling, catastrophe theory, and nonlinear regression are tackled in one article.

This Special Issue highlights achievements of complexity science in studying patterns of organization and change in human dynamics. It also highlights new challenges that lie ahead. First, it clarifies how complex systems present plural structural forms and varieties of organization and disorganization [15,16]. These varieties are frequently distributed even within any singular system, creating rugged dynamical landscapes. This applies more specifically to human systems, which are hybrid by default [17]. They present multiple scales and heterogeneous subsystems; synchronous and asynchronous interactions; stable, unstable, and metastable states; and localized and generalized dynamics. Therefore, considering the hyper-complexity of the human dynamical landscapes, empirical studies can use mixed methods with several different approaches (sometimes all at once). Accordingly, multiple and varied interventions can induce change or facilitate its natural evolution and emersion [18,19]. This can explain how multiple and different therapeutic techniques can produce similar (though not identical) outcomes in the clinical field. It is the good old equifinality principle of complex open systems still at work.

We hope that with this, new questions and new research might be incited, as Voltaire once suggested: "The most useful books are those in which the readers themselves supply half of the meaning".

Author Contributions: Writing—original draft preparation, F.O. and W.T.; writing—review and editing, F.O. and W.T. All authors have read and agreed to the published version of the manuscript.

Acknowledgments: We express our thanks to the authors of the above contributions, and the journal Entropy and MDPI for their support during this work.

Conflicts of Interest: The authors declare no conflict of interest.

References

1. Orsucci, F. Human Synchronization Maps—The Hybrid Consciousness of the Embodied Mind. *Entropy* **2021**, *23*, 1569. [CrossRef] [PubMed]
2. Tomashin, A.; Leonardi, G.; Wallot, S. Four Methods to Distinguish between Fractal Dimensions in Time Series through Recurrence Quantification Analysis. *Entropy* **2022**, *24*, 1314. [CrossRef] [PubMed]
3. Altmann, U.; Strauss, B.; Tschacher, W. Cross-Correlation- and Entropy-Based Measures of Movement Synchrony: Non-Convergence of Measures Leads to Different Associations with Depressive Symptoms. *Entropy* **2022**, *24*, 1307. [CrossRef] [PubMed]
4. Stamovlasis, D.; Giannouli, V.; Vaiopoulou, J.; Tsolaki, M. Catastrophe Theory Applied to Neuropsychological Data: Nonlinear Effects of Depression on Financial Capacity in Amnestic Mild Cognitive Impairment and Dementia. *Entropy* **2022**, *24*, 1089. [CrossRef] [PubMed]
5. Zubek, J.; Nagórska, E.; Komorowska-Mach, J.; Skowrońska, K.; Zieliński, K.; Rączaszek-Leonardi, J. Dynamics of Remote Communication: Movement Coordination in Video-Mediated and Face-to-Face Conversations. *Entropy* **2022**, *24*, 559. [CrossRef] [PubMed]
6. Laudańska, Z.; López Pérez, D.; Radkowska, A.; Babis, K.; Malinowska-Korczak, A.; Wallot, S.; Tomalski, P. Changes in the Complexity of Limb Movements during the First Year of Life across Different Tasks. *Entropy* **2022**, *24*, 552. [CrossRef] [PubMed]
7. Ganesh, K.; Gabora, L. A Dynamic Autocatalytic Network Model of Therapeutic Change. *Entropy* **2022**, *24*, 547. [CrossRef] [PubMed]
8. Avdi, E.; Paraskevopoulos, E.; Lagogianni, C.; Kartsidis, P.; Plaskasovitis, F. Studying Physiological Synchrony in Couple Therapy through Partial Directed Coherence: Associations with the Therapeutic Alliance and Meaning Construction. *Entropy* **2022**, *24*, 517. [CrossRef] [PubMed]
9. Nkomidio, A.M.; Ngamga, E.K.; Nbendjo, B.R.N.; Kurths, J.; Marwan, N. Recurrence-Based Synchronization Analysis of Weakly Coupled Bursting Neurons under External ELF Fields. *Entropy* **2022**, *24*, 235. [CrossRef] [PubMed]
10. Webber, C.L., Jr. Alternate Entropy Computations by Applying Recurrence Matrix Masking. *Entropy* **2022**, *24*, 16. [CrossRef] [PubMed]
11. Ciompi, L.; Tschacher, W. Affect-Logic, Embodiment, Synergetics, and the Free Energy Principle: New Approaches to the Understanding and Treatment of Schizophrenia. *Entropy* **2021**, *23*, 1619. [CrossRef] [PubMed]
12. Prinz, J.; Rafaeli, E.; Wasserheß, J.; Lutz, W. Clients' Emotional Experiences Tied to Therapist-Led (but Not Client-Led) Physiological Synchrony during Imagery Rescripting. *Entropy* **2021**, *23*, 1556. [CrossRef] [PubMed]
13. Gennaro, A.; Carola, V.; Ottaviani, C.; Pesca, C.; Palmieri, A.; Salvatore, S. Affective Saturation Index: A Lexical Measure of Affect. *Entropy* **2021**, *23*, 1421. [CrossRef] [PubMed]
14. Meier, D.; Tschacher, W. Beyond Dyadic Coupling: The Method of Multivariate Surrogate Synchrony (mv-SUSY). *Entropy* **2021**, *23*, 1385. [CrossRef] [PubMed]
15. Weaver, W. Science and complexity. In *Facets of Systems Science*; Springer: Cham, Switzerland, 1991; pp. 449–456. Available online: http://link.springer.com/chapter/10.1007/978-1-4899-0718-9_30 (accessed on 10 October 2022).
16. Schreiber, T. Interdisciplinary application of nonlinear time series methods. *Phys. Rep.* **1999**, *308*, 1–64. [CrossRef]
17. Orsucci, F. *Changing Mind Transitions in Natural and Artificial Environments*; World Scientific: Singapore, 2002.
18. Orsucci, F. Towards a meta-model of human change, from singularity to event horizon. *Chaos Complex. Lett.* **2015**, *9*, 107.
19. Tschacher, W.; Haken, H. *The Process of Psychotherapy: Causation and Chance*; Springer Nature: Cham, Switzerland, 2019. [CrossRef]

 MDPI

Perspective

Human Synchronization Maps—The Hybrid Consciousness of the Embodied Mind

Franco Orsucci [1,2]

[1] Department of Psychology, University College London, London WC1E 6BT, UK; f.orsucci@ucl.ac.uk
[2] Norfolk and Suffolk NHS Foundation Trust, Drayton High Road, Norwich NR6 5BE, UK

Abstract: We examine the theoretical implications of empirical studies developed over recent years. These experiments have explored the biosemiotic nature of communication streams from emotional neuroscience and embodied mind perspectives. Information combinatorics analysis enabled a deeper understanding of the coupling and decoupling dynamics of biosemiotics streams. We investigated intraindividual and interpersonal relations as coevolution dynamics of hybrid couplings, synchronizations, and desynchronizations. Cluster analysis and Markov chains produced evidence of chimaera states and phase transitions. A probabilistic and nondeterministic approach clarified the properties of these hybrid dynamics. Thus, multidimensional theoretical models can represent the hybrid nature of human interactions.

Keywords: synchronization; semiotics; information; cognitive neuroscience; psychotherapy; conversation; mapping; chimaera states; statistical dynamics; coupling

Citation: Orsucci, F. Human Synchronization Maps—The Hybrid Consciousness of the Embodied Mind. *Entropy* **2021**, *23*, 1569. https://doi.org/10.3390/e23121569

Academic Editor: Milan Paluš

Received: 11 October 2021
Accepted: 18 November 2021
Published: 25 November 2021

> Science is built up with facts, as a house is with stones.
> However, a collection of facts is no more a science than a heap of stones is a house.
>
> *Henri Poincaré, Science and Hypothesis*

1. Introduction. Complexity, Noise, and Orders

We will try to expand some theoretical outcomes of empirical and experimental research on human interactions published by our laboratories in recent years. We built an advanced multidimensional methodology for analyzing human dynamics, mainly focusing on synchronization in an embodied mind framework [1,2]. Patterns of synchronization form the foundations of the cognition [3,4] continuum between healthy and disease states [5]. Structural coupling and synchronization arise in human dynamics in many ways, including coordination in conversations: speech, movement, emotions, and physiology [6–12]. It is a partially self-contained setting and practices to observe and facilitate transformation in human conditions and relations. Psychotherapy has been described as "one of the most complex bio-psycho-social systems in which patterns of language, cognition, emotion, and behavior are formed and changed through the dynamics of therapist and patient interactions" [13,14]. Beyond clinical research, studying such an exceptional human dynamics environment can lead to a general model of human dynamics, comprehending the linguistic, behavioral, and physiological realms. The integration of communication, action, bodies, and environments highlights our embodied interactions' multimodality and parallel multiactivity [15,16].

We started by focusing our studies on language. Language study is scaled in complex structures: from informational systems to mesoscopic morphological patterns to semantic and narrative streams. In verbal interactions, voice tonality, volume, pitch, cadence, rhythm, and turn-taking are relevant. Shannon [17] built the foundations of the information theory of texts and speech. His less famous work on the prediction and entropy of printed English [18] is a resource for inspiring new research. It might be interesting to consider the distribution of information and organization in different living and nonliving systems in the same perspective. In this perspective, a graph proposed by Schreiber [19] mapped scattered areas of various forms of order, entropy and knowledge still interspersed with regions of the unknown, as in old charts. Following his mapping, we can find periodic and noisy oscillations, deterministic and stochastic areas of chaos, stochastic resonance, self-organized criticality, nonlinearity, or noise. Then, there are a few other islands where a connection between our knowledge models and real-world phenomena is yet to be well established. This kind of dynamical mapping might be synchronic and diachronic, in spatial distribution and time transitions.

The structure of different systems can be known and modified through the emergence of self-organization or by external actions, by casual or planned perturbations, including measurements. Some interactions can lead to coupling between systems, and if they repeat in time, they might produce forms of synchronization. Maturana and Varela [20] considered synchronization a form of structural coupling occurring when two systems repeatedly perturb each other. "Synchronization is a nonlinear phenomenon discovered at the beginning of the scientific revolution", and in its classical definition, synchronization refers to adjustment or entrainment in frequencies or phase of periodic oscillators due to weak interactions that lead to structural coordination between systems" [21]. This process can lead to the emergence of adaptive behavior between interacting systems. Pecora and Carroll [22], Ott, Grebogi and Yorke [23], and Pyragas [24] found that synchronization can be used to change the dynamic behavior of complex systems.

2. Materials and Methods. Biosemiotics Pattern Analysis

Our initial approach was different from most of the studies mentioned above. We chose a method, Recurrence Quantification Analysis–RQA [25,26], that does not generate any specific hypothesis on the form of data and does not need to consider time series produced by a dynamic system. Our primary aim was to build a statistical tool for reliable quantitative measures of the degree of organization (as expressed by the recurrence of patterns) of a flow of signs. We demonstrated how this could be performed with a relatively simple mathematical model. The analysis of the informational structure of a text (irrespective of its meaning) could unveil the hidden matching of patterns between two speaking persons. The hidden matching relates to the flow and forms of information linking partners in conversation. Through the phonetic configuration of speech, as represented in orthography, we can extract relevant patterns in the dynamic structures of human interactions.

We used Recurrence Quantification Analysis, a methodology that can reliably measure the recurrence of patterns, determinism, and entropy. Recurrence Plots (RP) were first pioneered in physics by Eckmann, Kamphorst and Ruelle [25,26]. Later, Webber, Zbilut, Giuliani and Marwan augmented this technique by identifying nonlinear variables for the quantitative assessment of RPs, thus creating RQA. Since then, RQA has been used in different areas, from molecular dynamics [27,28] to physiology [29] and bioinformatics [30–32]. In performing RQA, the original time series must be placed into an embedding matrix by converting the original n elements column vector correspondent to the symbol series into a p-dimensional matrix with columns as the original X_n series plus its lagged copies X_{n+1}, $X_{n+2}, \ldots, X_{n+1}$, while p is the embedding dimension. The quantification of recurrences is acquired by many different 'counts' of repetitions within the matrix.

While testing the robustness of this methodology on written language [33], we had to set to three (letters) for dimensional embedding, as this amplifies its sensitivity while avoiding noise from low-level statistical features (for example, asymmetrical distribution of couplets of letters). We might notice that a three-letter dimension represents a mesoscopic information level in natural language, just between single letters and whole words. We will later see the theoretical implications of this seemingly technical specification. Our time series analysis used RQA and CRQA (cross recurrence) to measure the synchronization in conversations as semiotic interactions. These informational patterns represent a preverbal and a-conscious communication channel revealed by the frequent emergence of patterns of prosodic structures (such as the musicality of phoneme sequences, stereotyped words, pauses and phrases).

Other independent centers started developing research on social and clinical interactions using a similar methodology based on recurrence analysis. For example, they studied postural or verbal time series of interpersonal coordination during conversations [34–36]. These studies usually took one type of time series (i.e., movement, speech, or physiology) while not considering the mutual influence between different kinds of interaction. However, as human relationship dynamics are naturally hybrid, one type of interaction can influence the coupling or uncoupling of the other streams: motor, semiotic or physiological.

3. Results. Hybrid Couplings and Synchronizations

Human interactions constantly involve multiple streams (language, movement, emotions) which undergo coupling, decoupling and synchronizing. These multiscale and hybrid interactions are better comprehended within the biosemiotics, embodied mind framework that we defined as Mind Force [37,38]. We built the empirical paradigm of this approach as a multidimensional analysis of speech and emotions in patients and therapists in psychotherapy [39]. We chose Galvanic Skin Response—GSR and verbal prosody, as both variables reveal, in different flows, the expression of emotions [40,41]. Our new experiments studied four signals: the therapist's speech transcription, the patient's speech transcription, the therapist's GSR, and the patient's GSR. We focused on how those four variables modulated, coupled, synchronized, or desynchronized with each other. First, we considered the combinatorics and patterns of letters (phonemes) and morphemes (the minor portion of words that communicate significance). As mentioned, we had established this methodology in previous studies, which validated robust informational measures of entropy and determinism. In this new study, we initially considered the synchronization with standard correlation coefficients of Principal Components Analysis. Afterwards, we clustered all four signals using k-means resulting in a model representing this complex system's phase space and state transitions. Then, using a Markov Transition Matrix (see Figure 1), we disclosed phase transition probabilities between linguistic and physiological time series.

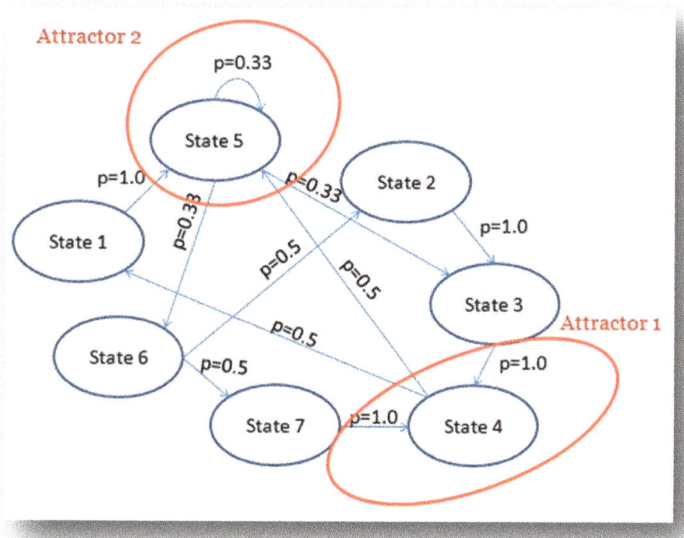

Figure 1. The Markov Transition Matrix map of attractors and phase transitions [1].

The complex dyadic system evolves between two attractors. In the first attractor, state four, the therapist strives to attune and entrain with the patient presenting low values of GSR recurrence and determinism. The therapist has high recurrence and determinism in prosody with repetitive semiotic patterns, perhaps to direct the patient's emotional expressions. The second attractor, at state five, is characterized by a medium level of GSR recurrence and determinism for both patient and therapist. We evidence semiotic medium recurrence and determinism for the therapist and low recurrence and determinism for the patient. Overall, this phase represents a state in which the patient's physiological anxiety becomes more manageable and linguistic expressions are more integrated. In short, while state four is an erratic phase of the interaction in which semiotics seems independent from passions, state five shows an integration. This sequence in human interactions is consistent with the literature in psychotherapy and neuroscience research on the embodied mind.

4. Discussion. The Chimera States in Human Interactions

This data analysis and mapping highlight dynamical landscapes of mixed states of coupling, with mixed zones of synchronization, noninteraction, and drift in uncoupling that can change over time. The Japanese physicist Yoshiki Kuramoto (1984b, 1984a) proposed a paradigmatic mathematical model to describe synchronization dynamics in a large set of coupled oscillators. The most frequent form of the model has the following equation:

$$\frac{d\theta_i}{dt} = \omega_i + \frac{K}{N}\sum_{j=1}^{N} sin(\theta_j - \theta_i),\ i = 1 \ldots .\ N, \tag{1}$$

where the system is formed of N limit-cycle oscillators with phase θ_i and coupling K.

Then, in November 2002, Yoshiki Kuramoto and Dorjsuren Battogtokh published the paper "Coexistence of Coherence and Incoherence in Nonlocally Coupled Phase Oscillators" [42,43]. They observed the coexistence of coherence and incoherence in a network of identical, nonlocally coupled, complex Ginzburg–Landau oscillators. While coupled nonidentical oscillators were known to exhibit mixed complex behavior (frequency locking, phase synchronization, partial synchronization, and incoherence), identical oscillators were supposed to either synchronize in phase or incoherently drift. They showed that oscillators that were identically coupled with similar natural frequencies could behave differently from one another for specific initial conditions. Some could synchronize while others remained incoherent in a stable state. They considered the following equation, which they called the nonlocally coupled complex Ginzburg–Landau equation:

$$\frac{\delta}{\delta t}\psi(x,t) = \omega(x) - \int G(x - x')\sin(\psi(x.t) - \psi(x',t) + \alpha)dx'$$

with $\omega(x) = \omega$ for all x.

Later, Abrams and Strogatz [44,45] named it a chimaera state, from the mythological Greek creature made up of parts of different animals and introduced some theoretical clarifications for such behavior. Finally, they studied the most straightforward system presenting a chimaera state, a ring of phase oscillators governed by:

$$\frac{\vartheta\phi}{\theta t}\omega - \int_{-\pi}^{\pi} G(x - x')\sin[\phi(x, xt) - \phi(x',t) + \alpha]dx\prime$$

Here, $\phi(x',t)$ is the phase of the oscillator at position x at time t. The space variable x runs from $-\pi$ to π with periodic boundary conditions. The frequency ω plays no role in the dynamics; one can set $\omega = 0$ by redefining $\phi \to \phi + \omega t$ without otherwise changing the form of the equation.

Chimaera states were later found in limit-cycle oscillators, chaotic oscillators, chaotic maps and in neuronal systems. In the beginning, chimaera patterns were observed in nonlocally coupled networks, but afterwards, these states were also found globally and locally (nearest neighbor) coupled networks and in modular networks [46,47]. The usage of Markov chains for mapping couplings and chimaera states was also explored ([48,49]. C.R. Laing studied chimaera state in heterogeneous networks, analyzing the influence of heterogeneous coupling strengths. Of further interest for human dynamics is the emergence of chimaera states in multiscale networks that result from the networking of different networks [50,51]. The ubiquity of chimaera mapping of synchronization and its different typologies extended its original definition to areas that might include nonidentical coupling oscillators in hybrid networks and multiscale networking of networks that were already known to present chimeralike dynamics before this definition started to be used.

Our studies' dynamic mapping of heterogeneous synchronization indicates that similar dynamics involving different brain areas related to emotional, motor, and verbal interactions co-occur. Cognitive tasks constantly require a balance between segregated and integrated neural processing with relevant consequences for cognitive performance. Segregation enables efficient computations in specialized brain regions, while integrated systems ensure coordinated, robust performance. Focused states tend to involve shorter, local connections, while integration largely relies on subcortical regions and cortical hubs with diverse connections to other brain regions [52,53]. "Recognizing chimaera dynamics can help to clarify the hybrid complexity of synchronization in critical cognitive states where a balance between integration and segregation is required for adaptive cognition and social interactions" [54]. Brain chimaera dynamics might also be related to different neuronal interactions mediated by different electrical or chemical synapses in the nervous system.

Further neural interactions involve neuromodulators and hormones, faster or slower action, and different time frames [55]. Various types of neural interaction are undoubtedly an additional factor in the emergence of chimaera dynamically states in human hybrid synchronization [51]. As separate regions interact to perform neurocognitive tasks, variable patterns of partial synchrony form chimaera states [3].

5. Conclusions: From Determinism to Statistical Dynamics

Human dynamics are so complex and prone to indeterminacy and randomness that even deterministic chaos might be considered, in many cases, as a reductionist simplification. Therefore, we might consider probabilistic models including elements of randomness. The previous study highlighted how biopsychosocial dynamics are hybrid, discontinuous, and have many degrees of freedom. We also highlighted how cluster analysis and Markov states could help to clarify the dynamics. However, our knowledge of the state of the systems is always incomplete, and some uncertainty is part of the game. While standard dynamics usually consider the behavior of a single state, statistical dynamics define the statistical ensemble as a probability cloud of the possible conditions in the system [56]. The ensemble probability can be interpreted in two main ways:

(a) Epistemic probability of all the possible states.
(b) Empirical probability in repeated experiments.

Following this perspective, we used a probabilistic approach in the study of empathy [57]. Empathy plays a significant role in human coordination, collaboration, and change, in human interactions. Most authors agree that forms of resonance in imitation, emotions, and communication are relevant factors of empathy. Following the expanding literature on relational physiology, we explored if empathy would present physiological evidence. We applied a Principal Component Analysis (PCA) on simultaneous GSR and HR signals from a patient-therapist dyad. PCA revealed a 'shared' component in signals, and two 'individual' components of independent correlation. Regression analysis showed that the shared component predicted a therapy outcome (R^2 = 0.28). We further examined the common component dynamics in a symbolic Markovian discrete model and cluster analysis.

Several studies on cognitive neuroscience [58,59] established statistical dynamics in biological systems focusing on the reciprocal correlations between system descriptors.

This scientific position focuses on the mesoscopic level [60,61], potentially expanding correlations among system variables. This is the midpoint between pure "bottom-up" and "top-down" approaches. The crucial role of mesoscopic dynamics was validated in our physiological analysis and semiotics, as highlighted by the robust evidence for our mesoscopic embedding in RQA since the first experiments. In this way, we focused on the level of morphemes as word subcomponents. Morphemes have meaning and grammatical functions. They can be decomposed into smaller morphemes without losing these two crucial properties. Morphemes can be considered as semiotic quanta of information in natural language, as they are the basic lexical item in a language. They are usually composed of more than one phoneme and several letters or informational units [62–64]. Therefore, we can consider morphemes as the information quanta structuring coupling and synchronization in natural human language.

We explored informational patterns in human interactions. We investigated intraindividual and interpersonal relations as coevolution dynamics of hybrid couplings, synchronizations, and desynchronization. Cluster analyses and Markov chains produced evidence of chimaera states and phase transitions. A probabilistic and nondeterministic approach can clarify relevant properties of human dynamics, focusing on the mesoscopic scale and statistical dynamics. Theoretical models of human interactions should be founded on the hybrid nature of human structural couplings.

Funding: Parts of this research have been funded by: 2012–2014 Grant, MURST (Italian Ministry for University, Research & Technology) with ENEA, Rome, Italy. 2005 Invited Lecturer fund, Max Planck Institute for Human Cognitive and Brain Sciences, Leipzig, Germany. 1999–2001 Research Project, EU ICT 11696, European Commission, Brussels, Belgium.

Institutional Review Board Statement: The study was conducted according to the guidelines of the Declaration of Helsinki and approved by the Institutional Review Board (or Ethics Committee) of the University of Siena and of ENEA, Rome, Italy.

Informed Consent Statement: Informed consent was obtained from all subjects involved in the study.

Conflicts of Interest: Parts of this research have been presented at the 8th Experimental Chaos Conference (Florence, June 2004), at the Tandem Workshop sponsored by the Max Planck Institute for Human Cognitive and Brain Sciences, University of Potsdam (Berlin, March 2005), at the International Conferences of the Society for Psychotherapy Research (Jerusalem 2016, Oxford 2017, Amsterdam 2018).

References

1. Orsucci, F.; Musmeci, N.; Aas, B.; Schiepek, G.; Reda, M.A.; Canestri, L.; de Felice, G. Synchronization analysis of language and physiology in human dyads. *Nonlinear Dyn. Psychol. Life Sci.* **2016**, *20*, 167–191.
2. Orsucci, F. Towards the integration of semiotic and physiological dynamics: From nonlinear dynamics to quantum fields. In *Selbstorganization–Ein Paradigma für die Humanwissenschaften*; Springer: Wiesbaden, Germany, 2020; pp. 153–175.
3. Bansal, K.; Garcia, J.O.; Tompson, S.H.; Verstynen, T.; Vettel, J.M.; Muldoon, S.F. Cognitive chimera states in human brain networks. *Sci. Adv.* **2019**, *5*, eaau8535. [CrossRef] [PubMed]
4. Shine, J.M.; Breakspear, M.; Bell, P.T.; Martens, K.A.E.; Shine, R.; Koyejo, O.; Sporns, O.; Poldrack, R.A. Human cognition involves the dynamic integration of neural activity and neuromodulatory systems. *Nat. Neurosci.* **2019**, *22*, 289–296. [CrossRef] [PubMed]
5. Hizanidis, J.; Kouvaris, N.E.; Zamora-López, G.; Díaz-Guilera, A.; & Antonopoulos, C.G. Chimera-like states in modular neural networks. *Sci. Rep.* **2016**, *6*, 1–11.
6. Glass, L. Synchronization and rhythmic processes in physiology. *Nature* **2001**, *410*, 277–284. [CrossRef]
7. Orsucci, F.; Giuliani, A.; Webber, C.; Fonagy, P. Combinatorics and synchronization in natural semiotics. *Phys. A Stat. Mech. Appl.* **2006**, *361*, 665–676. [CrossRef]
8. Orsucci, F.; Petrosino, R.; Paoloni, G.; Canestri, L.; Conte, E.; Reda, M.A.; Fulcheri, M. Prosody and synchronization in cognitive neuroscience. *EPJ Nonlinear Biomed. Phys.* **2013**, *1*, 1–11. [CrossRef]
9. Wiltshire, T.J.; Philipsen, J.S.; Trasmundi, S.B.; Jensen, T.W.; Steffensen, S.V. Interpersonal coordination dynamics in psychotherapy: A systematic review. *Cognit. Ther. Res.* **2020**, *44*, 752–773. [CrossRef]
10. Tschacher, W.; Meier, D. Physiological synchrony in psychotherapy sessions. *Psychother. Res.* **2020**, *30*, 558–573. [CrossRef]
11. Repp, B.H.; Su, Y.-H. Sensorimotor synchronization: A review of recent research (2006–2012). *Psychon. Bull. Rev.* **2013**, *20*, 403–452. [CrossRef]

12. Delaherche, E.; Chetouani, M.; Mahdhaoui, A.; Saint-Georges, C.; Viaux, S.; Cohen, D. Interpersonal synchrony: A survey of evaluation methods across disciplines. *IEEE Trans. Affect. Comput.* **2012**, *3*, 349–365. [CrossRef]
13. Schiepek, G.; Fricke, B.; Kaimer, P. Synergetics of psychotherapy. In *Self-Organization and Clinical Psychology*; Springer: Berlin/Heidelberg, Germany, 1992; pp. 239–267.
14. Gelo, O.C.; Salvatore, S. A dynamic systems approach to psychotherapy: A meta-theoretical framework for explaining psychotherapy change processes. *J. Couns. Psychol.* **2016**, *63*, 379. [CrossRef]
15. Mondada, L. Challenges of multimodality: Language and the body in social interaction. *J. Socioling.* **2016**, *20*, 336–366. [CrossRef]
16. Mondada, L. Contemporary issues in conversation analysis: Embodiment and materiality, multimodality and multisensoriality in social interaction. *J. Pragmat.* **2019**, *145*, 47–62. [CrossRef]
17. Shannon, C.E.; Weaver, W. *The Mathematical Theory of Communication*; University of Illinois Press: Champaign, IL, USA, 1948.
18. Shannon, C.E. Prediction and Entropy of Printed English. *Bell Syst. Tech. J.* **1951**, *30*, 50–64. [CrossRef]
19. Schreiber, T. Interdisciplinary application of nonlinear time series methods. *Phys. Rep.* **1999**, *308*, 1–64. [CrossRef]
20. Maturana, H.R.; Varela, F.J. *Autopoiesis and Cognition the Realisation of the Living*; D. Reidel Publishing Company: Dordrecht, The Netherlands, 1980.
21. Strogatz, S.H. *Sync the Emerging Science of Spontaneous Order*; Hyperion: New York, NY, USA, 2003.
22. Pecora, L.M.; Carroll, T.L. Synchronization in chaotic systems. *Phys. Rev. Lett.* **1990**, *64*, 821–824. [CrossRef] [PubMed]
23. Ott, E.; Grebogi, C.; Yorke, J.A. Controlling chaotic dynamical systems. In *Chaos: Soviet-American Perspective on Nonlinear Science*; American Institute of Physics: Melville, NY, USA, 1990; pp. 153–172.
24. Pyragas, K. Weak and strong synchronization of chaos. *Phys. Rev. E Stat. Phys.* **1996**, *54*, R4508–R4511. [CrossRef]
25. Eckmann, J.-P.; Kamphorst, S.O.; Ruelle, D. Recurrence plots of dynamical systems. *Europhys. Lett.* **1987**, *4*, 973. [CrossRef]
26. Webber, C.L.; Zbilut, J.P. Dynamical assessment of physiological systems and states using recurrence plot strategies. *J. Appl. Physiol.* **1994**, *76*, 965–973. [CrossRef]
27. Giuliani, A.; Benigni, R.; Zbilut, J.P.; Webber, C.L.; Sirabella, P.; Colosimo, A. Nonlinear signal analysis methods in the elucidation of protein sequence-structure relationships. *Chem. Rev.* **2002**, *102*, 1471–1492. [CrossRef] [PubMed]
28. Manetti, C.; Ceruso, M.A.; Giuliani, A.; Webber, C.L.; Zbilut, J.P. Recurrence quantification analysis in molecular dynamics. *Ann. N. Y. Acad. Sci.* **1999**, *879*, 258–266. [CrossRef] [PubMed]
29. Webber, C.L.; Zbilut, J.P. Recurrence quantification analysis of nonlinear dynamical systems. *Tutor. Contemp. Nonlinear Methods Behav. Sci.* **2005**, *94*, 26–94.
30. Marwan, N. A historical review of recurrence plots. *Eur. Phys. J. Spec. Top.* **2008**, *164*, 3–12. [CrossRef]
31. Webber, C.L.; Marwan, N.; Facchini, A.; Giuliani, A. Simpler methods do it better: Success of Recurrence Quantification Analysis as a general-purpose data analysis tool. *Phys. Lett. A* **2009**, *373*, 3753–3756. [CrossRef]
32. Webber, C.L.; Marwan, N. Recurrence Quantification Analysis. In *Theory and Best Practices*; Springer: New York, NY, USA, 2015.
33. Orsucci, F.; Walter, K.; Giuliani, A.; Webber, C.L., Jr.; Zbilut, J.P. Orthographic Structuring of Human Speech and Texts: Linguistic Application of Recurrence Quantification Analysis. *arXiv* **1997**, arXiv:cmp-lg/9712010.
34. Keller, E.; Tschacher, W. Prosodic and gestural expression of interactional agreement. In *Verbal and Nonverbal Communication Behaviors*; Springer: Berlin/Heidelberg, Germany, 2007; pp. 85–98.
35. Shockley, K.; Richardson, D.C.; Dale, R. Conversation and coordinative structures. *Top. Cogn. Sci.* **2009**, *1*, 305–319. [CrossRef]
36. Fusaroli, R.; Konvalinka, I.; Wallot, S. Analyzing Social Interactions: The Promises and Challenges of Using Cross Recurrence Quantification Analysis. In *Translational Recurrences*; Springer: Berlin/Heidelberg, Germany, 2014; pp. 137–155.
37. Orsucci, F. Mind force theory: Hyper-network dynamics in neuroscience. *Chaos Complex. Lett.* **2009**, *4*, 1–26.
38. Freeman, W.J. On the Nature and Neural Mechanisms of Mind Force. *Chaos Complex. Lett.* **2012**, *6*, 7.
39. Orsucci, F. *Human Dynamics: A Complexity Science Open Handbook*; Nova Science Publishers Inc.: New York, NY, USA, 2016.
40. Pichon, S.; Kell, C.A. Affective and sensorimotor components of emotional prosody generation. *J. Neurosci.* **2013**, *33*, 1640–1650. [CrossRef]
41. Koolagudi, S.G.; Rao, K.S. Emotion recognition from speech: A review. *Int. J. Speech Technol.* **2012**, *15*, 99–117. [CrossRef]
42. Kuramoto, Y.; Battogtokh, D. Coexistence of coherence and incoherence in nonlocally coupled phase oscillators. *arXiv* **2002**, arXiv:Cond-Mat/0210694.
43. Smirnov, L.; Osipov, G.; Pikovsky, A. Chimera patterns in the Kuramoto–Battogtokh model. *J. Phys. A Math. Theor.* **2017**, *50*, 08LT01. [CrossRef]
44. Abrams, D.M.; Strogatz, S.H. Chimera states for coupled oscillators. *Phys. Rev. Lett.* **2004**, *93*, 174102. [CrossRef] [PubMed]
45. Panaggio, M.J.; Abrams, D.M. Chimera states: Coexistence of coherence and incoherence in networks of coupled oscillators. *Nonlinearity* **2015**, *28*, R67. [CrossRef]
46. Schöll, E.; Zakharova, A.; Andrzejak, R.G. Chimera states in complex networks. *Front. Appl. Math. Stat.* **2019**, *5*, 62. [CrossRef]
47. Wang, Z.; Liu, Z. A brief review of chimera state in empirical brain networks. *Front. Physiol.* **2020**, *11*, 724. [CrossRef] [PubMed]
48. Cavers, M.; Vasudevan, K. Spatio-temporal complex Markov Chain (SCMC) model using directed graphs: Earthquake sequencing. *Pure Appl. Geophys.* **2015**, *172*, 225–241. [CrossRef]
49. Vasudevan, K.; Cavers, M.; Ware, A. Earthquake sequencing: Chimera states with Kuramoto model dynamics on directed graphs. *Nonlinear Process. Geophys.* **2015**, *22*, 499–512. [CrossRef]
50. Laing, C.R. Chimera states in heterogeneous networks. *Chaos Interdiscip. J. Nonlinear Sci.* **2009**, *19*, 013113. [CrossRef]

51. Makarov, V.V.; Kundu, S.; Kirsanov, D.V.; Frolov, N.S.; Maksimenko, V.A.; Ghosh, D.; Dana, S.K.; Hramov, A.E. Multiscale interaction promotes chimera states in complex networks. *Commun. Nonlinear Sci. Numer. Simul.* **2019**, *71*, 118–129. [CrossRef]
52. Liégeois, R.; Ziegler, E.; Phillips, C.; Geurts, P.; Gómez, F.; Bahri, M.A.; Yeo, B.T.; Soddu, A.; Vanhaudenhuyse, A.; Laureys, S.; et al. Cerebral functional connectivity periodically (de) synchronises with anatomical constraints. *Brain. Struct. Funct.* **2016**, *221*, 2985–2997. [CrossRef]
53. Shine, J.M. Neuromodulatory influences on integration and segregation in the brain. *Trends Cogn. Sci.* **2019**, *23*, 572–583. [CrossRef] [PubMed]
54. Chouzouris, T.; Omelchenko, I.; Zakharova, A.; Hlinka, J.; Jiruska, P.; Schöll, E. Chimera states in brain networks: Empirical neural vs. modular fractal connectivity. *Chaos: Interdiscip. J. Nonlinear Sci.* **2018**, *28*, 045112. [CrossRef]
55. Majhi, S.; Bera, B.K.; Ghosh, D.; Perc, M. Chimera states in neuronal networks: A review. *Phys. Life Rev.* **2019**, *28*, 100–121. [CrossRef]
56. Kolmogorov, A.N.; Bharucha-Reid, A.T. *Foundations of the theory of probability, Second English Edition*; Courier Dover Publications: Mineola, NY, USA, 2018.
57. Kleinbub, J.R.; Palmieri, A.; Orsucci, F.F.; Andreassi, S.; Musmeci, N.; Benelli, E.; Giuliani, A.; de Felice, G. Measuring empathy: A statistical physics grounded approach. *Phys. A Stat. Mech. Appl.* **2019**, *526*, 120979. [CrossRef]
58. Giuliani, A.; Tsuchiya, M.; Yoshikawa, K. Self-organization of genome expression from embryo to terminal cell fate: Single-cell statistical mechanics of biological regulation. *Entropy* **2018**, *20*, 13. [CrossRef]
59. Mojtahedi, M.; Skupin, A.; Zhou, J.; Castaño, I.G.; Leong-Quong, R.Y.; Chang, H.; Huang, S. Cell fate decision as high-dimensional critical state transition. *PLoS Biol.* **2016**, *14*, e2000640. [CrossRef] [PubMed]
60. Freeman, W.J. Mesoscopic neurodynamics: From neuron to brain. *J. Physiol. Paris* **2000**, *94*, 303–322. [CrossRef]
61. Giuliani, A. Networks as a privileged way to develop mesoscopic level approaches in systems biology. *Systems* **2014**, *2*, 237–242. [CrossRef]
62. Feldman, L.B. *Morphological Aspects of Language Processing*; Lawrence Erlbaum: Mahwah, NJ, USA, 1995.
63. Martinčić-Ipšić, S.; Margan, D.; Meštrović, A. Multilayer network of language: A unified framework for structural analysis of linguistic subsystems. *Phys. A Stat. Mech. Appl.* **2016**, *457*, 117–128. [CrossRef]
64. Martin, A.E.; Baggio, G. Modelling meaning composition from formalism to mechanism. *Phil. Trans. R. Soc.* **2020**, *375*. [CrossRef] [PubMed]

Article

Four Methods to Distinguish between Fractal Dimensions in Time Series through Recurrence Quantification Analysis

Alon Tomashin [1], Giuseppe Leonardi [2] and Sebastian Wallot [3,4,*]

1 The Gonda Multidisciplinary Brain Research Center, Bar-Ilan University, Ramat-Gan 5290002, Israel
2 Institute of Psychology, University of Economics and Human Sciences, 01-043 Warsaw, Poland
3 Institute for Sustainability Education and Psychology, Leuphana University of Lüneburg, 21335 Lüneburg, Germany
4 Department of Language and Literature, Max Planck Institute of Empirical Aesthetics, 60322 Frankfurt am Main, Germany
* Correspondence: sebastian.wallot@leuphana.de

Abstract: Fractal properties in time series of human behavior and physiology are quite ubiquitous, and several methods to capture such properties have been proposed in the past decades. Fractal properties are marked by similarities in statistical characteristics over time and space, and it has been suggested that such properties can be well-captured through recurrence quantification analysis. However, no methods to capture fractal fluctuations by means of recurrence-based methods have been developed yet. The present paper takes this suggestion as a point of departure to propose and test several approaches to quantifying fractal fluctuations in synthetic and empirical time-series data using recurrence-based analysis. We show that such measures can be extracted based on recurrence plots, and contrast the different approaches in terms of their accuracy and range of applicability.

Keywords: recurrence quantification analysis; fractals; monofractals; fractal time series

Citation: Tomashin, A.; Leonardi, G.; Wallot, S. Four Methods to Distinguish between Fractal Dimensions in Time Series through Recurrence Quantification Analysis. *Entropy* **2022**, *24*, 1314. https://doi.org/10.3390/e24091314

Academic Editors: Franco Orsucci and Wolfgang Tschacher

Received: 19 July 2022
Accepted: 13 September 2022
Published: 19 September 2022

1. Introduction

Since Gilden et al.'s [1] seminal paper, showing the presence of $1/f^{\alpha}$-fluctuations in human time estimation performance, a huge interest in the presence and meaning of fractal fluctuations in human behavior has emerged. On the one hand, fractal patterns have been found in virtually all aspects of human physiology and behavior across recent studies [2–11]. On the other hand, their meaning has been intensely discussed [12–19].

Through the same period, the development and refinement of different time-series analysis techniques gained momentum, so that fractal properties could be quantified with a variety of methods, based on the power spectrum of a time series [20], their standard deviation [21] or residual fluctuations [22]—each of which has particular advantages and downsides, as well as requirements for preprocessing [21,23]. This was of central importance, because methods that are suitable for special fractals, such as box counting, are not equally applicable to time-series data [24].

In the current paper, we want to present another way of quantifying fractal fluctuations in time-series data using recurrence quantification analysis [25,26]. Our motivation for the present work is two-fold: firstly, to extend the use of recurrence plot-based methods to capture fractal properties. This is something that recurrence plot-based analyses have not been capable of. Further, to pave the way to provide an easy-to-use tool to compare fractal dimensions of time series that are well-applicable to binary data, and in the future also to multidimensional time series using multidimensional recurrence plot methods [27]. As has been suggested elsewhere [28], fractal properties in time-series data can be well-captured by the concept of (imperfectly) recurring patterns over time, and this is—as the name implies—what recurrence quantification analysis is about. Specifically, Webber [28] encouraged researchers to explore RQA as a bridge to further understand fractal systems in various fields. However, Webber did not specify how to quantify fractal fluctuations by means of recurrence plots.

Hence, the aim of the present paper is to take this next step and propose, as well as compare, novel recurrence-based approaches that can be used to quantify fractal fluctuations. In the following, each approach is introduced, tested on synthetic data, and evaluated; in addition, a Matlab (The MathWorks, Inc., Natick, MA, USA) implementation of the approaches presented in this paper is available on GitHub: https://github.com/alontom/FARQA (Accessed on 19 July 2022, see Appendix A). Finally, we discuss the individual strengths and weaknesses of each approach and relate the results to those obtained from detrended fluctuation analysis (DFA; [22]), as DFA is one of the most widely used methods with accurate performance in capturing fractal fluctuations in time-series data [29–31].

2. Methods and Results

2.1. Synthetic Data

In this section, we show four new approaches to differentiate between the power-law scaling exponent (α; $1/f^{\alpha}$) based on several RQA properties. Each evaluation approach was applied to synthetic data consisting of 1026 data points with different fractal dimensions ranging from $\alpha = -1$ (antipersistent) to 2 (persistent) generated by 'power noise' function [32] using Matlab version 2021b (The MathWorks, Inc.). For every fractal dimension, 100 time series were generated under two conditions: idealized fractal time series and a noisy fractal time series (SNR 2:1). The noise component added was drawn from a normal distribution with 50% of the SD of the idealized fractal time series. We conducted RQA without embedding (delay and embedding parameters of 1, euclidean normalization of the phase space, and radius = 0.4) on the z-scored generated time series and utilized its properties to discriminate between signals with different $1/f$ values. As a benchmark to compare against, alongside the true predetermined α, we also subjected the data to detrended fluctuation analysis (DFA; [22]). In the next sections, we will describe each of the methods and present the results of their application. After that, we will apply the methods to empirical data of a time-estimation task. Finally, we will provide a summary of the strengths and weaknesses of each method and the intercorrelations of their results.

2.1.1. Detrended Fluctuations Analysis (DFA)

First, we tested the fractal properties of the dataset by applying a detrended fluctuation analysis [22]. To do so, we used the following DFA parameters: a minimum bin size of 10, a maximum of 510, linear detrending. The results are presented in Figure 1 and show that the Hurst exponents H estimated via the DFA scale well with the true α-values of the time series. In the absence of random noise, DFA distinguishes scaling relations well down to antipersistent fluctuations with $\alpha = -1$ (Figure 1, left panel, $R^2 = 0.997$). When noise is added, the capacity of DFA to distinguish among antipersistent was slightly compromised (Figure 1, right panel, $R^2 = 0.965$).

2.1.2. First Approach: Estimating Scaling Using the *SD* of *%REC* over a Range of Bin Sizes (*%REC SD*)

To capture the change in fluctuations with scale, the RP was split into bins of various sizes (powers of two). In each, we calculated the recurrence rate. Then, the *SD* of all the bins of the same size was computed, and we fitted a linear line to the log–log plot of the *SD* vs. the bin sizes. Figure 2 illustrates the approach.

The rationale behind this approach is that a time series of i.i.d. white noise will yield a recurrence plot that is statistically uniformly populated by mostly isolated recurrence points, while the correlation structure of persistent fluctuations will yield a more clustered, nonuniform distribution, and will hence lead to a slower increase in *SD* compared to the white noise case (Figure 3, $\alpha > 0$). However, antipersistent fluctuations tend to systematically decluster recurrences, and the result is likewise a relatively uniformly distributed recurrence plot (Figure 3, $\alpha = -1$).

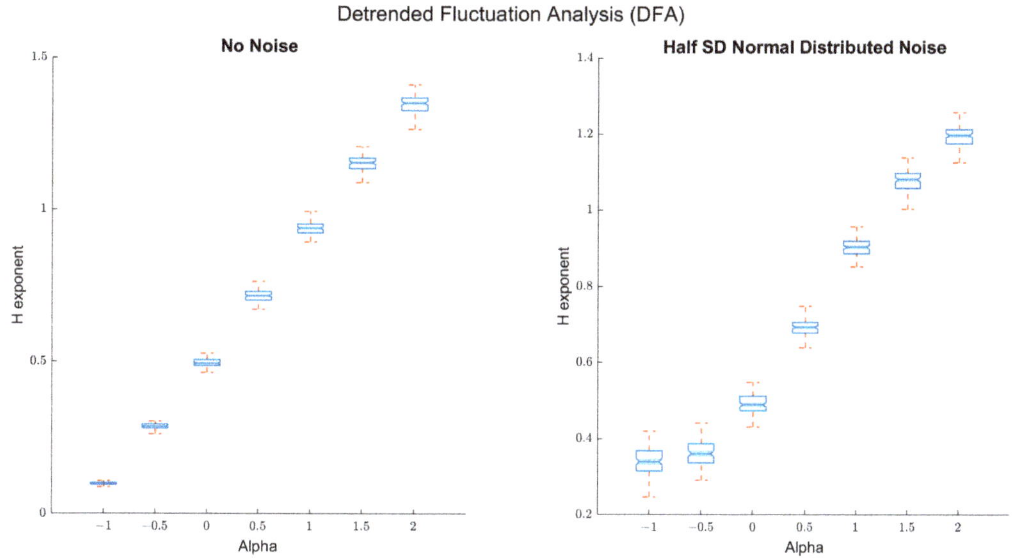

Detrended Fluctuation Analysis (DFA)

No Noise

Half SD Normal Distributed Noise

H exponent

Alpha

Figure 1. Detrended fluctuation analysis (DFA) results: Left panel: Box plots of the true α-values on the x-axis and the estimated Hurst exponents H on the y-axis from DFA. As can be seen, the DFA H scales well with the true alpha values down to antipersistent fluctuations ($\alpha = -1$). Right panel: Box plots of the true α-values on the x-axis and the estimated Hurst exponents H on the y-axis from DFA, when random noise is added (SNR = 2:1). DFA still scales well for persistent fluctuations with the true α-values, but is relatively less sensitive to distinguishing between different types of antipersistent fluctuations.

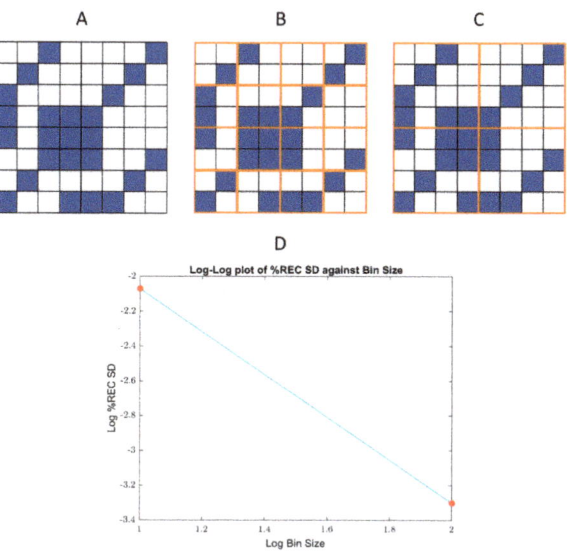

Figure 2. Demonstration of approach 1 over a simple recurrence plot: (**A**) A hypothetic 8 $\times$ 8 recurrence plot (RP) where blue squares stand for recurrence points and blank squares for nonrecurrent ones. In (**B**,**C**) the RP is split into bins of 2 and 4 (respectively, marked in brown). With approach 1, one finds the *%REC* in every bin and computes the *SD* between the recurrence percentages. Afterward, a linear trend is fitted to the log–log scaling plot (**D**) and the slope represents the scaling.

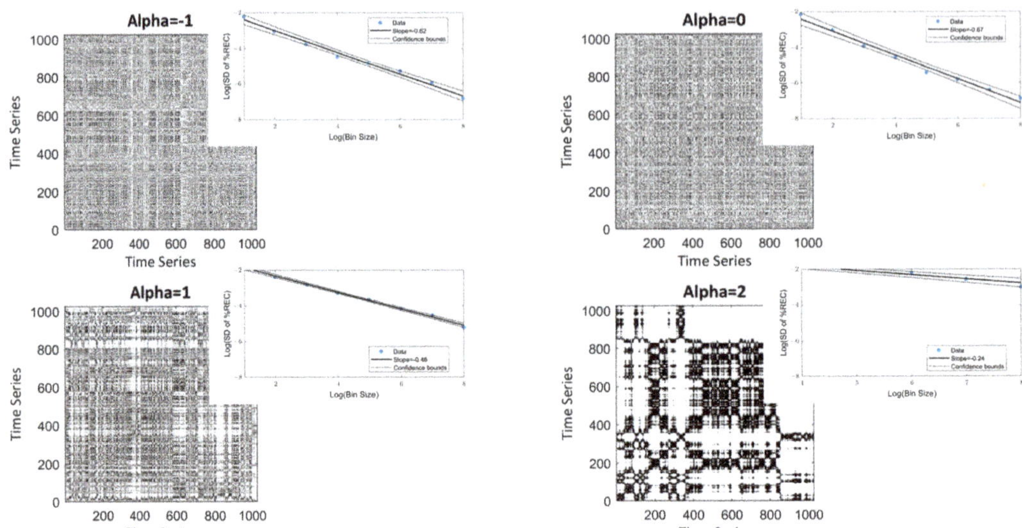

Figure 3. RP and scaling plots for different alpha values: Examples of univariate RP time series generated with different α-values, and scaling plots demonstrate the association of bin sizes and the *SD* of the recurrence rates between bins. As can be seen, from α = 0 the slope tends to decrease, suggesting a lower fractal dimension (i.e., higher α).

Figure 4 shows the model coefficients for each α (n = 100 simulations each). This method seems appropriate for distinguishing among persistent signals (α > 0) for the idealized, but also noise data. It does not work for antipersistent fluctuations (α < 0). Here, the method simply does not distinguish between α-values of 0 and −1. For the noisy time series, we fitted linear regressions between α values and the power-law coefficients separately for the $\alpha \geq 0$ (R^2 = 0.9) and $\alpha \leq 0$ (R^2 = 0.04), which support the above statement.

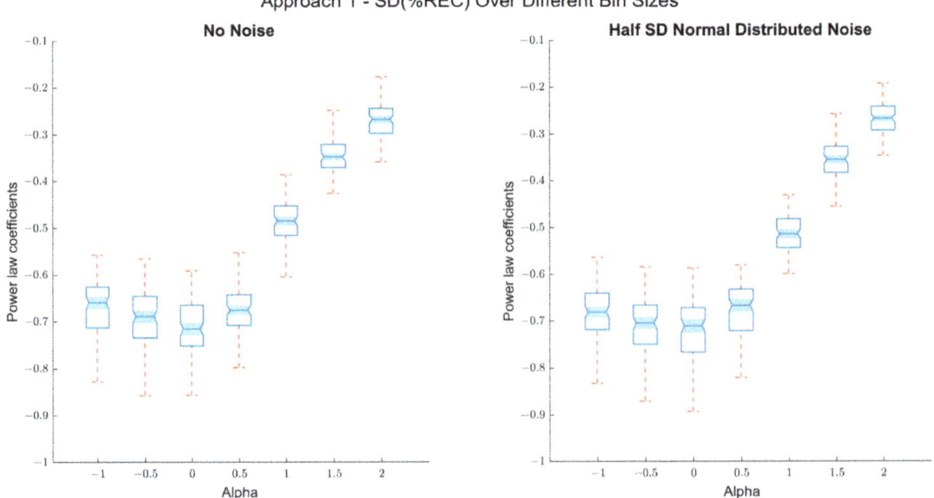

Figure 4. Results of approach 1 (*SD %REC*): Left panel: Box plots of the true alpha values on the x-axis and the power-law coefficients for the association of *SD* of *%REC* between the bins and bin

sizes on the y-axis. As observed, the coefficient scales well with the true alpha values for the persistent fluctuations ($\alpha > 0$). Right panel: Box plots of the true alpha values on the x-axis and the power-law coefficients for the association of *SD* of *%REC* between the bins and bin sizes on the y-axis, when random noise (SNR 2:1) is added. Still, the resulted coefficients scale well for persistent fluctuations with the true alpha values but are relatively insensitive to distinguishing between different types of antipersistent fluctuations.

2.1.3. Second Approach: Estimating Scaling Using Laminarity *(%LAM)*

For this approach, we simply calculated the percentage of recurrence points that have a vertical/horizontal neighbor (*%LAM*, laminarity; [33]) over the whole plot (Figure 5). The rationale behind this approach is somewhat similar to the first approach, which is that fractal fluctuations tend to be manifested by patches or squares in the recurrence plot (see Figure 3). Hence, *%LAM* would represent the persistence of the data well.

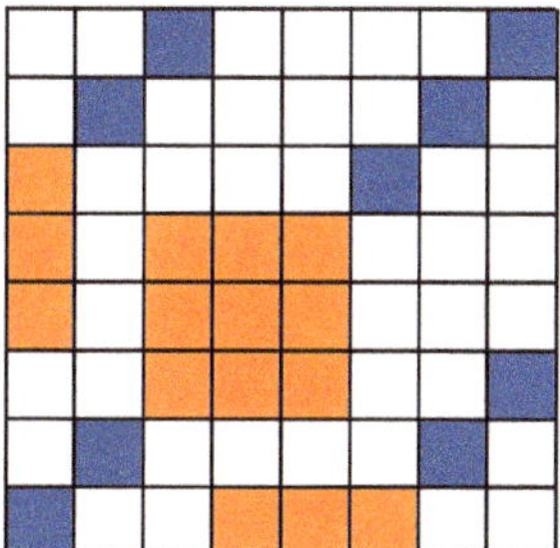

Figure 5. Quantifying laminarity: An 8 × 8 RP where colored squares represent recurrence points. The orange-filled recurrence points have a vertical/horizontal neighbor, while the blue squares do not. *%LAM* is the percentage of the recurrence points that have a vertical neighbor (orange) out of all the recurrence points (colored).

The results corroborate this: persistent fractal fluctuations lead to increased laminarity with and without noise (Figure 6). In addition, there was a tight connection between the *%LAM* values and the true α-values, marked by a high R^2 (0.96) quantifying correlation between α and *%LAM* for the noise condition.

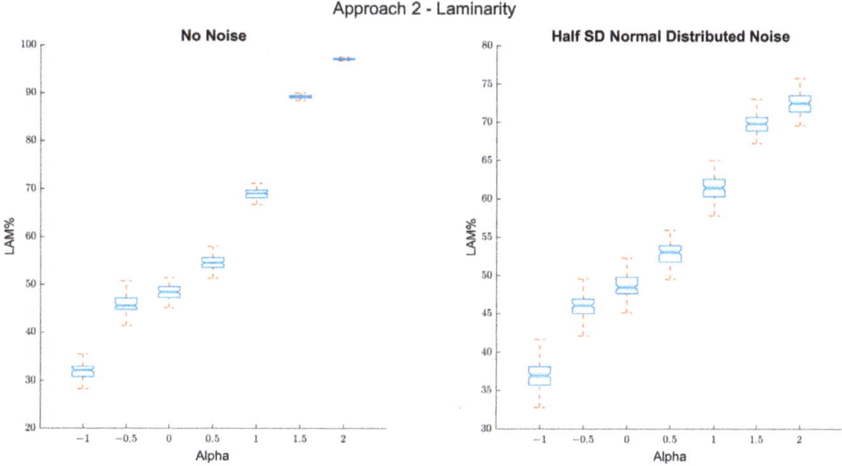

Figure 6. Results of approach 2 (laminarity). Left panel: box plots of the true alpha values on the x-axis and the *%LAM* on the y-axis. As can be seen, the coefficient scales well with the true α-values

for both persistent and antipersistent fluctuations ($\alpha > 0$). Right panel: box plots of the true α-values on the x-axis and the *%LAM* on the y-axis, when random noise (SNR 2:1) is added. Still, the resulted coefficients scale well for every alpha value ($-1 < \alpha < 2$).

While there is a mathematical relation between *%LAM* and autocorrelations in a time series, the method has a downside in that it does not capture scaling relations within the data per se, and hence represents more of a correlate of fractal fluctuations, albeit a very useful one.

2.1.4. Third Approach: Estimating Scaling Relations via Diagonal Recurrence Rates (*Diag %REC*)

The third approach is based on diagonal recurrence profiles of a time series. The diagonal recurrence profile quantifies the number of recurrences at different lags, similar to the autocorrelation function [34]. To obtain the diagonal recurrence profile, one simply counts the proportion of recurrence points in the off-diagonals towards the lower-right or lower-left of the recurrence plot and plots them as a function of distance from the main diagonal; that is, lag [35]. Figure 7 illustrates the computation of the diagonal recurrence profile.

Figure 7. Approaches 3 and 4—diagonals in RP: An 8 × 8 RP presents the diagonal lines from 0 (main diagonal) to 7; due to the univariate RP's symmetrical characteristic, only the bottom triangle was used. In approach 3, we counted the recurrence points in each diagonal and divided them by the diagonal's length. Additionally, approach 4 utilizes the ratio of recurrence percentage between every two subsequent diagonal lines. Both approaches focused on the middle diagonals to avoid the main diagonal's 100% recurrence points and the short diagonals towards the edges of the recurrence plot.

The rationale behind the approach is that the diagonal recurrence profile is a model-free type of autocorrelation [33,36], and hence captures the magnitude of autocorrelation at different lags, which is related to fractal fluctuations in a time series [37]. Accordingly, a scaling relation between the logarithm of the recurrence rate and the logarithm of the diagonal number (reflecting the frequency spectra) should be related to fractal scaling. Here, a sharper negative slope indicates dominance of lower frequencies. Hence, contrasting the previous approaches, a lower power-law coefficient evidence a more persistent fluctuation. Correspondingly to spectral scaling analysis, this method yielded a scaling exponent of 0 to white noise ($\alpha = 0$)—a benchmark to determine whether the time series is persistent, random, or antipersistent.

As can be seen in Figure 8, this approach distinguishes comparatively well between the different exponents for persistent fluctuations, with and without noise, but is less sensitive to the antipersistent fluctuations (however, the exponents are still increasing with decreasing negative alpha-values). Moreover, the relation to the true α-values appears strong for this range, even with noise ($R^2 = 0.88$).

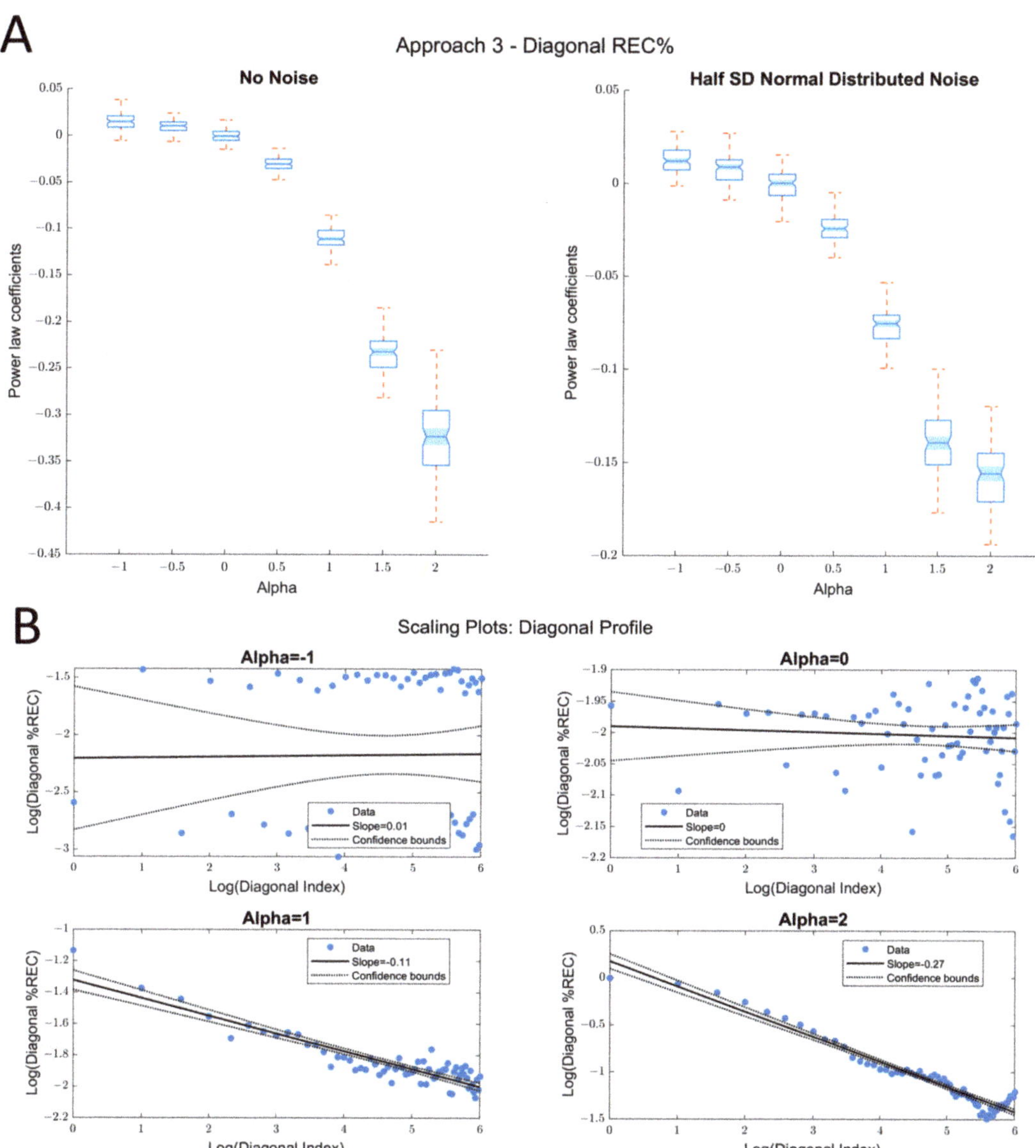

Figure 8. Results of approach 3 (*Diag %REC*). (**A**) Left panel: box plots of the true alpha values on the x-axis and the power-law coefficients for the association of diagonal *%REC* and the diagonal index (distance from the main diagonal) on the y-axis. As can be seen, the coefficient scales well with the true α-values for both persistent and antipersistent fluctuations ($-1 < \alpha < 2$). Right panel: box plots of the true alpha values on the x-axis and the power-law coefficients for the association of diagonal *%REC* and the diagonal index on the y-axis when random noise (SNR = 2:1) is added. Still, the resulted coefficients scale well with the true α-values for persistent and antipersistent fluctuations, but are somewhat less sensitive to distinguishing between different types of antipersistent fluctuations ($\alpha < 0$). (**B**) Example of scaling plots demonstrating the association of diagonal *%REC* and diagonal index for different α values.

Another version of this approach is derived from Zbilut and Marwan's [38] proposal, which applied the Wiener–Khinchin theorem [38] to the analysis of diagonal recurrence profiles. They show that one can detect (nonlinear) periodicities by applying a Fourier transform to the diagonal recurrence profile of an RP (Figure 7). Just as with the raw diagonal recurrence profile, we fitted a linear trend line to the log–log plot power spectrum (obtained via the Fourier Transform) of the diagonal recurrence profile (Figure 9). The results were similar to what we observed for the raw diagonal recurrence profile in that the method distinguished between persistent (R^2 = 0.68, $\alpha \geq 0$) fluctuations. However, the standard errors were higher, and the method did not capture antipersistent fluctuations (R^2 = 0.002, $\alpha \leq 0$).

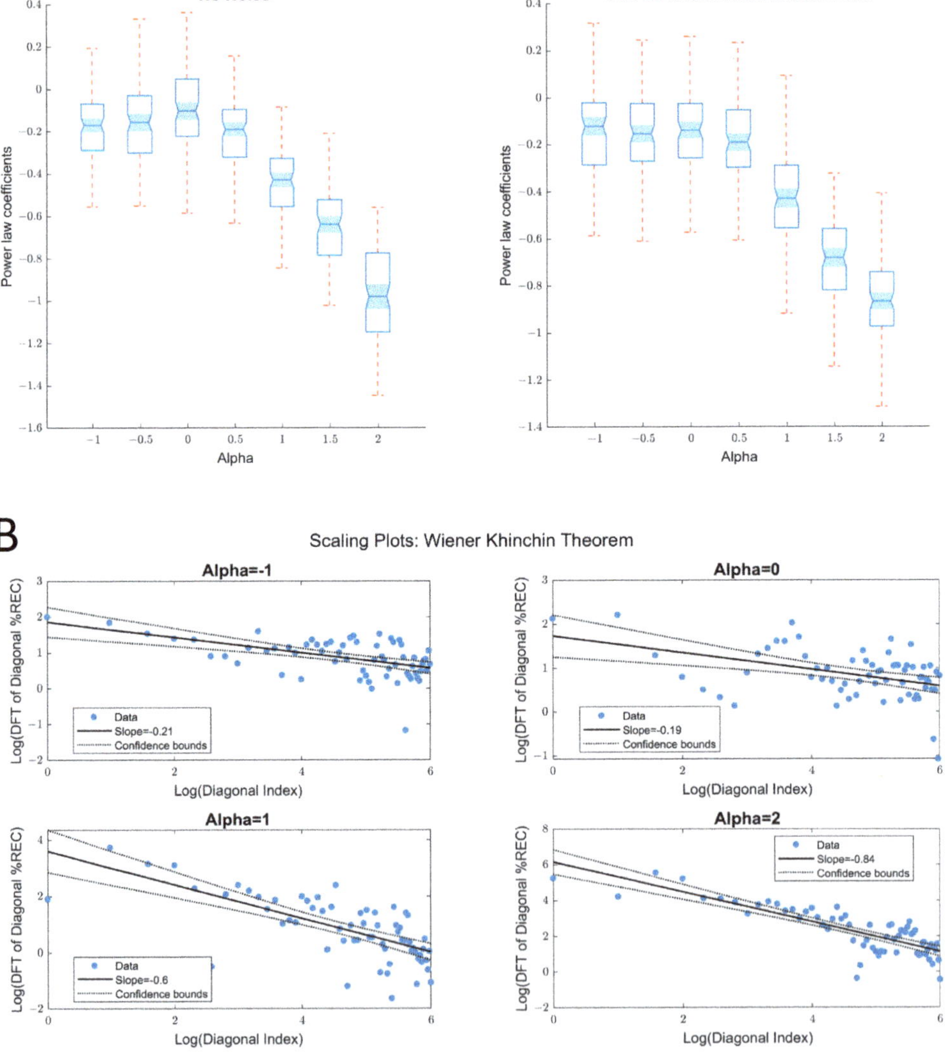

Figure 9. Results of Wiener–Khinchin theorem. (**A**) Left panel: box plots of the true α-values on the x-axis and the power-law coefficients for the association of FT of the diagonal $\%REC$ and the diagonal

index on the y-axis. As can be seen, the coefficient scales well with the true α-values for the persistent fluctuations ($\alpha > 0$). Right panel: box plots of the true α-values on the x-axis and the power-law coefficients for the association of FT of the diagonal %*REC* and the diagonal index on the y-axis, when random noise (SNR = 2:1) is added. The relation of the resulting coefficients to the true α-values is not as good for persistent fluctuations (cannot differentiate $\alpha = 0$ and 0.5) and is relatively insensitive to distinguishing between different types of antipersistent fluctuations. (**B**) Example of scaling plots demonstrating the association of FT of the diagonal %*REC* and diagonal index for different α values.

2.1.5. Fourth Approach: Consecutive Diagonals Recurrence Ratio *(Diag ratio)*

Until this point, the analysis techniques were more effective for persistence signals and did not distinguish between antipersistent signals well. Approach number four solves this issue to some degree. Similar to the third approach, we utilized the recurrence percentage of the diagonal lines. Here, however, we calculate the ratio between each couple of consecutive diagonal lines (Figure 7). The rationale behind the approach is that antipersistent fluctuations will tend to yield oscillations at high frequencies, and the ratio of recurrence rate of adjacent diagonals in the recurrence plot will capture the magnitude of such oscillations. Just as with the laminarity measure, however, this method is more of a correlate of antipersistent fractal scaling, and does not capture scaling properties directly.

As seen in Figure 10, with this measure, we can differentiate negative α-values (antipersistent) from $\alpha = 0$, both with and without external noise. However, the method does not distinguish between the different alpha values of the persistent fluctuations.

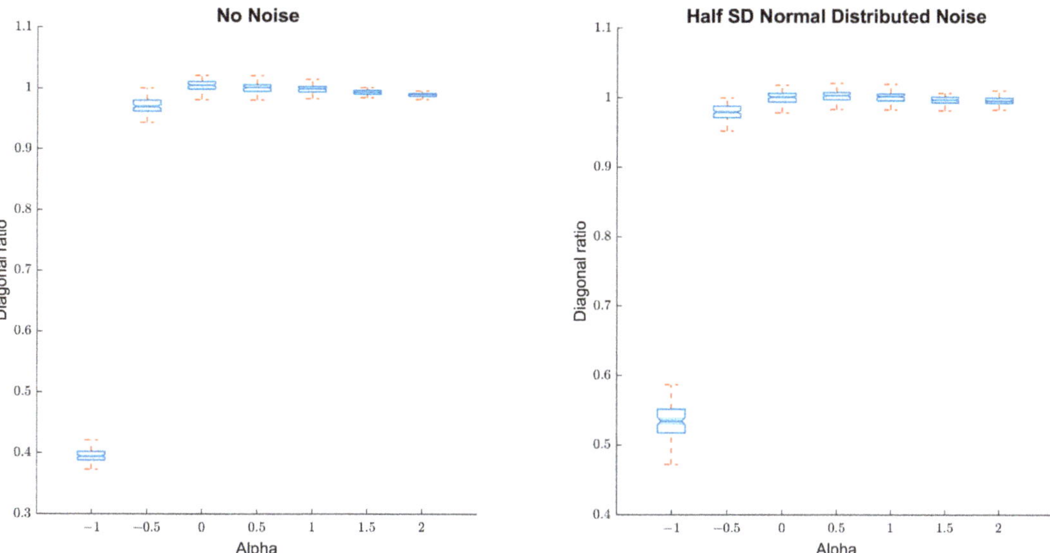

Figure 10. Results of approach 4 (consecutive diagonals %*REC* ratio). Left panel: box plots of the true alpha values on the x-axis and the mean ratio between subsequent diagonals' %*REC* on the y-axis. As can be seen, the coefficient scales well with the true α-values for antipersistent fluctuations ($\alpha < 0$) and converges to 1 from $\alpha = 0$. Right panel: box plots of the true α-values on the x-axis and the mean ratio between subsequent diagonals' %*REC* on the y-axis when random noise (SNR 2:1) is added. Still, the resulted coefficients scale well for negative alpha value ($R^2 = 0.79$).

2.2. Empirical Example

The approaches were tested on a dataset of a tapping experiment during which participants listened to a certain beat and were then instructed to tap according to the tempo they had heard. Under one of the two within-participant conditions, participants received visual feedback on every trial to help them align their tapping performance with the target tempo, while in the other condition no such feedback was provided. The sample was comprised of 36 time series from 18 participants with at about 1000 tapping intervals per time series.

Drawing on previous research on cognitive processes, we expected the time series to show persistent fractal fluctuation. Moreover, previous research showed that receiving feedback would reduce long-range dependencies in the data related to cognitive-motor processes of timing, and hence yield a more random ('whiter') noise manifested by a lower α exponent [39]. Our findings, displayed in Figure 11, support these expectations in several ways. Firstly, a negative power-law coefficient in approach 3 along with a ~1 ratio between subsequent diagonals (approach 4) indicate a persistent fluctuation in both conditions and is supported by a Hurst exponent $1.0 > H > 0.5$, suggesting a pinkish noise. Further, approaches 1–3, as well as Wiener–Khinchin theorem's results, imply a lower α-exponent for the feedback condition (see Table 1). While *SD %REC* and *%LAM* exhibit it by presenting a higher clustering characteristic for the no-feedback condition, *Diag %REC* and the Wiener–Khinchin theorem display it with a stronger lower frequency dominance when no feedback is given.

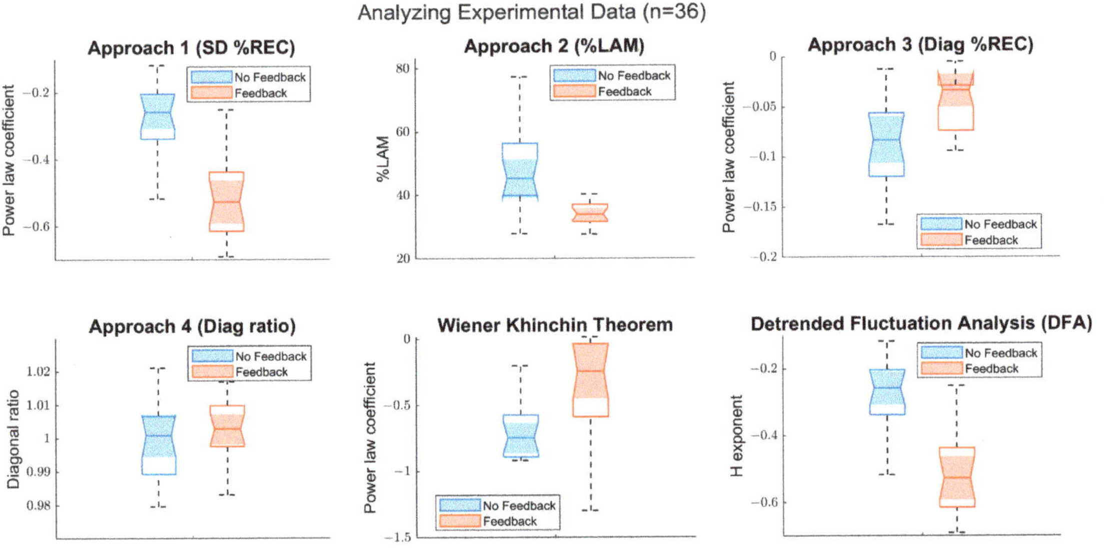

Figure 11. Box plots illustrating the outcomes of feedback and no-feedback conditions: Six sets of box plots represent a comparison between the outcomes of each approach for feedback (orange) and no-feedback (blue) conditions. While the frame of the boxplot is defined by the interquartile range, the notch represents a 95% confidence interval and the whiskers show the maximum and the minimum of each distribution (except outliers). As expected, due to its persistent noise characteristics ($\alpha > 0$), behavioral data would be appropriately analyzed by approaches 1–3 but not approach 4. Approaches 1 and 2, as well as DFA, yield higher results for the no-feedback condition, indicating a larger α, meaning a more persistent behavior. Likewise, approach 3 and Wiener–Khinchin theorem suggest a lower frequency dominancy in the no-feedback condition.

Table 1. Paired *t*-test results comparing the outcomes of feedback and no-feedback conditions.

Approach	*t*	*df*	*p*
1—*SD %REC*	−5.12	17	>0.001
2—*%LAM*	−3.33	17	0.004
3—*Diag %REC*	3.43	17	0.003
4—*Diag ratio*	0.82	17	0.42
Wiener–Khinchin theorem	3.68	17	0.002
DFA	−4.08	17	>0.001

2.3. *Comparison of the Approaches*

To evaluate the presented approaches in relation to the true alpha values of the generated time series, we focus on three main parameters: (a) fractal dimension range, (b) sensitivity to noise, and (c) summary of the quantitative relation to the true alpha values. Furthermore, we investigated their applicability to empirical behavioral data.

2.3.1. Range

As presented above, approaches 1 (*SD %REC*) and the Wiener–Khinchin-based analysis are sensitive to persistent fluctuations. Conversely, approach 4 (*Diag ratio*) differentiates only antipersistent fluctuations, whereas approaches 2 and 3 (*%LAM*, *Diag %REC*) are applicable throughout the whole tested range ($-1 < \alpha < 2$), like DFA. Hence, with no estimation of the time series' fractal dimension, one should conduct an analysis according to approaches 2 or 3, otherwise the researcher might prefer to pick the analysis technique that best fits his data's characteristics. On a similar note, one can try to detect whether there are persistent fluctuations using approach 4, which yields a ~1 ratio for $\alpha \geq 0$.

2.3.2. Robustness to Noise

Most of the analysis techniques that were applied were robust to noise. Except for the Wiener–Khinchin theorem approach, the rest distinguished between α-values within their range comparably with and without noise. Nevertheless, antipersistent fluctuations were less distinguishable by both approach 3 and DFA when i.i.d. noise (SNR = 2:1) was introduced.

2.3.3. Quantitative Relation to True Alpha Values

Table 2 provides a summary of the R^2-values that capture the relation between the true α-values and the estimated parameters of the different approaches, separately for persistent and antipersistent fluctuations. As DFA is the gold standard for fractal analyses in time-series methods, the comparison of the recurrence-based approaches to DFA is of particular interest here. Comparing the likelihoods of the linear models of each of our four approaches to DFA, we found that the association between the true α values and Hurst exponent is significantly stronger than in almost every other method ($\alpha < 0.05$). On the contrary, approaches 2 and 4 yielded significantly higher association (than DFA) with the true α values for antipersistent fluctuation when noise is introduced, but somewhat below DFA under the no-noise condition. Nevertheless, approaches 1–3 show similar R^2 to DFA when analyzing persistent noise. It has to be kept in mind that the sample sizes here are quite large, and tests of significance are of limited value in this case.

2.3.4. Applicability

All approaches were found applicable to behavioral data and concluded conformally despite the small sample size. The utilized data were most likely to behave persistently and hence were out of the fourth approach range. Yet, we suggest using approach 4 to confirm whether the time series is persistent or not (persistent fluctuations are indicated by a 1:1 ratio between subsequent diagonals). Table 3 provides an overview of the R^2 for the different approaches (including DFA) when comparing the two time-estimation groups (i.e., with and without feedback). In a model comparison, approaches 2–4 were less predictive than DFA ($\alpha < 0.05$), while approach 1 was not significantly lower.

Table 2. Comparison of approaches on simulated data.

Approach	R^2—*Persistent (No Noise)*	R^2—*Antipersistent (No Noise)*	R^2—*Persistent (with Noise)*	R^2—*Antipersistent (with Noise)*
1—*SD %REC*	0.9	0.08	0.9	0.04
2—*%LAM*	0.97	0.82	0.95	0.81
3—*Diag %REC*	0.93	0.38	0.93	0.33
4—*Diag ratio*	0.33	0.79	0.06	0.78
Wiener–Khinchin theorem	0.7	0.01	0.68	0.002
DFA	0.99	0.99	0.98	0.68

Table 3. Comparison of approaches on empirical data.

Approach	R^2—*with* vs. *without Feedback*
1—*SD %REC*	0.33
2—*%LAM*	0.11
3—*Diag %REC*	0.25
4—*Diag ratio*	0
Wiener–Khinchin theorem	0.23
DFA	0.36

3. Conclusions

In the current paper, we presented and compared several recurrence-based approaches to quantify the strength of monofractal autocorrelations in time-series data. This is a major step forward for integrating the quantification of scaling properties into recurrence quantification analysis, as previous research has suggested that such analyses are theoretically possible (e.g., [28]), but did not point to concrete means for how to deduce such properties. The proposed methods differ in quality, as well as in the range of applicability to particular types of colored noise, as we have shown on synthetic and empirical data. Based on our results, we recommend using approaches 3 and 4 to determine whether the data are persistent, antipersistent, or white noise. Then, approaches 1, 2, and 3 would be suitable to compare the fractal dimensionality of persistent data, while approaches 2 and 4 would fit antipersistent time series.

Thus, the present work lays the foundations for integrating fractal analysis into an RQA framework, and defining appropriate recurrence-based quantifies. Moreover, these methods might be amenable to quantifying time-dependent fractal fluctuations of not only univariate time series, but also strange attractor profiles, which possess fractal properties and are readily analyzable within the framework of recurrence quantification analysis [28]. In the future, these methods could be extended to capturing fractal dimensions in multidimensional systems via multidimensional recurrence quantification analysis. In addition, an evaluation and adaptation of these approaches to multifractals would be valuable [40,41].

Author Contributions: Conceptualization, A.T. and S.W.; methodology, A.T.; formal analysis, A.T. and S.W.; investigation, A.T., S.W. and G.L.; writing—original draft preparation, A.T., S.W. and G.L.; writing—review and editing, A.T., S.W. and G.L.; All authors have read and agreed to the published version of the manuscript.

Funding: S.W. acknowledges funding from the German Science Foundation (DFG; grant number 442405852).

Institutional Review Board Statement: The collection of the timing data was approved by the institutional review board of Leuphana University of Lüneburg ("EB-Antrag_202202-03-Wallot_Timing Distraction").

Data Availability Statement: The data is available upon request from the corresponding author. Also, a Matlab function (The MathWorks, Inc.) to perform the four analyses is available on GitHub, see Appendix A.

Acknowledgments: We thank Stine Hollah and Gerke Feindt for collecting the timing data presented in this manuscript. S.W. acknowledges funding by the German Research Foundation (DFG; grant numbers 442405852 and 442405919).

Conflicts of Interest: The authors declare no conflict of interest.

Appendix A

A Matlab (The MathWorks, Inc.) implementation of the four approaches is available on GitHub: https://github.com/alontom/FARQA. Accessed on 19 July 2022.

References

1. Gilden, D.L.; Thornton, T.; Mallon, M.W. 1/f noise in human cognition. *Science* **1995**, *267*, 1837–1839. [CrossRef] [PubMed]
2. Delignières, D.; Fortes, M.; Ninot, G. The fractal dynamics of self-esteem and physical self. *Nonlinear Dyn. Psychol. Life Sci.* **2004**, *8*, 479–510.
3. Goldberger, A.L.; Amaral, L.A.N.; Hausdorff, J.M.; Ivanov, P.C.; Peng, C.K.; Stanley, H.E. Fractal dynamics in physiology: Alterations with disease and aging. *Proc. Natl. Acad. Sci. USA* **2002**, *99* (Suppl. 1), 2466–2472. [CrossRef] [PubMed]
4. Kello, C.T.; Anderson, G.G.; Holden, J.G.; van Orden, G.C. The Pervasiveness of 1/f Scaling in Speech Reflects the Metastable Basis of Cognition. *Cogn. Sci.* **2008**, *32*, 1217–1231. [CrossRef]
5. Miller, K.J.; Sorensen, L.B.; Ojemann, J.G.; den Nijs, M. Power-Law Scaling in the Brain Surface Electric Potential. *PLOS Comput. Biol.* **2009**, *5*, e1000609. [CrossRef]
6. Nobukawa, S.; Yamanishi, T.; Nishimura, H.; Wada, Y.; Kikuchi, M.; Takahashi, T. Atypical temporal-scale-specific fractal changes in Alzheimer's disease EEG and their relevance to cognitive decline. *Cogn. Neurodyn.* **2019**, *13*, 1–11. [CrossRef]
7. Shelhamer, M.; Joiner, W.M. Saccades exhibit abrupt transition between reactive and predictive, predictive saccade sequences have long-term correlations. *J. Neurophysiol.* **2003**, *90*, 2763–2769. [CrossRef]
8. Wallot, S.; Coey, C.A.; Richardson, M.J. Cue predictability changes scaling in eye-movement fluctuations. *Atten. Percept. Psychophys.* **2015**, *77*, 2169–2180. [CrossRef]
9. Wijnants, M.L.; Cox, R.F.A.; Hasselman, F.; Bosman, A.M.T.; van Orden, G. A trade-off study revealing nested timescales of constraint. *Front. Physiol.* **2012**, *3*, 116. [CrossRef]
10. Wijnants, M.L.; Hasselman, F.; Cox, R.F.A.; Bosman, A.M.T.; van Orden, G. An interaction-dominant perspective on reading fluency and dyslexia. *Ann. Dyslexia* **2012**, *62*, 100–119. [CrossRef]
11. Wiltshire, T.J.; Euler, M.J.; McKinney, T.L.; Butner, J.E. Changes in dimensionality and fractal scaling suggest soft-assembled dynamics in human EEG. *Front. Physiol.* **2017**, *8*, 633. [CrossRef] [PubMed]
12. Delignières, D.; Marmelat, V. Fractal Fluctuations and Complexity: Current Debates and Future Challenges. *Crit. Rev. Biomed. Eng.* **2012**, *40*, 485–500. [CrossRef] [PubMed]
13. Farrell, S.; Wagenmakers, E.J.; Ratcliff, R. 1/f noise in human cognition: Is it ubiquitous, and what does it mean? *Psychon. Bull. Rev.* **2006**, *13*, 737–741. [CrossRef] [PubMed]
14. Holden, J.G.; Choi, I.; Amazeen, P.G.; van Orden, G. Fractal 1/f dynamics suggest entanglement of measurement and human performance. *J. Exp. Psychol. Hum. Percept. Perform.* **2011**, *37*, 935–948. [CrossRef]
15. Kelty-Stephen, D.G.; Wallot, S. Multifractality Versus (Mono-) Fractality as Evidence of Nonlinear Interactions Across Timescales: Disentangling the Belief in Nonlinearity From the Diagnosis of Nonlinearity in Empirical Data. *Ecol. Psychol.* **2017**, *29*, 259–299. [CrossRef]
16. Kloos, H.; van Orden, G. Voluntary Behavior in Cognitive and Motor Tasks. *Mind Matter* **2010**, *8*, 19–43.
17. Van Orden, G.C.; Holden, J.G.; Turvey, M.T. Self-organization of cognitive performance. *J. Exp. Psychol. Gen.* **2003**, *132*, 331–350. [CrossRef]
18. Wagenmakers, E.J.; Farrell, S.; Ratcliff, R. Estimation and interpretation of $1/f\alpha$ noise in human cognition. *Psychon. Bull. Rev.* **2004**, *11*, 579. [CrossRef]
19. Wagenmakers, E.J.; Farrell, S.; Ratcliff, R. Human Cognition and a Pile of Sand: A Discussion on Serial Correlations and Self-Organized Criticality. *J. Exp. Psychol. Gen.* **2005**, *134*, 108. [CrossRef]
20. Gilden, D.L. Cognitive emissions of 1/f noise. *Psychol. Rev.* **2001**, *108*, 33–56. [CrossRef]
21. Holden, J. Gauging the fractal dimension of response times from cognitive tasks. *Contemp. Nonlinear Methods Behav. Sci. A Webbook Tutor, 1* **2005**, *1*, 267–318.
22. Peng, C.K.; Havlin, S.; Stanley, H.E.; Goldberger, A.L. Quantification of scaling exponents and crossover phenomena in nonstationary heartbeat time series. *Chaos Interdiscip. J. Nonlinear Sci.* **1998**, *5*, 82. [CrossRef] [PubMed]
23. Riley, M.A.; Bonnette, S.; Kuznetsov, N.; Wallot, S.; Gao, J. A tutorial introduction to adaptive fractal analysis. *Front. Physiol.* **2012**, *3*, 371. [CrossRef] [PubMed]
24. Pilgrim, I.; Taylor, R.P. Fractal Analysis of Time-Series Data Sets: Methods and Challenges. In *Fractal Analysis*; Ouadfeul, S., Ed.; IntechOpen: London, UK, 2019; pp. 5–30.
25. Webber, C.L.; Zbilut, J.P. Dynamical assessment of physiological systems and states using recurrence plot strategies. *J. Appl. Physiol.* **1994**, *76*, 965–973. [CrossRef] [PubMed]

26. Zbilut, J.P.; Webber, C.L. Embeddings and delays as derived from quantification of recurrence plots. *Phys. Lett. A* **1992**, *171*, 199–203. [CrossRef]
27. Wallot, S.; Roepstorff, A.; Mønster, D. Multidimensional recurrence quantification analysis (MdRQA) for the analysis of multidimensional time-series: A software implementation in MATLAB and its application to group-level data in joint action. *Front. Psychol.* **2016**, *7*, 1835. [CrossRef]
28. Webber, C.L. Recurrence quantification of fractal structures. *Front. Physiol.* **2012**, *3*, 382. [CrossRef]
29. Hu, K.; Ivanov, P.C.; Chen, Z.; Carpena, P.; Stanley, H.E. Effect of trends on detrended fluctuation analysis. *Phys. Rev. E* **2001**, *64*, 011114. [CrossRef]
30. Phinyomark, A.; Larracy, R.; Scheme, E. Fractal Analysis of Human Gait Variability via Stride Interval Time Series. *Front. Physiol.* **2020**, *11*, 333. [CrossRef]
31. Ravi, D.K.; Marmelat, V.; Taylor, W.R.; Newell, K.M.; Stergiou, N.; Singh, N.B. Assessing the temporal organization of walking variability: A systematic review and consensus guidelines on detrended fluctuation analysis. *Front. Physiol.* **2020**, *11*, 562. [CrossRef]
32. Little, M.A.; Mcsharry, P.E.; Roberts, S.J.; Ae Costello, D.; Moroz, I.M. Exploiting Nonlinear Recurrence and Fractal Scaling Properties for Voice Disorder Detection. *Nat. Preced.* **2007**. [CrossRef]
33. Marwan, N.; Carmen Romano, M.; Thiel, M.; Kurths, J. Recurrence plots for the analysis of complex systems. *Phys. Rep.* **2007**, *438*, 237–329. [CrossRef]
34. Dale, R.; Warlaumont, A.S.; Richardson, D.C. Nominal cross recurrence as a generalized lag sequential analysis for behavioral streams. *Int. J. Bifurc. Chaos* **2011**, *21*, 1153–1161. [CrossRef]
35. Wallot, S.; Leonardi, G. Analyzing multivariate dynamics using cross-recurrence quantification analysis (CRQA), diagonal-cross-recurrence profiles (DCRP), and multidimensional recurrence quantification analysis (MdRQA)—A tutorial in R. *Front. Psychol.* **2018**, *9*, 2232. [CrossRef]
36. Richardson, D.C.; Dale, R. Looking to understand: The coupling between speakers' and listeners' eye movements and its relationship to discourse comprehension. *Cogn. Sci.* **2005**, *29*, 1045–1060. [CrossRef]
37. Granger, C.W.J.; Joyeux, R. An Introduction To Long-Memory Time Series Models And Fractional Differencing. *J. Time Ser. Anal.* **1980**, *1*, 15–29. [CrossRef]
38. Zbilut, J.P.; Marwan, N. The Wiener-Khinchin theorem and recurrence quantification. *Phys. Lett. Sect. A Gen. At. Solid State Phys.* **2008**, *372*, 6622–6626. [CrossRef]
39. Kuznetsov, N.A.; Wallot, S. Effects of accuracy feedback on fractal characteristics of time estimation. *Front. Integr. Neurosci.* **2011**, *5*, 62. [CrossRef]
40. Dixon, J.A.; Holden, J.G.; Mirman, D.; Stephen, D.G. Multifractal Dynamics in the Emergence of Cognitive Structure. *Top. Cogn. Sci.* **2012**, *4*, 51–62. [CrossRef]
41. Kelty-Stephen, D.G.; Palatinus, K.; Saltzman, E.; Dixon, J.A. A Tutorial on Multifractality, Cascades, and Interactivity for Empirical Time Series in Ecological Science. *Ecol. Psychol.* **2013**, *25*, 1–62. [CrossRef]

Article

Cross-Correlation- and Entropy-Based Measures of Movement Synchrony: Non-Convergence of Measures Leads to Different Associations with Depressive Symptoms

Uwe Altmann [1,*], Bernhard Strauss [1] and Wolfgang Tschacher [2]

1 Institute of Psychosocial Medicine, Psychotherapy and Psycho-Oncology, Jena University Hospital, D-07743 Jena, Germany
2 Department of Experimental Psychology, University Hospital of Psychiatry and Psychotherapy, CH-3060 Bern, Switzerland
* Correspondence: uwe.altmann@med.uni-jena.de

Abstract: Background: Several algorithms have been proposed to quantify synchronization. However, little is known about their convergent and predictive validity. Methods: The sample included 30 persons who completed a manualized interview focusing on psychosomatic symptoms. The intensity of body motions was measured using motion-energy analysis. We computed several measures of movement synchrony based on the time series of the interviewer and participant: mutual information, windowed cross-recurrence analysis, cross-correlation, rMEA, SUSY, SUCO, WCLC–PP and WCLR–PP. Depressive symptoms were assessed with the Patient Health Questionnaire (PHQ9). Results: According to the explorative factor analyses, all the variants of cross-correlation and all the measures of SUSY, SUCO and rMEA–WCC led to similar synchrony measures and could be assigned to the same factor. All the mutual-information measures, rMEA–WCLC, WCLC–PP–F, WCLC–PP–R2, WCLR–PP–F, and WinCRQA–DET loaded on the second factor. Depressive symptoms correlated negatively with WCLC–PP–F and WCLR–PP–F and positively with rMEA–WCC, SUCO–ES–CO, and MI–Z. Conclusion: More standardization efforts are needed because different synchrony measures have little convergent validity, which can lead to contradictory conclusions concerning associations between depressive symptoms and movement synchrony using the same dataset.

Keywords: nonverbal communication; movement synchrony; time-series analysis; validity; depression

Citation: Altmann, U.; Strauss, B.; Tschacher, W. Cross-Correlation- and Entropy-Based Measures of Movement Synchrony: Non-Convergence of Measures Leads to Different Associations with Depressive Symptoms. *Entropy* **2022**, *24*, 1307. https://doi.org/10.3390/e24091307

Academic Editor: Philip Broadbridge

Received: 16 July 2022
Accepted: 7 September 2022
Published: 15 September 2022

Publisher's Note: MDPI stays neutral with regard to jurisdictional claims in published maps and institutional affiliations.

1. Introduction

Processes relevant in psychotherapy can be located on different time scales ranging from neuronal processes that change within milliseconds, to affective and interpersonal processes representing single sessions, to between-session changes of mood stages [1–3]. Both bottom-up effects (e.g., when patient–therapist interactions have impacts on patient's mood) and top-down effects (e.g., mood affecting the kind of interpersonal interaction) are assumed [1]. The core of psychotherapy process is generally considered to rest in the exchanges between the patient and therapist, which consist of verbal–semantic and nonverbal aspects.

The nonverbal interaction of patient and therapist may be understood as a coupled dynamical system [4–8]. Each sub-system (the patient's or the therapist's) obeys its own eigen-dynamics and coupling dynamics. The eigen-dynamic is constituted by an actor's ability to perceive and process information and act accordingly (see Figure 1 left). The coupling dynamic refers to the mutual influence between the patient and therapist, may be asymmetrical (e.g., the therapist affecting the patient's state more than vice versa) and may change during the interpersonal interaction (e.g., at the beginning of a session, the patient influences the therapist, whereas at the end the influence is reversed). There are two different understandings of coupling dynamics. One is that the degree of coupling changes more or less smoothly over time [9–11], the other regards the coupling dynamics as an on–off process whereby phases with no or very weak coupling (no synchronization visible) may alternate with strongly coupled phases [5,8]. The former dynamics may be assumed in more stationary processes (e.g., physiological data), whereas the latter on–off dynamics occur in behavioral processes with non-stationary bursts (e.g., movement activity). Phases of strong coupling are characterized by a synchronization of sub-system states and are hence called synchronization intervals [4,12] or mimicry episodes [13]. The person who acts as the driver during the coupling is called leader, and the person who follows/imitates is the driven (see Figure 1 right).

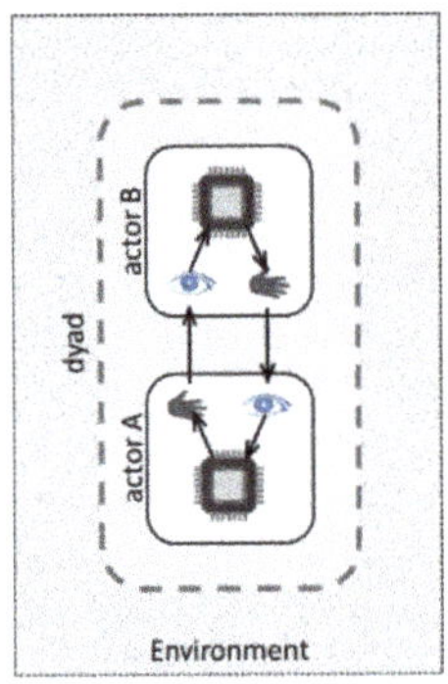

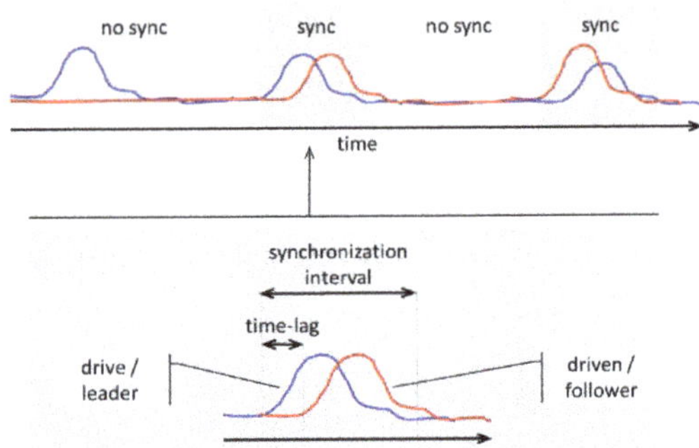

Figure 1. Schematic illustration of a dyad as coupled dynamical system (**left**) and hypothetical motion activity of two interactants with synchronization intervals (**right**).

1.1. Synchronization in Patient–Therapist Interaction

Psychotherapy research has investigated the synchronization of physiological parameters [14,15], body movements [5,11], facial expressions [16], prosodic cues [17–19], and language style [20,21]. Many studies have investigated the relationship between (nonverbal) synchronization and therapeutic success. According to the systematic review of Wiltshire et al. [22], physiological synchrony was most frequently correlated with empathy and language, vocal synchrony with therapeutic alliance, and movement synchrony with psychotherapy outcomes. This review supported the InSync model of psychotherapy [3,23], which assumes that (nonverbal) synchrony in patient–therapist interaction affects the emotion regulation of patients (as a top-down effect at medium/tonic to longer/chronic time scales) as well as the therapeutic relationship and, as a consequence, therapeutic success (bottom-up effect at tonic and fast/phasic time scales).

Other psychotherapy studies have investigated (nonverbal) synchrony in interpersonal interaction as a diagnostic feature of mental disorders. Multiple studies suggested, for example, that attenuated nonverbal synchrony was linked with depressive symptoms [16,24–26]. These findings correspond with neurophysiological [27,28] and interpersonal models of depression [29]. The former explains changes in emotion regulation and interpersonal interaction (e.g., less smiling or movements) by disorder-related changes in neurophysiological processes (e.g., dysfunctions in the left frontal hemisphere of the brain) [30,31]. Interpersonal models of depression [29] assume that depressed persons induce a negative mood in their conversation partners, thus provoking negative responses, which in turn confirm the negative expectations of the depressed person. Accordingly, in interviews with depressed patients, the interviewers synchronize their nonverbal behavior less often.

It may be noted, however, that the findings are not homogeneous. Some studies did not find significant associations between synchrony and therapeutic outcome [17,32], or even reported that higher synchrony was related to higher symptom levels [33,34]. This was also true for synchrony as a diagnostic feature, e.g., when more vocal synchrony was correlated with higher anxiety symptoms in the study of [35].

From a methodological point of view, an explanation of the conflicting results may be that nonverbal synchrony was measured differently. Thus, researchers may have addressed different aspects, or even different concepts, of synchrony, which may have resulted in varying correlations between synchrony and therapeutic outcome as well as symptom load [36]. This unsettled state of research has motivated the present study on the validity of different synchronization measures.

1.2. Measures of Synchronization and Its Convergent Validity

Various statistical methods may be used to estimate the average degree of coupling (e.g., [14,37]) or identify synchronization intervals (e.g., [12]). Despite multiple overviews of methods applied in psychotherapy research [10,36,38] so far, there are only few studies on the validity of synchrony measures.

First, it should be noted that synchrony measures depend on the parameter settings of algorithms. Ramseyer and Tschacher [39], Schoenherr et al. [40] and Behrens et al. [41] applied windowed cross-lagged correlation algorithms multiple times to the same bivariate time series and varied parameters such as degree of smoothing, window size and maximum time lag. Among other things, they showed that smaller windows [39–41], non-transformed movement data and slight smoothing [40] lead to higher synchronization values and higher correlations between synchrony and therapeutic alliance, respectively [39]. The application of a pseudo-synchrony approach [42] also affects the measured synchrony. For each real-world time-series pair, Moulder et al. [43] generated multiple artificial time-series pairs by (a) shuffling participants, (b) shuffling time-series segments within a dyad, (c) shuffling measurement points within a dyad and (d) reversing one of the time series in the pair. They found that the decision as to whether synchrony was present in a time-series pair strongly depended on the applied shuffling method. All these findings imply that the validity of synchrony measures depends on the parameter settings of an algorithm.

Regarding the validity of synchrony measures, one should distinguish different kinds of validity. Predictive validity is present when a synchrony measure predicts an external criterion in accordance with theoretical expectations, as was investigated by [36,39,44,45]. Feniger-Schaal, Schoenherr, Altmann and Strauss [44] applied windowed cross-lagged correlation (WCLC) with peak picking (PP) by [12] to motion time series that were captured in a "mirror game". In the first phase of the mirror game, the study assistant mirrored the movements of a participant. In the second phase, the leading–following roles were switched. In the third phase, these roles were not predetermined. In concordance with instructions, the algorithm measured more synchrony with the participant leading in the first study phase, and more synchrony with the assistant leading in the second phase. In the study of Luehof [45], WCLC with PP by [9], windowed cross-lagged regression (WCLR) with PP by [4,12], and recurrence quantification analysis (RQA) were able to discriminate between interviews with rapport-trained interviewers and control interviewers, finding more synchrony with the trained interviewers. The WCLR–PP showed the best discrimination. Schoenherr et al. [36] used therapeutic success as the criterion to be predicted by synchrony. They found that only windowed cross-correlation (WCC), WCLC–PP and WCLR–PP correlated significantly in the expected direction with therapeutic success.

Schoenherr, Paulick, Deisenhofer, Schwartz, Rubel, Lutz, Strauss and Altmann [40] studied the criterion validity of synchronization measures using artificially generated time-series pairs that contained a single synchronization interval. The WCLC–PP and WCLR–PP by [4,12] were applied to each time-series pair and correct identifications of the synchronization interval (the criterion) were counted. The best concordance in terms of the average Cohen's κ was observed for both WCLC–PP and WCLR–PP with window widths of 3 and 5 s, non-transformed movement data and slight smoothing. When applying the algorithms to real motion time series with isolated synchronization intervals (no other movement activity before or after the synchronization interval), the identification rate varied between moderate and substantial Cohen's κ, depending on the parameter settings.

Validity defined as congruence between different measures was investigated by Schoenherr, Paulick, Worrack, Strauss, Rubel, Schwartz, Deisenhofer, Lutz and Altmann [36], Luehof [45] and Tschacher and Meier [14]. They applied multiple algorithms to naturalistic bivariate time series and determined convergent validity by the correlations between different synchrony measures. In their study of physiological synchrony, Tschacher and Meier [14] found little or no inter-correlations between SUSY-ES_{abs}, SUSY-ES_{noabs} and the SUCO algorithm. In a study of movement synchrony, Luehof [45] found no concordance between the synchrony measures of WCLC–PP by [9], WCLR–PP by [4,12], and recurrence quantification analysis (RQA). In the study of Schoenherr, Paulick, Worrack, Strauss, Rubel, Schwartz, Deisenhofer, Lutz and Altmann [36], cross-lagged correlation (CLC), cross-lagged regression (CLR), windowed cross-correlation (WCC), windowed cross-lagged correlation (WCLC) by [37], WCLC by [32], WCLC–PP and WCLR–PP by [4,12], and cross-recurrence quantification analysis (CRQA) by [46] were conducted. The correlation between two synchrony measures ranged from not present ($r \approx 0$) to almost perfect ($r \approx 1$). In a subsequent exploratory factor analysis, CLC, WCLC by [37], and WCLC by [32] formed a factor of highly correlated synchrony measures. The second factor loaded on average cross-correlation within the synchronization intervals assessed with WCLC–PP and WCLR–PP by [12]. The third factor consisted of non-linear synchrony such as CRQA and the frequency of synchrony of WCLC–PP and WCLR–PP by [12]. Schoenherr et al. [36] concluded that the examined algorithms did not measure the same kind of synchrony and that different measures predicted different effects on therapeutic outcome.

1.3. Research Question

The present study explored the convergent validity and predictive validity of cross-correlation- and entropy-based measures of movement synchrony. We used data from a pilot study on nonverbal communication in depressive patients and healthy controls [16,47,48]. The primary study focused on the evaluation of feasibility of recruitment, assessment procedures, automatic coding of nonverbal behavior and provided first empirical results on the differences between patients with depression and healthy controls in terms of body motion, facial expressions and prosody. In the present secondary analysis, we addressed a methodological question: the validity of movement synchronization measures. For this purpose, multiple algorithms measuring synchronization were applied to motion times series of participants and interviewers. The convergent validity was examined by correlations between the synchrony measures. According to [36], we assume weak convergent validity in terms of low correlations between different synchrony measures, and that some measures can be assigned to different facets of synchrony measures. As in [36], we conducted an exploratory factor analysis to identify the facets of synchronization. Due to the fact that the distribution of synchrony measures is non-normal, we conducted a minimum rank factor analysis, which is more appropriate for non-normally distributed data. The predictive validity was investigated by comparing the synchrony measures in patients with major depression and in healthy controls as well as by the correlation between synchrony measures and symptom load, which was assessed with questionnaires. According to the literature mentioned above, movement synchrony was expected to be lower in the interview dyads with depressive patients.

To our knowledge, the present study is the second peer-reviewed study on the convergent validity of synchronization measures applied in clinical research. In the first study [36], the synchronization of the patient and psychotherapist in an early therapy session was investigated. In comparison to [36], we applied additional synchronization measures, especially measures based on information theory, and the homogeneity of the interactions was given (manual-guided interviews rather than therapy sessions addressing patient-specific conversation topics in [36]), and predicted criterion and synchrony were much closer in terms of time (the criterion—depression—was assessed before the interviews rather than measured weeks after the sessions, as was true for the criterion—reduction of interpersonal problems—in [36]).

2. Materials and Methods

2.1. Sample of Participants

The sample included 15 inpatients with major depression and 15 healthy controls matched by age and gender, thus groups did not differ regarding mean age and gender distribution. The age range was 20 to 30 years. Of the 30 participants 40% were female. Table 1 gives a short description of both groups, which showed no group differences regarding further socio-demographic characteristics. Patients with depression reported higher degrees of depressive and anxiety symptoms. For a detailed description of inclusion criteria, recruitment, and group characteristics, see the primary study [16].

Table 1. Description of included study subjects.

	All ($N_{\text{Persons}} = 30$)	Healthy Controls ($N_{\text{Persons}} = 15$)	Depressive Patients ($N_{\text{Persons}} = 15$)	p-Value
		Socio-demographic characteristics		
Age in years	25.2 (3.14)	25.5 (3.25)	24.9 (3.10)	0.6091
Gender				1.0000
Male	18 (60.0%)	9 (60.0%)	9 (60.0%)	
Female	12 (40.0%)	6 (40.0%)	6 (40.0%)	
Education				0.1686
No high-school degree	6 (20.0%)	1 (6.67%)	5 (33.3%)	
High-school degree	24 (80.0%)	14 (93.3%)	10 (66.7%)	
Partner status				0.6817
Without partner	22 (73.3%)	10 (66.7%)	12 (80.0%)	
In steady relationship	8 (26.7%)	5 (33.3%)	3 (20.0%)	
		Questionnaires (pre interview)		
Depressive symptoms (PHQ9)	9.43 (7.10)	3.40 (2.44)	15.5 (4.52)	<0.0001
Anxiety symptoms (GAD7)	6.80 (5.76)	1.73 (1.71)	11.9 (3.29)	<0.0001

Note: For continuous variables average and standard deviation are reported and for categorical variables frequency and percentage. For categorical variables a chi-squared or exact Fisher test was applied (the latter, when one or more expected cell frequencies were less than 5). For continuous variables we used a t-test or Kruskall–Wallis test (the latter for non-normally distributed data). For more details see [16]. N denotes the number of persons.

2.2. Instruments

Prior to the interviews, several questionnaires were administered. We assessed the degree of depressive symptoms using the corresponding scale of the Patient Health Questionnaire (PHQ9) [49]. A sum score of 0–4 is interpreted as no or minimal, 5–9 as mild, 10–14 as moderate, and 15–27 as severe depressive symptoms. The degree of anxiety symptoms was measured with the Generalized Anxiety Disorder Scale (GAD7) [49]. The values 0–4 are interpreted as no or minimal, 5–9 as mild, 10–14 as moderate, and 15–21 as severe anxiety symptoms. Both sum scores have an acceptable internal consistency (Cronbach's $\alpha > 0.8$). Further questionnaires were assessed in the primary study but not used in the present study.

2.3. Interviews of Patients and Controls

The focus of the interviews was on somatic complaints, which may be present in healthy participants as well, similar to anamnestic interviews regarding somatoform disorders (SCID-I, section G) [50]. Example questions were: "Have you been ill during the last few years?", or "Have you had any significant problems with headaches?". At the beginning of interviews, the interviewer asked warm-up questions (e.g., "Did you find your way to the site easily?") to allow the interviewee to become accustomed to the recording situation (cameras, etc.).

We used two cameras to record a frontal view of each person. Both recordings were subsequently synchronized by means of a film clapperboard and merged into a split-screen video. Interviews were held in a neutral counseling room where the interviewee and interviewer sat across each other at a table on identical chairs. The interviews were conducted by two female medical students (age ~25 years) in their senior semester. Both interviewers were trained and instructed to adopt a professional and neutral style. Further details on interviews and video recording are reported in [16,48].

2.4. Coding of Motor Activity during the Interviews

Using the interview videos, the motor activity of the interviewees and interviewers was captured using motion-energy analysis, or MEA [51]. We applied the MATLAB© scripts developed by Altmann [4,12], where regions of interest (ROI) can be drawn by hand [5] (free download at https://github.com/10101-00001/MEA, accessed on 15 July 2022). To capture motor activity, the MEA considers all changes of subsequent (t; t + 1) video frames of the recording. First, for each person, a ROI is defined that covers the region in which movements are visible. Inside the ROI, those pixels are counted whose grayscale values change substantially from t to t + 1 (cut-off value: 12 of 256 grayscale degrees). The number of such pixels defines the motion energy of the respective person's ROI at t. For each of the 30 interview videos, we thus generated a bivariate time series that represents in detail (25 measures per second) the visible movement activity (movement of torso, arm, hands, and head of each interlocutor were aggregated to one measure of individual motion energy).

After the MEA, we applied several pre-processing steps. First, each time series was standardized by the size of the corresponding ROI and multiplied by 100. Accordingly, the values of time series ranged from zero (no motion) to 100 (entire ROI activated). Finally, all the time series were smoothed using a moving median with a bandwidth of five frames.

Figure 2 shows, as an example, one pair of motion-energy time series to which the aforementioned preprocessing steps were applied. The time axis is given in frames (25 frames = 1 s). Some algorithms analyze the time series window-wise, e.g., in rMEA und WinCRQA, 1500 frames, or in WCLR–PP, 125 frames. To illustrate what amount of motion activity was captured during such intervals, in Figure 2 we plotted examples of time series segments with length 1500 frames and 125 frames, respectively.

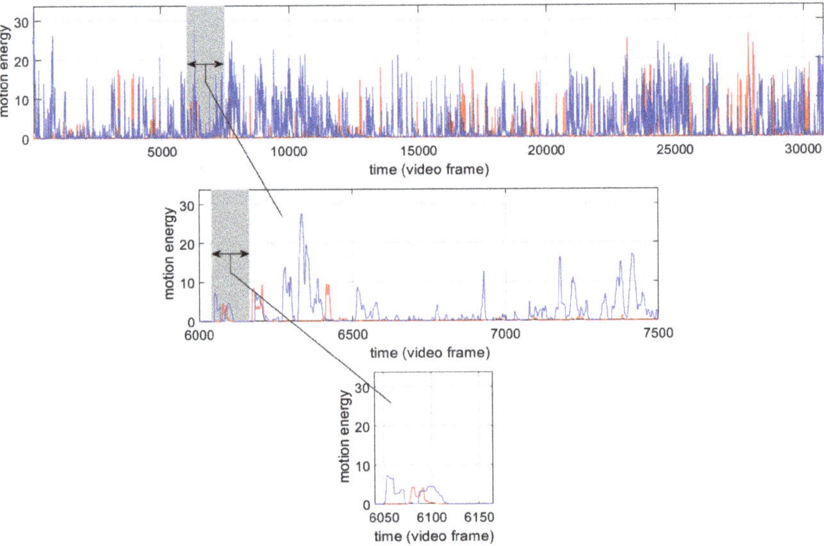

Figure 2. Example of an examined pair of standardized motion-energy time series (participant: red line; interviewer: blue line; 1st row: entire time series; 2nd row: time-series segment with length of 1500 frames; 3rd row: time-series segment with length of 125 frames).

The length of the time series ranged from 10,325 to 42,250 frames (from 413 to 1850 s, respectively; median = 855 s). The interviews of the patients lasted longer than those of controls ($\text{median}_{\text{Patients}}$ = 1276 s, $\text{median}_{\text{Controls}}$ = 759 s, median test $p = 0.001$).

2.5. Measures of Movement Synchrony

The 30 bivariate time series originating from the interviews served as the data input for the synchronization measures that we wished to assess. Features of the measures provided by the algorithms introduced below are summarized in Table 2.

2.5.1. Cross-Correlation

The "simplest" measure of movement synchrony is the cross-correlation (CC) of both time series describing the movements of interlocutors. Please note that in this approach no time lag between both time series is modeled.

Some research has considered the sign problem when the averages of cross-correlations are computed: For example, a time series may include sections with $r = 0.5$ and the same number of sections with $r = -0.5$, so that the aggregated cross-correlation is zero, leading to the conclusion that on average there is no interrelatedness, or no synchrony. To avoid this problem, prior to aggregation, the signs of cross-correlations may be removed by calculating the absolute values (e.g., [37,42,55]), or the coefficient of determination (squaring the correlations) may be used (e.g., [12]). In the latter approach, large cross-correlations will be weighted higher. Furthermore, sometimes Fisher's Z-transformed correlation is considered, because then values are approximately normally distributed. Since a consensus has not been reached, we considered all the options in the present study: raw values of cross-correlations (including negative and positive values when averaging; CC–raw), the absolute values (CC–abs), Fisher's Z-transformed (CC–Z), and squaring of cross-correlations (CC–R2).

2.5.2. rMEA

The R package rMEA [37,52] is based on the work of Ramseyer and Tschacher [11,42] and includes motion capture via MEA as well as (a variant of) windowed cross-lagged correlation (WCLC) to compute the averages of local cross-correlations. Similar to the approach of Boker, Rotondo, Xu and King [9], local associations of both time series are quantified by the cross-lagged correlations of time-series segments—so-called windows. When starting the algorithm, the user has to define the window size (default value: $b = 60$ s, i.e. $60 \cdot 25 = 1500$ respective time points when the video frame rate per second is 25) as well as the maximum time lag (default value: $\tau_{max} = 5$ s, 125 respective time points) which defines the range of the considered time lags. The calculation of WCLC contains three steps. First, a cross-correlation (time lag $\tau = 0$) for a pair of windows with the same starting point (t) is computed. Second, the start position of the reference window is kept constant, whereas the start position of the second window is shifted up to the maximum time lag. In the third phase, the position of the reference window is shifted with an increment of 30 s (default value). Then, the algorithm repeats step 1 and 2 for this start position of the reference window. The result is a matrix whose columns refer to the time lag ($-\tau_{max}, \ldots, 0, \ldots, \tau_{max}$) and rows to the start position of the reference window ($1, \ldots, L - b + 1$; L: time-series length, b: window width). The values of this matrix are the Fisher's Z-transformed coefficients of WCLC (default setting). Before applying the transformation, the absolute values of all cross-correlations are computed to remove the signs (default setting).

Table 2. Features of synchronization measures.

Algorithm/Package	Method	Global or Local Synchrony Estimation	Time Lag	Significance Test	Sign of Correlations	Peak Picking	Output/Synchrony Scores
CC	Cross-correlation	Global	No	No	Positive and negative values	no	Cross-correlation (CC–raw) Fisher's Z-transformed CC (CC–Z) Absolut values of CC (CC–abs) R square of CC (CC–R2)
rMEA by Kleinbub and Ramseyer [52]	Windowed cross-lagged correlation, Fisher's Z-transformed	Local	Yes	Sample shuffling	Absolute values	no	Windowed cross-correlation without time lags (rMEA–WCC) Windowed cross-lagged correlation with time lags (rMEA–WCLC)
SUSY by Tschacher and Haken [53]	Windowed cross-lagged correlation, Fisher's Z-transformed	Local	Yes	Segment shuffling	Absolute values (SUSY–ES_{abs}); positive and negative (SUSY–ES_{noabs})	no	global effect size of absolute values of cross-correlations (SUSY–ES_{abs}) global effect size of cross-correlations (SUSY–ES_{noabs})
SUCO by Tschacher and Meier [14]	Windowed regression, Fisher's Z-transformed	Local	No	Segment shuffling	Absolute values (SUCO–ES_{abs}); positive and negative (SUCO–CO, SUCO–ES–CO)	no	global concordance value (SUCO–CO) global effect size of regressions (SUCO–ES_{abs}) global effect size of concordance values (SUCO–ES–CO)
WCLC–PP by Altmann [4,12]	Windowed cross-lagged correlation and peak-picking algorithm	Local	Yes	R squared difference test with $\alpha = 0.001$	Squared (positive)	yes	Frequency of synchrony relative to conversation duration (WCLC–PP–F), average R2 of WCLC within the synchronization intervals (WCLC–PP–R2)
WCLR–PP by Altmann [4,12]	Windowed cross-lagged regression and peak-picking algorithm	Local	Yes	R squared difference test with $\alpha = 0.001$	Squared (positive)	yes	Frequency of synchrony relative to conversation duration (WCLR–PP–F), average R2 of WCLR within the synchronization intervals (WCLR–PP–R2)
MI by Pardy [54]	Information theory	Global	No	No	not applicable	no	mutual information (MI–raw) Jackknife bias corrected MI (BCMI) (MI–COR) Z-scores of BCMI (MI–Z)
WinCRQA by Coco and Dale [46]	Windowed cross-recurrence	Local	Yes	No	not applicable	no	Recurrence rate in % (WinCRQA–RR) Determination rate in % (WinCRQA–DET) Normalized entropy (WinCRQA–ENTR)

Note: CC: cross-correlation; rMEA: R package motion-energy analysis; SUSY: surrogate synchrony; SUCO: surrogate concordance; WCLC: windowed cross-lagged correlation; WCLR: windowed cross-lagged regression; PP: peak picking; R2: squared correlation; MI: mutual information; WinCRQA: windowed cross-recurrence quantification analysis.

In the present study we considered two measures of the degree of synchrony provided by the rMEA package: First, the average windowed cross-correlations (step 1 above; rMEA–WCC), and second, the average windowed cross-lagged correlations (step 3 above; rMEA–WCLC). The former only includes values of the column of the WCLC matrix referring to $\tau = 0$, whereas the latter considers all columns. In contrast to Boker, Rotondo, Xu and King [9] and Altmann [4,12] there is no selection of WCLC maxima, thus no application of a peak-picking algorithm.

Due to the fact that noise and non-stationarity can cause cross-lagged correlation, a pseudo-synchrony approach [42,56] is conducted in the next step. The corresponding bootstrap algorithm randomly recombines the time series of person A and person B of another interview multiple times (default value: 100 times) and each time computes the WCLC. Thus, the surrogate generation is based on person shuffling. In this way, a statistic is produced to test whether the present WCLCs are different from the expected value of a random distribution of WCLC values.

2.5.3. SUSY

Surrogate synchrony (SUSY) [53,55] is based on the cross-correlation function of dyadic time series (the algorithm can be used online: https://www.embodiment.ch/, accessed on 15 July 2022). The cross-correlations are computed across a range of lags L (here $-5\text{ s} \leq L \leq 5\text{ s}$). The cross-correlation function is computed segment-wise, i.e., separately in all non-overlapping segments of the time series (here segment-size = 30 s). It may be noted that terminology differs in the rMEA package, where "window" is used to denote segments. All cross-correlations are transformed using Fisher's Z-transformation to allow the aggregation of cross-correlations. The synchrony of any segment is then defined as the mean of all (lagged) cross-correlations of this segment, and the synchrony of the time series as the mean of segment means. Aggregation is performed using either absolute Z cross-correlations (Zabs) or the original, negative or positive, cross-correlations (Znoabs). The reason for taking the absolutes of correlations is that one may define synchrony irrespective of the direction of coupling, which may be negative ("anti-phase") or positive ("in-phase"); in Zabs, both are collapsed into one signature of synchrony. Znoabs differentiates in-phase from anti-phase coupling. Then surrogate tests are performed to establish a control condition for the Zabs and Znoabs values of each dyad. Surrogate time series in SUSY are generated by randomly shuffling the sequence of segments, independently for each dyad member (surrogate generation by segment shuffling). From a dyadic time series with n segments, $n(n-1)$ surrogates can be produced. In the present analysis, all $n(n-1)$ respective surrogates were used. The surrogate step produces effect sizes (ES) as the final signatures of synchrony in SUSY, namely $SUSY\text{–}ES_{\text{abs}}$ and $SUSY\text{–}ES_{\text{noabs}}$. $SUSY\text{–}ES_{\text{abs}}$ is derived using the mean surrogate Zabs and their standard distribution: $SUSY\text{–}ES_{\text{abs}}$ = (Zabs − Zabs-surr)/SD(Zabs-surr). $SUSY\text{–}ES_{\text{noabs}}$ is computed analogously. The leading–following relationships of synchrony may further be operationalized by differentiating between positive and negative lags L.

2.5.4. SUCO

Surrogate concordance (SUCO; online access https://www.embodiment.ch/, accessed on 15 July 2022) [14,55] is based on the correlations of the local slopes of dyadic time series (A,B). The slopes are determined by least-squares regression in windows (here, window size was 3 s) of the time series, and the time series are again partitioned into segments of 30 s duration as in SUSY. The linear slopes are computed inside the first window of segment i, the window is then shifted by an increment of 1 s and the slopes are again computed; thus, overlapping windows are used. This is repeated until all windows in segment i are covered. The slopes in segment i of time series A are Pearson-correlated with those of the same segment of B. The resulting correlation ri denotes the relation between A's and B's slopes in segment i of the time series. This is performed in all segments of the time series A and B. The absolute values of all correlations ri are Fisher's Z-transformed and aggregated, yielding Z′abs (with high comma to distinguish from SUSY). Segment-wise shuffling is used, as in SUSY, to create surrogate time series, yielding the effect size of Z′abs, labeled $SUCO\text{–}ES_{\text{abs}}$. The concordance index (*SUCO–CO*) across all segments of the client–therapist interaction is defined by the natural logarithm of the sum of all positive correlations *ri* divided by the absolute value of the sum of all negative correlations *ri*, as previously suggested by Marci and Orr [57]. Using surrogate analysis, an effect size *SUCO–ES–CO* is computed by standardizing the concordance index by the mean and standard deviation of the concordance indexes of surrogate data, in analogy to the procedure in SUSY. The leading–following relationships of concordance synchrony are operationalized by shifting of A's windows with respect to B prior to computing *ri*, yet such lags were not computed in the present analyses.

2.5.5. WCLC–PP and WCLR–PP

The algorithm by Altmann [4,12] (download at https://github.com/10101-00001/sync_ident, accessed on 15 July 2022) combines three approaches: First, the computation of local associations proposed by Boker, Rotondo, Xu and King [9], Ramseyer and Tschacher [11] and Watanabe [58]; second, the reduction of auto-correlation bias [59] by a regression approach, e.g., as performed by Gottman and Ringland [60]; and third, the differentiation between significant and non-significant local associations and their selection by a peak-picking algorithm as proposed by Boker, Rotondo, Xu and King [9]. The algorithm is based on the assumptions that in interpersonal interaction, phases of synchrony (high degree of coupling) alternate with phases of non-synchrony (no coupling), and that within a phase of synchrony the data are sufficiently described by cross-lagged regression. The validity and high detection rate of the algorithm has been shown in multiple studies [36,40,44,45].

The computation consists of three steps. First, the local associations are computed. This can be performed with windowed cross-lagged correlation (WCLC) or windowed cross-lagged regression (WCLR). Similar to rMEA and SUSY, time-lagged windows of both time series are considered. In WCLR, for each start position of a reference window (e.g., of person A) and each possible time lag (τ), two cross-lagged regressions are applied. In model 1, the window of person A beginning at $t + \tau$ is predicted by the window of person A beginning at t. This means that only the auto-correlation is modeled. However, in model 2, the window of person A beginning at $t + \tau$ is predicted by a window of person A beginning at t (corresponding to the auto-correlation) as well as a window of person B beginning at t (corresponding to the cross-correlation). Then, the coefficient of determination is computed based on both models: $\Delta R^2_{t,\tau} = R^2_{M2,t,\tau} - R^2_{M1,t,\tau}$ (note: M1: model 1; M2: model 2; t: start position of window; τ: time lag between "action" and "response"). $\Delta R^2_{t,\tau}$ quantifies the proportion of variance of window A at t, which is explained by cross-lagged correlation with time lag τ and which is unbiased by the auto-correlation with time lag τ. The procedure described above is conducted for each position of the reference window ($t \in \{1, \ldots, L - b + 1\}$; L: length of time series; b: window width) and each time lag of interest ($\tau \in \{-\tau_{max}, \ldots, \tau_{max}\}$). All resulting $\Delta R^2_{t,\tau}$ values are stored in a matrix (so-called R square matrix; [12]). Similar to rMEA, the column refers to the time lag ($-\tau_{max}, \ldots,\ 0,\ \ldots, \tau_{max}$) and the row to the start position of the reference window ($1, \ldots, L - b + 1$; L: time-series length; b: window width). However, the values of matrix ($\Delta R^2_{t,\tau}$) are not correlation coefficients but the proportion of explained variance by cross-correlation adjusted by auto-correlation with the same time lag (also called R square or coefficient of determination). In contrast to WCLR, the WCLC by Altmann [4,12] estimates the local associations between two time-series windows with cross-lagged correlations. They can be confounded by auto-correlation. However, the process is similar: The correlations computed for windows that are time-lagged and "moved" over the time axis. The result is also a R square matrix. Its elements ($R^2_{t,\tau}$) are the squared windowed cross-lagged correlations at a specific start position of reference window (t) and a specific time lag (τ) of the interlocutor's window.

In the second step of the analysis, the R square matrix is analyzed by a peak-picking algorithm (abbreviation: PP) [4,12]. For each start position of reference window at t, local maxima of $\Delta R^2_{t,\tau}$ ($R^2_{t,\tau}$) values are detected (for illustrations see [12]). Next, local maxima with equal time lag and directly consecutive in time are combined into intervals. When a start position of the reference window is part of multiple intervals, then the interval with the largest average $\Delta R^2_{t,\tau}$ is selected. These selected intervals are called synchronization intervals [12]. The output of the peak-picking algorithm is a list of synchronization intervals (abbreviation: LOSI). Based on this list, an interpersonal interaction can be described as a process where phases of movement synchronization (synchronization intervals with a high degree of cross-lagged correlation) alternate with phases without synchronization (without significant cross-lagged correlation).

In the last step of WCLC–PP and WCLR–PP, various synchronization measures can be quantified based on the LOSI. In the present study, we considered the frequency of movement synchrony defined as the proportion of synchronization intervals in relation to the duration of the time-series length (WCLC–PP–F and WCLR–PP–F) and the average interrelatedness of both time series within the synchronization intervals quantified by the average R square of the synchronization intervals (WCLC–PP–R2 and WCLR–PP–R2).

Before starting WCLC–PP and WCLR–PP, some parameter values have to be defined. In the present study, the window width was 125 time points (5 s), the R^2 cut-off was 0.25 (both values suggested by the simulation study of [40]), the increment was one frame (resulting in overlapping windows), and the maximum time lag τ_{max} = 125 frames (both recommended by [5,8]). According to the simulation study of [40], in the LOSI we considered only synchronization intervals with $average(\Delta R^2) > 0.25$, which led to better identification rates and lower false positives.

2.5.6. Mutual Information

Mutual information (MI) [61] quantifies the amount of information that is shared by two random variables and uses this as a measure of dependence. Shannon information is closely linked with entropy [62]. In other terms, MI is the joint distribution of both time series related to the marginal distributions of both time series under the assumption of independence. In contrast to cross-correlation, which assumes continuous or interval-scaled time series, mutual information can only be computed for categorical variables or continuous variables binned into categories. A further difference is that MI does not rest on the assumption of linear dependencies between time series.

In the present study, we estimated MI using the R package mpmi (command cmi.pw) [54] which automatically calculates a vector of smoothing bandwidths for each of the dyadic time series. It uses a kernel-smoothing approach to estimate the joint distribution and both marginal distributions. The package provides three measures: an (uncorrected) raw value of MI (MI–raw), a Jackknife bias-corrected MI (MI–cor), and a Z-score of bias-corrected MI that provides a statistic for the null-hypothesis that the bias-corrected MI is zero (MI–Z).

2.5.7. Recurrence Techniques

Cross-recurrence quantification analysis (CRQA) [63–65] is based on a state–space approach. Given time series of two coupled dynamical systems, in the first step, recurrence techniques identify the time points when both systems are in the same state (e.g., both interlocutors smile). This includes simultaneous and time-lagged states. The information is stored in the recurrence matrix (illustrated as a recurrence plot). In continuous data (e.g., movement intensity), a distance measure has to be defined (usually Euclidian distance) and a recurrence threshold (radius: ε) has to be specified. Instead of same states, the simultaneous and time-lagged similarity of continuous state parameters is identified ($\epsilon <||x_t - x_{t+\tau}||$).

The values of the recurrence matrix are zero or one, depending on the similarity of values (in categorical time series, identity of values) at a specific time point of the reference time series and time lag of the interlocutors' time series. Based on the recurrence matrix, various parameters describing aspects of coupling can be computed, e.g., the percentage of recurrence points in the recurrence plot (recurrence rate: RR in %), the percentage of recurrence points forming diagonal lines (percentage of determinism: DET in %) or the Shannon information entropy of the diagonal line length longer than the minimum line (entropy: ENTR; entropy normalized by number of diagonal lines in the recurrence plot: rENTR) [46,66]. Of these measures, WinCRQA–DET is often used as a synchrony measure, e.g., in [40,67,68]. Please note that as in other algorithms, the result of the recurrence analysis depends on the parameter values, especially on the recurrence threshold [69].

In the present study, we conducted the windowed cross-recurrence quantification analysis (R command: wincrqa) implemented in the R package CRQA [46,66]. We transformed all the time series to the unit-interval and used a Euclidian distance with $\varepsilon = 0.05$ as recurrence threshold. The embedding dimension was three. As in rMEA, the window width was 1500 frames (60 s), the overlap of windows was 750 and the maximal time lag was 125 (5 s). The algorithms provided various outcome parameters for each window (e.g., RR, DET and ENTR). To obtain a measure for the entire conversation, we averaged these values over all the windows.

2.6. Statistical Analysis of Synchrony Measures

After the video recording of the 30 interviews and the measurement of motion energy during the interviews using the MEA, we applied the listed algorithms on the motion-energy time series to quantify synchrony. We created a data matrix in which a column refers to a specific synchrony measure and a line to an interview. Based on this table we investigated the validity of synchrony measures.

First, the convergent validity of synchronization measures was examined by Pearson and Spearman correlations. Thus, we assumed that all synchrony measures correlated with each other. According to Cohen [70], $r > 0.1$ can be interpreted as small, $r > 0.3$ as moderate and $r > 0.5$ as a large effect.

Due to the findings of [36], we explored facets of synchrony using factor analysis. To determine the number of extracted factors, we applied a parallel test with 100 bootstraps. We computed an exploratory factor analysis (EFA) with a maximum likelihood estimator (ML) as well as a minimum rank factor analysis (MINRANK), which is more appropriate in non-normally distributed data. In both factor analyses, the factors were allowed to correlate (oblimin rotation). An acceptable model fit is given when the root-mean-square error of approximation (RMSEA) is <0.08 and the Tucker Lewis Index (TLI) is >0.9.

Next, the predictive validity of synchronization measures was examined. We assumed that in dyads of patients with major depression, less movement synchronization would be observed than in the dyads of healthy controls. The synchronization measures of both groups were compared using the Kruskall–Wallis tests. In significant group differences, we reported Hedges g as an effect size measure. According to Cohen [70], $g > 0.2$ can be interpreted as small, $g > 0.5$ as moderate and $g > 0.8$ as a large effect.

Furthermore, for the predictive validity we assumed that a higher symptom load (assessed with PHQ9 and GAD7) would be related to less synchronization observed in the interviews. We computed Pearson and Spearman correlations.

3. Results

First, we investigated the convergent validity with correlations between different synchronization measures (Table 3). The three measures based on mutual information were highly interrelated (all Pearson $r > 0.845$, $p < 0.01$). This also holds for the three measures of the SUCO approach (all $r > 0.718$, all $p < 0.05$). Moderate correlations were found for both measures of rMEA ($r = 0.685$, $p < 0.05$), both measures of SUSY ($r = 0.502$, $p < 0.05$), both measures of WCLC–PP ($r = 0.51$, $p < 0.05$), and the four variants of cross-correlation (all $r > 0.69$, all $p < 0.05$).

Table 3. Pearson correlations (lower left triangle) and Spearman correlations (upper right triangle) of synchronization measures.

	1. CC–raw	2. CC–abs	3. CC–Z	4. CC–R2	5. rMEA–WCC	6. rMEA–WCLC	7. SUSY–ES_{abs}	8. SUSY–ES_{noabs}	9. SUCO–CI	10. SUCO–ES_{abs}	11. SUCO–ES–CI	12. WCLC–PP–F	13. WCLC–PP–R2	14. WCLR–PP–F	15. WCLR–PP–R2	16. MI–raw	17. MI–cor	18. MI–Z	19. Win-CRQA–RR	20. Win-CRQA–DET	21. Win-CRQA–ENTR
1		0.22	1.00 *	0.22	0.48 *	0.22	0.64 *	0.78 *	0.70 *	0.56 *	0.73 *	−0.17	0.04	−0.17	0.06	0.02	−0.01	−0.12	−0.2	0.02	0.18
2	0.69 *		0.22	1.00 *	0.72 *	0.57 *	0.46 *	0.19	0.27	0.25	0.23	−0.25	0.23	−0.26	0.06	0.03	0	0.07	−0.24	−0.18	0.07
3	1.00 *	0.69 *		0.22	0.48 *	0.22	0.64 *	0.78 *	0.70 *	0.56 *	0.73 *	−0.17	0.04	−0.17	0.06	0.02	−0.01	−0.12	−0.2	0.02	0.18
4	0.73 *	0.95 *	0.74 *		0.72 *	0.57 *	0.46 *	0.19	0.27	0.25	0.23	−0.25	0.23	−0.26	0.06	0.03	0	0.07	−0.24	−0.18	0.07
5	0.73 *	0.88 *	0.73 *	0.82 *		0.75 *	0.63 *	0.35	0.50 *	0.56 *	0.53 *	−0.1	0.39 *	−0.16	0.18	0.22	0.18	0.28	−0.32	−0.24	−0.1
6	0.21	0.56 *	0.21	0.44 *	0.68 *		0.33	0.08	0.22	0.33	0.2	0	0.57 *	−0.09	0.09	0.43 *	0.39 *	0.49 *	−0.33	−0.33	−0.1
7	0.72 *	0.66 *	0.72 *	0.64 *	0.74 *	0.40 *		0.57 *	0.52 *	0.54 *	0.54 *	−0.13	0.07	−0.13	0.09	−0.09	−0.12	−0.09	−0.12	0.01	0.17
8	0.74 *	0.43 *	0.74 *	0.50 *	0.40 *	0.02	0.50 *		0.66 *	0.50 *	0.60 *	0	0.12	−0.02	0.16	−0.06	−0.05	−0.28	−0.08	0.22	0.3
9	0.83 *	0.64 *	0.83 *	0.68 *	0.72 *	0.25	0.64 *	0.57 *		0.57 *	0.96 *	−0.12	0.2	−0.07	0.29	−0.02	−0.02	−0.01	−0.02	0.14	0.13
10	0.70 *	0.61 *	0.70 *	0.59 *	0.72 *	0.3	0.64 *	0.51 *	0.72 *		0.62 *	0	0.31	−0.14	0.35	0.05	0.04	0.05	−0.04	0.14	0.22
11	0.82 *	0.62 *	0.82 *	0.64 *	0.73 *	0.21	0.66 *	0.53 *	0.96 *	0.75 *		−0.19	0.14	−0.15	0.2	−0.1	−0.12	−0.03	−0.08	0.14	0.16
12	−0.3	−0.25	−0.3	−0.32	−0.16	0.31	−0.15	−0.01	−0.16	−0.17	−0.29		0.44 *	0.87 *	0.03	0.1	0.06	0.09	−0.46 *	−0.44 *	−0.32
13	0.17	0.32	0.17	0.32	0.38 *	0.65 *	0.15	0.14	0.36	0.26	0.21	0.51 *		0.31	0.48 *	0.29	0.29	0.3	−0.31	−0.27	0.02
14	−0.28	−0.31	−0.28	−0.40 *	−0.21	0.21	−0.15	0.02	−0.11	−0.2	−0.24	0.93 *	0.39 *		0.02	0	−0.01	−0.01	−0.37 *	−0.43 *	−0.28
15	0.14	0.19	0.14	0.21	0.22	0.25	0.08	0.12	0.34	0.3	0.21	0.32	0.66 *	0.35		0.2	0.23	0.07	0.2	0.16	0.22
16	−0.01	−0.06	−0.02	−0.14	0.13	0.53 *	0	−0.16	0.02	0.1	−0.11	0.28	0.49 *	0.26	0.35		0.98 *	0.80 *	−0.18	−0.38 *	−0.17
17	0	−0.07	−0.01	−0.14	0.1	0.47 *	0	−0.16	0.01	0.1	−0.12	0.22	0.46 *	0.22	0.39 *	0.99 *		0.74 *	−0.09	−0.28	−0.15
18	−0.08	−0.05	−0.08	−0.12	0.16	0.58 *	0	−0.26	0.01	0.04	−0.05	0.27	0.44 *	0.27	0.27	0.89 *	0.84 *		−0.3	−0.58 *	−0.25
19	−0.06	−0.15	−0.06	−0.07	−0.24	−0.41 *	−0.09	−0.08	0.03	−0.01	−0.03	−0.44 *	−0.22	−0.31	0.19	−0.16	−0.06	−0.23		0.83 *	−0.05
20	0.16	−0.07	0.16	0.09	−0.19	−0.51 *	0.01	0.2	0.18	0.07	0.17	−0.49 *	−0.24	−0.41 *	0.07	−0.33	−0.22	−0.43 *	0.84 *		0.13
21	0.26	0.23	0.26	0.26	0.1	−0.21	0.28	0.28	0.22	0.37 *	0.28	−0.41 *	−0.12	−0.40 *	0.02	−0.16	−0.15	−0.24	0.04	0.12	

Note: $N_{dyads} = 30$; * $p < 0.05$; CC: cross-correlation; SUSY: surrogate synchrony by Tschacher and Haken [53]; SUCO: surrogate concordance by Tschacher and Meier [14]; rMEA: R package for motion-energy analysis by Kleinbub and Ramseyer [37]; WCLC–PP and WCLR–PP: windowed cross-lagged correlation and windowed cross-lagged regression with subsequent peak picking by Altmann [4,12]; MI: mutual information by Pardy [54]; WinCRQA: windowed cross-recurrence quantification analysis by Coco and Dale [46], and Coco, Mønster, Leonardi, Dale and Wallot [66].

Contrary to our expectation, no single synchrony measure significantly correlated with all other synchrony measures. Synchrony quantified as cross-correlation (CC–raw, CC–abs, CC–Z, and CC–R2) correlated with synchrony measures of SUSY, SUCO and rMEA package moderately (most Pearson $r > 0.5$). In contrast, the synchrony measures of WinCRQA correlated negatively with rMEA–WCLC (e.g., Pearson r(WinCRQA–RR, rMEA–WCLC) = −0.41, $p < 0.05$), WCLC–PP–F (e.g., r(WinCRQA–DET, WCLC–PP–F) = −0.49, $p < 0.05$), WCLR–PP–F (e.g., r(WinCRQA–DET, WCLR–PP–F) = −0.41, $p < 0.05$) and mutual information (e.g., r(WinCRQA–DET, MI–Z) = −0.43, $p < 0.05$).

The parallel test suggested for EFA and the minimum rank factor analysis that two factors best describe the considered movement-synchrony measures. The loadings of both factor analyses were similar (Table 4). The variants of cross-correlation, all measures of the SUSY package, all measures of the SUCO package, and the rMEA–WCC formed a factor. The indicators of the second factor were all variants of mutual information and rMEA–WCLC. In the minimum rank factor analysis, WCLC–PP–F, WCLC–PP–R2, WCLR–PP–F, and WinCRQA–DET were also assigned to the second factor. WCLR–PP–R2, WinCRQA–RR, and WinCRQA–ENTR had low loadings (<0.5) and were not assigned to either factor. rMEA–WCLC, WCLC–PP–F, and WCLR–PP–F showed large cross-loadings (>0.3). Accordingly, in both factor analyses the model fit described by RMSEA and TLI was not acceptable.

Table 4. Loadings of exploratory factor analysis with maximum likelihood estimator ("ML") and minimum rank factor analysis ("MINRANK").

	ML		MINRANK	
	Factor 1	**Factor 2**	**Factor 1**	**Factor 2**
CC–raw	**1.00**	0.01	**0.93**	−0.09
CC–abs	**0.70**	−0.05	**0.86**	0.05
CC–Z	**1.00**	0.01	**0.93**	−0.09
CC–R2	**0.74**	−0.12	**0.88**	−0.06
rMEA–WCC	**0.74**	0.15	**0.88**	0.24
rMEA–WCLC	0.23	**0.53**	*0.38*	**0.74**
SUSY–ES_{abs}	**0.73**	0.02	**0.80**	0.02
SUSY–ES_{noabs}	**0.73**	−0.14	**0.66**	−0.14
SUCO–CI	**0.84**	0.04	**0.88**	0.03
SUCO–ES_{abs}	**0.71**	0.11	**0.80**	0.06
SUCO–ES–CI	**0.81**	−0.04	**0.87**	−0.03
WCLC–PP–F	−0.30	0.27	−0.30	**0.68**
WCLC–PP–R2	0.18	0.49	0.28	**0.72**
WCLR–PP–F	−0.28	0.26	*−0.31*	**0.62**
WCLR–PP–R2	0.15	0.37	0.24	0.45
MI–raw	−0.00	**1.00**	−0.04	**0.81**
MI–cor	0.00	**0.99**	−0.04	**0.75**
MI–Z	−0.07	**0.89**	−0.06	**0.82**
WinCRQA–RR	−0.07	−0.15	−0.05	−0.46
WinCRQA–DET	0.16	−0.31	0.12	−0.63
WinCRQA–ENTR	0.26	−0.16	0.34	−0.34
Variance explained by factor	32.8%	18.0%	37.4%	23.2%
Correlation of both factors	−0.02		0.04	
RMSR	0.16		0.12	
RMSEA	0.266		0.301	
TLI	0.367		0.189	

Note: $N_{dyads} = 30$; oblimin rotation; loadings > 0.5 marked bold and cross-loadings > 0.3 italic; CC: cross-correlation; SUSY: surrogate synchrony by Tschacher and Haken [53]; SUCO: surrogate concordance by Tschacher and Meier [14]; rMEA: R package for motion-energy analysis by Kleinbub and Ramseyer [37]; WCLC–PP and WCLR–PP: windowed cross-lagged correlation and windowed cross-lagged regression with subsequent peak picking by Altmann [4,12]; MI: mutual information by Pardy [54]; WinCRQA: windowed cross-recurrence quantification analysis by Coco and Dale [46], Coco, Mønster, Leonardi, Dale and Wallot [66]; RMSR: root mean square of the residuals; RMSEA: root-mean-square error of approximation; TLI: Tucker Lewis Index of factoring reliability.

Next, we examined the predictive validity based on the criterion whether the synchronization measures predicted the assignment into the group of healthy controls or of depressed patients. Table 5 reports the group averages of the different synchronization measures as well as the p-value of group mean comparisons. When measuring synchrony with rMEA–WCC ($g_{Hedges} = 0.838$, $p = 0.0274$), SUCO–ES–CO ($g_{Hedges} = 0.771$, $p = 0.0473$), and MI–Z ($g_{Hedges} = 0.882$, $p = 0.0197$), we found that patients with depression had a higher degree of synchrony (in terms of interrelatedness) than the healthy controls. Such an association at a trend level was also found for SUSY–ES$_{abs}$ ($g_{Hedges} = 0.620$, $p = 0.0918$) and SUCO–ES$_{abs}$ ($g_{Hedges} = 0.664$, $p = 0.0754$). In contrast, WCLC–PP–F and WCLR–PP–F (measuring the frequency of synchronization intervals) indicated that patients with depression synchronized less often than healthy controls (WCLC–PP–F: $g_{Hedges} = -1.03$, $p = 0.008$ and WCLR–PP–F: $g_{Hedges} = -0.994$, $p = 0.0114$). All other synchrony measures were unrelated to group assignment.

Table 5. Average synchronization depending on group assignment (averages and standard deviations, the p-value to Kruskall–Wallis test) and Spearman correlations (r) between symptoms and synchrony scores using the entire sample.

	Entire Sample	Healthy Controls	Depressive Patients	Group Comparison	r with	r with
	$N = 30$	$N = 15$	$N = 15$	p-Value	*PHQ9*	*GAD7*
CC–raw	0.02 (0.09)	0.00 (0.06)	0.04 (0.10)	0.1677	0.24	0.39 *
CC–abs	0.06 (0.06)	0.05 (0.03)	0.08 (0.07)	0.1617	0.28	0.39 *
CC–Z	0.02 (0.09)	0.00 (0.06)	0.04 (0.10)	0.1654	0.24	0.39 *
CC–R2	0.01 (0.01)	0.00 (0.00)	0.01 (0.02)	0.1402	0.29	0.39 *
rMEA–WCC	0.11 (0.04)	0.09 (0.03)	0.13 (0.05)	0.0274	0.49 *	0.60 *
rMEA–WCLC	0.09 (0.02)	0.09 (0.02)	0.09 (0.01)	0.1835	0.29	0.36 *
SUSY–ES$_{abs}$	0.59 (1.00)	0.28 (0.92)	0.90 (1.00)	0.0918	0.33	0.43 *
SUSY–ES$_{noabs}$	−2.60 (9.02)	−1.61 (4.66)	−3.59 (12.0)	0.5588	−0.19	−0.02
SUCO–CI	0.39 (0.77)	0.21 (0.62)	0.58 (0.88)	0.1989	0.36	0.43 *
SUCO–ES$_{abs}$	0.97 (1.92)	0.34 (1.25)	1.60 (2.29)	0.0753	0.46 *	0.49 *
SUCO–ES–CI	0.93 (1.48)	0.37 (1.06)	1.48 (1.67)	0.0473	0.49 *	0.56 *
WCLC–PP–F	0.41 (0.11)	0.46 (0.07)	0.36 (0.11)	0.0081	−0.43 *	−0.40 *
WCLC–PP–R2	0.40 (0.02)	0.40 (0.02)	0.40 (0.02)	0.5902	−0.02	−0.02
WCLR–PP–F	0.45 (0.09)	0.49 (0.05)	0.41 (0.11)	0.0114	−0.47 *	−0.39 *
WCLR–PP–R2	0.43 (0.02)	0.43 (0.02)	0.43 (0.03)	0.8538	0.01	0.04
MI–raw	0.70 (0.26)	0.64 (0.27)	0.75 (0.26)	0.2537	0.21	0.23
MI–cor	0.55 (0.21)	0.51 (0.22)	0.60 (0.20)	0.2508	0.20	0.22
MI–Z	64.3 (21.5)	55.3 (17.2)	73.3 (22.1)	0.0197	0.39 *	0.38 *
WinCRQA–RR	41.7 (10.7)	41.4 (12.9)	42.1 (8.32)	0.8581	0.03	−0.06
WinCRQA–DET	99.3 (0.45)	99.3 (0.49)	99.3 (0.42)	0.9594	0.03	−0.06
WinCRQA–ENTR	0.70 (0.02)	0.69 (0.02)	0.70 (0.02)	0.2987	0.19	0.14

Note: * $p < 0.05$; CC: cross-correlation, SUSY: surrogate synchrony by Tschacher and Haken [53]; SUCO: surrogate concordance by Tschacher and Meier [14]; rMEA: R package for motion-energy analysis by Kleinbub and Ramseyer [37]; WCLC–PP and WCLR–PP: windowed cross-lagged correlation and windowed cross-lagged regression with subsequent peak picking by Altmann [4,12]; MI: mutual information by Pardy [54]; WinCRQA: windowed cross-recurrence quantification analysis by Coco and Dale [46], Coco, Mønster, Leonardi, Dale and Wallot [66]; PHQ9: Depression Module of Patient Health Questionnaire; GAD7: Generalized Anxiety Disorder Scale.

To test predictive validity, we also examined the correlation between the degree of symptom load and synchronization measures (see Table 5). Similar to the group comparison, we found that rMEA–WCC (Spearman $r = 0.49$, $p < 0.05$), SUCO–ES$_{abs}$ ($r = 0.46$, $p < 0.05$), SUCO–ES–CI ($r = 0.49$, $p < 0.05$), and MI–Z ($r = 0.390$, $p < 0.05$) correlated with the degree of depressive symptoms (PHQ9 sum-score) in terms of more depression leading to more synchrony. In contrast, the significant correlation coefficients of WCLC–PP–F ($r = -0.43$, $p < 0.05$) and WCLR–PP–F ($r = -0.47$, $p < 0.05$) suggested that more depression is related to less synchronization. Regarding the degree of anxiety (GAD7 sum-score), we found more significant correlations than for depressive symptoms (see Table 5). Many of these correlations between anxiety symptoms and synchrony were larger than the corresponding correlations between depressive symptoms and the synchrony measure (e.g., GAD7 and rMEA–CC: $r = 0.600$ versus PHQ9 and rMEA–CC: $r = 0.46$). In contrast, the correlation between the frequency measures of synchronization and anxiety symptoms were lower than the corresponding correlation between synchrony and depressive symptoms (e.g., GAD7 and WCLR–PP–F: $r = -0.39$ versus PHQ9 and WCLR–PP–F: $r = -0.47$).

4. Discussion

Nonverbal interpersonal interaction can be regarded as a complex dynamical system as it comprises a high number of elements, considers changes in time depending on external parameters, and may form temporary self-organized patterns that decrease the initially high entropy of these systems. One such pattern that has received considerable attention in recent social and clinical psychology is movement synchrony. Sequences of movement synchronization defined as temporally coordinated motor activity are characterized by a reduced degree of complexity and entropy, respectively, and a high degree of interrelatedness between participants and their behavior. Currently, several synchrony measures are available, some based on information theory (e.g., mutual information) and some on cross-correlation (e.g., cross-lagged correlation or windowed cross-lagged correlation). Whereas developers (or users) claim that their algorithms actually measure "synchrony", there is as yet very little simulation or empirical evidence regarding the validity of synchrony measures, with few exceptions [40,55]. The present study therefore investigated two aspects of the validity of movement-synchrony measures: convergent validity and predictive validity. We applied several algorithms to the same dataset of 30 bivariate time series that represented the motor activity of both the interviewer and interviewee during clinical interviews on somatic complaints. From each interview video, bivariate motion time series were derived. Using these time series, we computed multiple synchronization measures and investigated the correlations between different measures (convergent validity). We also explored which synchrony measure predicted whether the interviewee belonged to the depression group (predictive validity).

4.1. Convergent Validity

Regarding the convergent validity, we found that synchrony measures originating from the same algorithmic approach were moderately to highly related. For instance, the three measures of mutual information of the R package mpmi [54] correlated highly among each other. The same was true to a moderate degree for measures of the SUCO algorithm [53], the rMEA package [37], and WCLC–PP [4].

When considering measures originating from different algorithms, their convergent validity (their correlation) varied considerably. The largest correlation was observed between CC–raw and SUCO–ES_{abs} (Spearman $r = 0.78$, $p < 0.05$). Many correlations, however, were insignificant and some were significant and negative, e.g., the correlation between MI–Z and WinCRQA–DET (Spearman $r = -0.58$, $p < 0.05$) or between WCLC–PP–F and WinCRQA–RR (Spearman $r = -0.46$, $p < 0.05$). When analyzing different aspects or facets of synchrony, research should consider synchrony measures of different algorithms instead of different measures of the same algorithm.

In detail, there are differences to other studies. In the study of Schoenherr, Paulick, Worrack, Strauss, Rubel, Schwartz, Deisenhofer, Lutz and Altmann [36], the correlation between rMEA–WCLC and WCLC–PP–F was higher (Pearson $r = 0.55$, $p < 0.05$, see ([36], Table 3, lower left triangle)) than in our study (Pearson $r = 0.31$, not significant). The same holds for the correlation between WinCRQA–RR and WCLC–PP–F (Pearson $r = 0.769$, $p < 0.05$, in ([36], Table 3, lower left triangle) versus $r = -0.44$, $p < 0.05$, in our study). Furthermore, our correlations between different synchrony measures did not correspond with the findings of Luehof [45] and Tschacher and Meier [14]. Depending on the kind of physiological data, Tschacher and Meier [14] found little or no inter-correlations between SUSY–ES_{abs}, SUSY–ES_{noabs} and SUCO. Yet in the present study of body movements and their synchronization, these measures correlated to a moderate amount (all Pearson correlations $r > 0.5$, all $p < 0.05$, see Table 2). Luehof [45] investigated body movements in interviews and quantified movement synchrony with CRQA and WCLR–PP. The correlation between CRQA–DET and WCLR–PP–F was $r = -0.01$ ([45] Table 4.40), whereas in the present study WinCRQA–DET and WCLR–PP–F correlated with $r = -0.41$ ($p < 0.05$, see Table 2, lower left triangle). However, it should be noted that in the discussed studies, different parameter settings (e.g., window size) were applied, especially in the recurrence techniques. Therefore, for each algorithm recommendations and guide lines for parameter settings should be developed that can be applied across future studies [40]. Another explanation for the heterogeneity may be that the kind of interaction (interviews versus psychotherapy sessions) and/or the kind of data (cyclic physiological time series versus movement time series characterized by bursts) affect the convergence of synchronization measures. Future studies should therefore test the convergent validity of synchronization measures with multiple and diverse datasets.

Next, we systematized the included synchrony measures using a data-driven approach: Factor analyses suggested two facets of synchrony. Indicators of the first factor were rMEA–WCC, all the variants of CC, and all the measures of SUSY and SUCO. These measures were based on cross-correlations and did not consider a specific time lag between the time series. All the MI measures, rMEA–WCLC, WCLC–PP–F, WCLC–PP–R2, WCLR–PP–F, and WinCRQA–DET loaded on the second factor (when applying a minimum rank factor analysis). MI and WinCRQA–DET are based on information theory and quantify a non-linear relationship in continuous data. The other synchrony measures of this factor use cross-lagged correlations (cross-lagged regression) to quantify a linear relationship between the time series. It should be noted that Schoenherr et al. [36] found a three-factor structure in EFA. The difference to our study may rest in that different synchronization measures were investigated and there was a small dataset in the present study. However, consistent with Schoenherr et al. [36], WCLC–PP–F, WCLR–PP–F, and WinCRQA–DET were assigned to the same factor.

In sum, we agree with Schoenherr et al. [36] by concluding that the convergent validity across the considered algorithmic approaches is insufficient, if present at all. While the mathematical justifications of all the approaches we tested here are clearly given, the quantifications of synchrony they are offering are in most cases only loosely connected. The factor analyses in Schoenherr et al. [36] and in the present study both suggest the presence of multiple facets of synchrony, where one facet appears to summarize coupling in terms of cross-correlation approaches, and the other relates to the frequency of synchronization intervals and the information-theory-based measures.

Further research is needed that can differentiate these synchrony aspects from one another. It would be straightforward to implement large studies with simulated datasets of pairs of time series that represent clear types of coupling between the respective pairs. The coupling may be locally restricted or globally present throughout the time series, coupling may be linear or nonlinear, and time series may be auto-correlated and stationary or not [71]. Such studies can ultimately elucidate which synchrony aspect is recognized by which algorithm. In addition, it would be possible to tailor the parameter settings of the algorithms to serve recognition.

A critical point to discuss is the convergent validity itself. Our study revealed that the absolute value of cross-correlation (CC–abs) was moderately to highly correlated with all the measures of rMEA, SUSY and SUCO (all Pearson $r > 0.43$, see Table 3). Accordingly, these measures formed a separate facet of synchrony in the factor analysis. The cross-correlation is one of simplest measures of synchrony by computing the linear relationship between two time series, not considering any time lag and without segmentation (as in windowed cross-correlations). The benefits of the more sophisticated algorithms rMEA, SUSY, and SUCO lie in the inclusion of surrogate testing that allows the computation of effect sizes and significance even in single-case time series. It remains to be seen how the various correlation-based algorithms fare in heterogeneous and non-stationary data. On the other hand, the measures that assess the frequency of synchronization intervals (WCLC–PP–F and WCLR–PP–F) were related only to cross-recurrence measures (WinCRQA) whereby the signs of correlations were negative (both Pearson $r \approx -0.4$, see Table 3). The question is whether the validity is given when a measure appears somewhat idiosyncratic; future research should explore in which conditions and in what kind of data the two facets of synchrony may collapse into one factor.

Interestingly, a similar situation regarding convergent validity was present in the measurement of adult attachment [71]. Possibly, the phenomenon of interest itself may have multiple aspects (facets) that are not related in a linear manner and may be measured currently only by one specific instrument (algorithm). Further methodological research is necessary to build bridges between these facets of synchrony, e.g., by developing further instruments (algorithms) or investigating non-linear relationships between the facets of synchrony.

4.2. Predictive Validity

Second, we studied the predictive validity based on the assumption that the presence of major depression as well as the degree of symptom load should result in a lower degree of synchrony and fewer synchronization intervals, respectively. In the present naturalistic dataset, more than half of the considered synchronization measures did not correlate with the degree of depressive symptoms, e.g., rMEA–WCLC, all the variants of CC, all the SUSY, and all the WinCRQA measures (see Table 5). The only synchrony measures that corresponded with our hypothesis were WCLC–PP–F and WCLR–PP–F. There was a negative correlation between these synchrony measures and depressive symptoms (Spearman r(WCLC–PP–F, PHQ9) = −0.43, r(WCLC–PP–F, PHQ9) = −0.47, respectively, both $p < 0.05$). In contrast to our assumption, rMEA–WCC, SUCO–ES–CO, and MI–Z showed positive correlations with depressive symptoms (all Spearman $r > 0.46$, all $p < 0.05$). These synchrony measures suggested that interpersonal interactions with depressed patients are characterized by a higher degree of movement synchrony. These results correspond with [36], who studied predictive validity based on psychotherapy data, finding inconsistent correlations with improvement of interpersonal problems in psychotherapy.

A possible explanation is that the algorithms measure different aspects of movement synchrony, which then correlate differently with depressive symptoms. WCLC–PP–F and WCLR–PP–F measure the frequency of synchronization intervals whereas rMEA–WCC, SUCO–ES–CO, and MI–Z quantify the degree of interrelatedness of both time series. Nevertheless, the present study revealed that in the diagnostic of depression, synchronization measures can lead to contrary conclusions (depressed synchronized less than control versus depressed synchronized more than controls). This raises the problem that the results of different synchrony studies cannot be aggregated when different measures have been used. A solution may be to measure movement synchrony with multiple algorithms, for example, when the relationship between depressive symptoms and synchronization is investigated. This would be comparable to studies on the efficacy of psychological treatment, in which both primary and secondary outcomes are assessed.

4.3. Limitations

Our sample of interview videos (bivariate time series) was rather small. Accordingly, the statistical analysis had low statistical power with limited generalizability. Future studies on the validity of synchronization measures should investigate large and diverse samples (e.g., free communication, structured interviews, and psychotherapy sessions) and consider time series related to different behavior modalities (e.g., movement synchrony and facial synchrony) and different contexts (e.g., mirror game and interviews). The present study investigated only movement synchrony in structured interviews.

Previous studies [39–41,43] showed in various algorithms that synchronization measures depend on the parameter settings. In the present study, each algorithm was applied with default parameter values recommended by the authors of the algorithms. Possibly, the convergence of synchrony measures depends on equal settings of corresponding parameters. In [36], WinCRQA and WCLR–PP were conducted with a window size of 5 frames (5 s). The correlation of the resulting synchrony measures was $r = 0.777$ ([36], Table 3). In the present study, the window size of WCLR–PP was 125 frames and the window size of WinCRQA was 1500 frames. Both synchrony measures correlated with $r = -0.41$ (see Table 3).

Study designs must be discussed, too. In the present study, we did not control the amount of synchrony in the experimental condition (patients versus controls) so that the "true" synchrony or a proxy for that is not known. Our analysis of predictive validity rested on the assumption that psychopathological symptom load should be linked to movement synchrony during interviews on somatic complaints. There is some plausibility for this assumption; yet it may be also true that both groups of participants were equally synchronized, as the topic of somatic complaints is an engaging topic for depressive as well as healthy interviewees. Additionally, as we discussed previously, the convergent validity of published findings on psychopathology and synchrony is not yet sufficiently robust because these findings originated from differing algorithms and differing parameter settings. Thus, a possible conclusion is that it is too early to study predictive validity; the (convergent) validity of the synchrony measures must be established in the first place.

At the very least, further studies building on the present one are necessary in the field of synchronization research to clarify especially convergent, but also predictive validity. On top of incorporating simulated data with known types of synchronized coupling (in order to analyze convergent validity) [40], experimental data with covert instructions for participants to synchronize (or not) [44,45], and sensitivity analyses on parameter settings and their influences [39,40,43] must be performed.

5. Conclusions

To date, only a few comparisons between synchrony measures deriving from different algorithms (frequency-, correlation-, information-based) have been performed systematically. Only recently and in the field of physics have such comparisons been performed on a large scale [72]. In the present study, we pursued a similar goal using a small naturalistic dataset that comprises psychological interaction processes.

Our study revealed that the convergent validity of synchronization measures applied in clinical research range from non-existent to very good. As expected, factor analyses suggested that the different convergence of the measures can be explained by the presence of facets: on the one hand cross-correlation measures and on the other hand measures based on information theory or describing the frequency of synchronization intervals. Moreover, patients with depression and healthy controls can be distinguished by many synchrony measures—which suggests predictive validity. However, some measures suggested that patients and interviewer synchronize less often than dyads with controls, whereas other measures suggested the opposite.

We believe the present study is a promising starting point for addressing the important question of what psychological meaning may reside in synchronization measures. Given the increasing number of synchrony studies in clinical, social, and developmental psychology, these are also pressing open questions in the light of what has been called the "replication crisis" in psychology and medicine.

Author Contributions: Conceptualization, U.A. and W.T.; methodology, U.A. and W.T.; data curation, U.A.; formal analysis, U.A.; writing—original draft preparation, U.A.; writing—review and editing, U.A., B.S. and W.T.; visualization, U.A.; supervision, B.S. All authors have read and agreed to the published version of the manuscript.

Funding: This research received no external funding.

Institutional Review Board Statement: The primary study that generated the movement time series was conducted in accordance with the World Medical Association Declaration of Helsinki. The ethical approval was obtained from the Ethics Committee of Jena University Hospital, Jena, Germany (ID 5043-01/17).

Informed Consent Statement: Before inclusion, all participants provided informed consent. Their participation was voluntary and unpaid.

Data Availability Statement: The data that support the findings of this study are available on request from the corresponding author, U.A.

Conflicts of Interest: The authors declare no conflict of interest.

References

1. Altmann, U.; Hermkes, R.; Alisch, L.-M. Analysis of nonverbal involvement in dyadic interactions. In *Verbal and Nonverbal Communication Behaviours*; Esposito, A., Faundez-Zanuy, M., Keller, E., Marinaro, M., Eds.; Springer: Berlin/Heidelberg, Germany, 2007; pp. 37–50. [CrossRef]
2. Orlinsky, D.E.; Ronnestad, M.H.; Willutzki, U. Fifty years of psychotherapy process-outcome research: Continuity and change. In *Bergin and Garfield's Handbook of Psychotherapy and Behavior Change*, 5th ed.; Lambert, M.J., Ed.; Wiley: Hoboken, NJ, USA, 2004; Volume 5, pp. 307–389.
3. Koole, S.L.; Tschacher, W. Synchrony in Psychotherapy: A Review and an Integrative Framework for the Therapeutic Alliance. *Front. Psychol.* **2016**, *7*, 862. [CrossRef] [PubMed]
4. Altmann, U. *Synchronisation Nonverbalen Verhaltens [Synchronization of Nonverbal Behavior]*; Springer: Berlin/Heidelberg, Germany, 2013.
5. Altmann, U.; Schoenherr, D.; Paulick, J.; Deisenhofer, A.-K.; Schwartz, B.; Rubel, J.A.; Stangier, U.; Lutz, W.; Strauss, B. Associations between movement synchrony and outcome in patients with social anxiety disorder: Evidence for treatment specific effects. *Psychother. Res.* **2020**, *30*, 574–590. [CrossRef] [PubMed]
6. Oullier, O.; de Guzman, G.C.; Jantzen, K.J.; Lagarde, J.; Kelso, J.A.S. Social coordination dynamics: Measuring human bonding. *Soc. Neurosci.* **2008**, *3*, 178–192. [CrossRef]
7. Schmidt, R.C.; O'Brien, B. Evaluating the dynamics of unintended interpersonal coordination. *Ecol. Psychol.* **1997**, *9*, 189–206. [CrossRef]
8. Schoenherr, D.; Paulick, J.; Strauss, B.; Deisenhofer, A.-K.; Schwartz, B.; Rubel, J.; Lutz, W.; Stangier, U.; Altmann, U. Nonverbal synchrony predicts premature termination of psychotherapy for social phobic patients. *Psychotherapy* **2019**, *56*, 503–513. [CrossRef] [PubMed]
9. Boker, S.M.; Rotondo, J.L.; Xu, M.; King, K. Windowed cross-correlation and peak picking for the analysis of variability in the association between behavioral time series. *Psychol. Methods* **2002**, *7*, 338–355. [CrossRef]
10. Delaherche, E.; Chetouani, M.; Mahdhaoui, A.; Saint-Georges, C.; Viaux, S.; Cohen, D. Interpersonal synchrony: A survey of evaluation methods across disciplines. *IEEE Trans. Affect. Comput.* **2012**, *3*, 349–365. [CrossRef]
11. Ramseyer, F.; Tschacher, W. Nonverbal synchrony in psychotherapy: Coordinated body movement reflects relationship quality and outcome. *J. Consult. Clin. Psychol.* **2011**, *79*, 284–295. [CrossRef]
12. Altmann, U. Investigation of movement synchrony using windowed cross-lagged regression. In *Analysis of Verbal and Nonverbal Communication and Enactment: The Processing Issue*; Esposito, A., Vinciarelli, A., Vicsi, K., Pelachaud, C., Nijholt, A., Eds.; Springer: Berlin/Heidelberg, Germany, 2011; pp. 344–354. [CrossRef]
13. Bilakhia, S.; Petridis, S.; Nijholt, A.; Pantic, M. The MAHNOB Mimicry Database: A database of naturalistic human interactions. *Pattern Recognit. Lett.* **2015**, *66*, 52–61. [CrossRef]
14. Tschacher, W.; Meier, D. Physiological synchrony in psychotherapy sessions. *Psychother. Res.* **2020**, *30*, 558–573. [CrossRef]
15. Kleinbub, J.R. State of the art of Interpersonal Physiology in Psychotherapy: A systematic review. *Front. Psychol.* **2017**, *8*, 2053. [CrossRef] [PubMed]
16. Altmann, U.; Brümmel, M.; Meier, J.; Strauss, B. Movement synchrony and facial synchrony as diagnostic features of depression: A pilot study. *J. Nerv. Ment. Dis.* **2021**, *209*, 128–136. [CrossRef] [PubMed]
17. Gaume, J.; Hallgren, K.A.; Clair, C.; Schmid Mast, M.; Carrard, V.; Atkins, D.C. Modeling empathy as synchrony in clinician and patient vocally encoded emotional arousal: A failure to replicate. *J. Couns. Psychol.* **2019**, *66*, 341–350. [CrossRef] [PubMed]
18. Imel, Z.E.; Barco, J.S.; Brown, H.J.; Baucom, B.R.; Baer, J.S.; Kircher, J.C.; Atkins, D.C. The association of therapist empathy and synchrony in vocally encoded arousal. *J. Couns. Psychol.* **2014**, *61*, 146–153. [CrossRef]
19. Schoenherr, D.; Strauss, B.; Stangier, U.; Altmann, U. The influence of vocal synchrony on outcome and attachment anxiety/avoidance in treatments of social anxiety disorder. *Psychotherapy* **2021**, *58*, 510–522. [CrossRef] [PubMed]
20. Aafjes-van Doorn, K.; Müller-Frommeyer, L. Reciprocal language style matching in psychotherapy research. *Couns. Psychother. Res.* **2020**, *20*, 449–455. [CrossRef]
21. Borelli, J.L.; Sohn, L.; Wang, B.A.; Hong, K.; DeCoste, C.; Suchman, N.E. Therapist–client language matching: Initial promise as a measure of therapist–client relationship quality. *Psychoanal. Psychol.* **2019**, *36*, 9. [CrossRef]
22. Wiltshire, T.J.; Philipsen, J.S.; Trasmundi, S.B.; Jensen, T.W.; Steffensen, S.V. Interpersonal Coordination Dynamics in Psychotherapy: A Systematic Review. *Cogn. Ther. Res.* **2020**, *44*, 752–773. [CrossRef]
23. Koole, S.L.; Atzil Slonim, D.; Butler, E.A.; Dikker, S.; Tschacher, W.; Wilderjans, T. In Sync with Your Shrink: Grounding psychotherapy in interpersonal synchrony. In *Applications of Social Psychology: How Social Psychology can Contribute to the Solution of Real-World Problems*; Forgas, J., Crano, W., Fiedler, K., Eds.; Routledge: New York, NY, USA, 2020; pp. 161–184.

24. Kupper, Z.; Ramseyer, F.; Hoffmann, H.; Tschacher, W. Nonverbal synchrony in social interactions of patients with schizophrenia indicates socio-communicative deficits. *PLoS ONE* **2015**, *10*, e0145882. [CrossRef]
25. Paulick, J.; Rubel, J.; Deisenhofer, A.-K.; Schwartz, B.; Thielemann, D.; Altmann, U.; Boyle, K.; Strauss, B.; Lutz, W. Diagnostic features of nonverbal synchrony in psychotherapy: Comparing depression and anxiety. *Cogn. Ther. Res.* **2018**, *42*, 539–551. [CrossRef]
26. Scherer, S.; Hammal, Z.; Yang, Y.; Morency, L.-P.; Cohn, J.F. Dyadic behavior analysis in depression severity assessment interviews. In Proceedings of the 16th International Conference on Multimodal Interaction, Istanbul, Turkey, 12–16 November 2014; pp. 112–119.
27. Kaltenboeck, A.; Harmer, C. The neuroscience of depressive disorders: A brief review of the past and some considerations about the future. *Brain Neurosci. Adv.* **2018**, *2*, 2398212818799269. [CrossRef] [PubMed]
28. Shenal, B.V.; Harrison, D.W.; Demaree, H.A. The neuropsychology of depression: A literature review and preliminary model. *Neuropsychol. Rev.* **2003**, *13*, 33–42. [CrossRef] [PubMed]
29. Hames, J.L.; Hagan, C.R.; Joiner, T.E. Interpersonal Processes in Depression. *Annu. Rev. Clin. Psychol.* **2013**, *9*, 355–377. [CrossRef] [PubMed]
30. Balsters, M.J.H.; Krahmer, E.J.; Swerts, M.G.J.; Vingerhoets, A.J.J.M. Verbal and nonverbal correlates for depression: A review. *Curr. Psychiatry Rev.* **2012**, *8*, 227–234. [CrossRef]
31. Burton, C.; McKinstry, B.; Szentagotai Tătar, A.; Serrano-Blanco, A.; Pagliari, C.; Wolters, M. Activity monitoring in patients with depression: A systematic review. *J. Affect. Disord.* **2013**, *145*, 21–28. [CrossRef]
32. Paulick, J.; Deisenhofer, A.-K.; Ramseyer, F.; Tschacher, W.; Boyle, K.; Rubel, J.; Lutz, W. Nonverbal Synchrony: A new approach to better understand psychotherapeutic processes and drop-out. *J. Psychother. Integr.* **2018**, *28*, 367–384. [CrossRef]
33. Babl, A. *Automatically Detected Nonverbal Synchrony between Patients and Therapists in Psychotherapeutic Dyads Assessed with Microsoft Kinect*; University of Bern: Bern, Switzerland, 2016.
34. Reich, C.M.; Berman, J.S.; Dale, R.; Levitt, H.M. Vocal synchrony in psychotherapy. *J. Soc. Clin. Psychol.* **2014**, *33*, 481–494. [CrossRef]
35. Schoenherr, D.; Strauss, B.; Paulick, J.; Deisenhofer, A.-K.; Schwartz, B.; Rubel, J.; Boyle, K.; Lutz, W.; Stangier, U.; Altmann, U. Movement synchrony and attachment related anxiety and avoidance in social anxiety disorder. *J. Psychother. Integr.* **2021**, *31*, 163–179. [CrossRef]
36. Schoenherr, D.; Paulick, J.; Worrack, S.; Strauss, B.; Rubel, J.; Schwartz, B.; Deisenhofer, A.-K.; Lutz, W.; Altmann, U. Quantification of nonverbal synchrony using linear time series analysis methods: Convergent validity of different methods. *Behav. Res. Methods* **2019**, *51*, 361–383. [CrossRef]
37. Kleinbub, J.R.; Ramseyer, F.T. rMEA: An R package to assess nonverbal synchronization in Motion Energy Analysis time-series. *Psychother. Res.* **2020**, *31*, 817–830. [CrossRef]
38. Gates, K.M.; Liu, S. Methods for Quantifying Patterns of Dynamic Interactions in Dyads. *Assessment* **2016**, *23*, 459–471. [CrossRef] [PubMed]
39. Ramseyer, F.; Tschacher, W. Movement Coordination in Psychotherapy: Synchrony of Hand Movements is Associated with Session Outcome. A Single-Case Study. *Nonlinear Dyn. Psychol. Life Sci.* **2016**, *20*, 145–166.
40. Schoenherr, D.; Paulick, J.; Deisenhofer, A.-K.; Schwartz, B.; Rubel, J.; Lutz, W.; Strauss, B.; Altmann, U. Identification of movement synchrony: Validation of time series analysis methods. *PLoS ONE* **2019**, *14*, e0211494. [CrossRef]
41. Behrens, F.; Moulder, R.; Boker, S.; Kret, M. Quantifying physiological synchrony through windowed cross-correlation analysis: Statistical and theoretical considerations. *BioRxiv* **2020**. [CrossRef]
42. Ramseyer, F.; Tschacher, W. Nonverbal Synchrony or Random Coincidence? How to Tell the Difference. In *Development of Multimodal Interfaces: Active Listening and Synchrony*; Esposito, A., Campbell, N., Vogel, C., Hussain, A., Nijholt, A., Eds.; Springer: Berlin/Heidelberg, Germany, 2010; Volume 5967, pp. 182–196.
43. Moulder, R.G.; Boker, S.M.; Ramseyer, F.; Tschacher, W. Determining synchrony between behavioral time series: An application of surrogate data generation for establishing falsifiable null-hypotheses. *Psychol. Methods* **2018**, *23*, 757–773. [CrossRef]
44. Feniger-Schaal, R.; Schoenherr, D.; Altmann, U.; Strauss, B. Movement synchrony in the Mirror Game. *J. Nonverbal Behav.* **2021**, *45*, 107–126. [CrossRef]
45. Luehof, S. *Automatic Analysis of Synchrony in Dyadic Interviews*; University of Utrecht: Utrecht, The Netherlands, 2019.
46. Coco, M.I.; Dale, R. Cross-recurrence quantification analysis of categorical and continuous time series: An R package. *Front. Psychol.* **2014**, *5*, 510. [CrossRef]
47. Shugaley, A.; Altmann, U.; Brümmel, M.; Meier, J.; Strauß, B.; Schönherr, D. Der Klang der Depression. Zusammenhang zwischen Depressivität und paraverbalen Merkmalen während der Anamnese depressiver Patient_innen und gesunden Probanden. *Psychotherapeut* **2021**, *67*, 158–165. [CrossRef]
48. Altmann, U.; Knitter, L.A.; Meier, J.; Brümmel, M.; Strauß, B. Nonverbale Korrelate depressiver Störungen: Eine Pilotstudie. *Z. Für Klin. Psychol. Psychother.* **2020**, *49*, 231–240. [CrossRef]
49. Löwe, B.; Spitzer, R.; Zipfel, S.; Herzog, W. *Gesundheitsfragebogen für Patienten (PHQ D)*; Komplettversion und Kurzform, Testmappe mit Manual, Fragebögen, Schablonen; Pfizer: Karlsruhe, Germany, 2002.

50. Wittchen, H.-U.; Wunderlich, U.; Gruschwitz, S.; Zaudig, M. *SKID I. Strukturiertes Klinisches Interview für DSM-IV. Achse I: Psychische Störungen. Interviewheft und Beurteilungsheft [SCID I. Strutured Clinical Interview for DSM-IV. Axis I: Mental Disorder. Manual]*; Hogrefe: Goettingen, Germany, 1997.
51. Grammer, K.; Honda, M.; Juette, A.; Schmitt, A. Fuzziness of nonverbal courtship communication unblurred by motion energy detection. *J. Personal. Soc. Psychol.* **1999**, *77*, 487–508. [CrossRef]
52. Kleinbub, J.R.; Ramseyer, F. R Package 'rMEA' Version 1.0.0—Synchrony in Motion Energy Analysis (MEA) Time-Series. 2018. Available online: https://cran.r-project.org/package=rMEA (accessed on 15 July 2022).
53. Tschacher, W.; Haken, H. *The Process of Psychotherapy: Causation and Chance*; Springer Nature: Cham, Switzerland, 2019. [CrossRef]
54. Pardy, C. mpmi: Mixed-Pair Mutual Information Estimators, 0.43.1. 2020. Available online: https://cran.r-project.org/package=mpmi (accessed on 15 July 2022).
55. Meier, D.; Tschacher, W. Beyond Dyadic Coupling: The Method of Multivariate Surrogate Synchrony (mv-SUSY). *Entropy* **2021**, *23*, 1385. [CrossRef] [PubMed]
56. Bernieri, F.J.; Reznick, J.S.; Rosenthal, R. Synchrony, pseudosynchrony, and dissynchrony: Measuring the entrainment process in mother-infant interactions. *J. Personal. Soc. Psychol.* **1988**, *54*, 243–253. [CrossRef]
57. Marci, C.D.; Orr, S.P. The effect of emotional distance on psychophysiologic concordance and perceived empathy between patient and interviewer. *Appl. Psychophysiol. Biofeedback* **2006**, *31*, 115–128. [CrossRef] [PubMed]
58. Watanabe, T. A study of motion-voice synchronization. *Bull. JSME* **1983**, *26*, 2244–2250. [CrossRef]
59. Dean, R.T.; Dunsmuir, W.T.M. Dangers and uses of cross-correlation in analyzing time series in perception, performance, movement, and neuroscience: The importance of constructing transfer function autoregressive models. *Behav. Res. Methods* **2016**, *48*, 783–802. [CrossRef]
60. Gottman, J.M.; Ringland, J.T. The analysis of dominance and bidirectionality in social development. *Child Dev.* **1981**, *52*, 393–412. [CrossRef]
61. Shannon, C.E. A Mathematical Theory of Communication. *Bell Syst. Tech. J.* **1948**, *27*, 623–666. [CrossRef]
62. Verdú, S. Empirical Estimation of Information Measures: A Literature Guide. *Entropy* **2019**, *21*, 720. [CrossRef]
63. Eckmann, J.-P.; Kamphorst, S.O.; Ruelle, D. Recurrence Plots of Dynamical Systems. *Europhys. Lett.* **1987**, *4*, 973–977. [CrossRef]
64. Fusaroli, R.; Konvalinka, I.; Wallot, S. Analyzing social interactions: The promises and challenges of using cross recurrence quantification analysis. In *Translational Recurrences. From Mathematical Theory to Real-World Applications*; Marwan, N., Riley, M., Giuliani, A., Webber, C.L., Jr., Eds.; Springer: Berlin/Heidelberg, Germany, 2014; Volume 103, pp. 137–155.
65. Wallot, S.; Leonardi, G. Analyzing multivariate dynamics using cross-recurrence quantification analysis (CRQA), diagonal-cross-recurrence profiles (DCRQ), and multidimensional recurrence quantification analysis (MDRQA)—A tutorial in R. *Front. Psychol.* **2018**, *9*, 2232. [CrossRef]
66. Coco, M.I.; Mønster, D.; Leonardi, G.; Dale, R.; Wallot, S. Unidimensional and multidimensional methods for Recurrence Quantification Analysis with crqa. *arXiv* **2020**, arXiv:2006.01954. [CrossRef]
67. Kodama, K.; Tanaka, S.; Shimizu, D.; Hori, K.; Matsui, H. Heart rate synchrony in psychological counseling: A case study. *Psychology* **2018**, *9*, 1858–1874. [CrossRef]
68. Konvalinka, I.; Xygalatas, D.; Bulbulia, J.; Schjødt, U.; Jegindø, E.-M.; Wallot, S.; Van Orden, G.; Roepstorff, A. Synchronized arousal between performers and related spectators in a fire-walking ritual. *Proc. Natl. Acad. Sci. USA* **2011**, *108*, 8514–8519. [CrossRef] [PubMed]
69. Marwan, N. How to avoid potential pitfalls in recurrence plot based data analysis. *Int. J. Bifurc. Chaos* **2011**, *21*, 1003–1017. [CrossRef]
70. Cohen, J. *Statistical Power Analysis for The Behavioral Sciences*, 2nd ed.; Erlbaum Associates: Hillsdale, NJ, USA, 1988.
71. Strauss, B.; Altmann, U.; Schönherr, D.; Schurig, S.; Singh, S.; Petrowski, K. Is there an elephant in the room? A study of convergences and divergences of adult attachment measures commonly used in clinical studies. *Psychother. Res.* **2022**, *32*, 695–709. [CrossRef]
72. Cliff, O.M.; Lizier, J.T.; Tsuchiya, N.; Fulcher, B.D. Unifying Pairwise Interactions in Complex Dynamics. *arXiv* **2022**, arXiv:2201.11941. [CrossRef]

Article

Catastrophe Theory Applied to Neuropsychological Data: Nonlinear Effects of Depression on Financial Capacity in Amnestic Mild Cognitive Impairment and Dementia

Dimitrios Stamovlasis [1,*], Vaitsa Giannouli [2], Julie Vaiopoulou [3,4] and Magda Tsolaki [2,5,6]

1 School of Philosophy and Education, Aristotle University of Thessaloniki, 54124 Thessaloniki, Greece
2 1st Department of Neurology, School of Medicine, Faculty of Health Sciences, Aristotle University of Thessaloniki, 54634 Thessaloniki, Greece
3 Department of Education, University of Nicosia, Nicosia 2417, Cyprus
4 School of Psychology, Aristotle University of Thessaloniki, 54124 Thessaloniki, Greece
5 Alzheimer Hellas, 54643 Thessaloniki, Greece
6 Laboratory of Neurodegenerative Diseases, Center for Interdisciplinary Research and Innovation (CIRI-AUTh), Balkan Center, Buildings A & B, Thessaloniki, Aristotle University of Thessaloniki, 10th km Thessaloniki-Thermi Rd, P.O. Box 8318, 54124 Thessaloniki, Greece
* Correspondence: stadi@edlit.auth.gr

Citation: Stamovlasis, D.; Giannouli, V.; Vaiopoulou, J.; Tsolaki, M. Catastrophe Theory Applied to Neuropsychological Data: Nonlinear Effects of Depression on Financial Capacity in Amnestic Mild Cognitive Impairment and Dementia. *Entropy* **2022**, *24*, 1089. https://doi.org/10.3390/e24081089

Academic Editors: Franco Orsucci and Wolfgang Tschacher

Received: 14 July 2022
Accepted: 5 August 2022
Published: 7 August 2022

Abstract: Financial incapacity is one of the cognitive deficits observed in amnestic mild cognitive impairment and dementia, while the combined interference of depression remains unexplored. The objective of this research is to investigate and propose a nonlinear model that explains empirical data better than ordinary linear ones and elucidates the role of depression. Four hundred eighteen (418) participants with a diagnosis of amnestic MCI with varying levels of depression were examined with the *Geriatric Depression Scale* (GDS-15), the *Functional Rating Scale for Symptoms of Dementia* (FRSSD), and the *Legal Capacity for Property Law Transactions Assessment Scale* (LCPLTAS). Cusp catastrophe analysis was applied to the data, which suggested that the nonlinear model was superior to the linear and logistic alternatives, demonstrating depression contributes to a bifurcation effect. Depressive symptomatology induces nonlinear effects, that is, beyond a threshold value sudden decline in financial capacity is observed. Implications for theory and practice are discussed.

Keywords: cusp catastrophe; complexity; nonlinear dynamics; financial capacity; amnestic mild cognitive impairment; depressive symptoms

1. Introduction

1.1. The Psychocognitive Framework

This section focuses on explaining the psychocognitive framework that hosts the present investigation, which has a predominately methodological orientation. Thus, from the existing extensive literature on dementia, amnestic mild cognitive impairment, and the factors involved in defining, measuring, and treating them, only the most relevant pieces will be cited, which will adequately familiarize the reader with the phenomenon under study.

An issue that has become crucial in modern societies, since it relates to legal implications, is the assessment of the financial capacity of older adults with psychocognitive problems [1–3]. The matter concerns a number of specialists, including not only clinical neuropsychologists and forensic psychiatrists, but also judges and lawyers, while there is an increasing theoretical interest in proposing models for describing and predicting empirical results. Although there is not a consensus among researchers about defining and measuring financial capacity [4–6], a predominant model (Marson's model) [7] proposes an effective way to deal with the multidimensionality of the latent variable in question, and it is acceptable for the legal systems in most countries. Conceptually, the model includes two components: the first encompasses the financial activity of a general domain of functioning, and the second takes into consideration specific financial abilities tasks. This general model has inspired potential endeavors for developing assessment instruments in different countries, given that the underlying process is culture-specific [8]. The interest in developing such tools aims at their implementation in clinical assessments, as direct measurements of relevant neuropsychological deficits. Analogous endeavors have been realized in other domains by examining different mental resources, such as memory skills or verbal fluency [9,10]. Given the abovementioned legal implications, the central idea continues to inspire a growing concern, specifically for the financial capacity, since it has been proven highly susceptible to Alzheimer's disease (AD) and related disorders [11]. In this direction, lately, a financial capacity test for the Greek population, namely the *Legal Capacity for Property Law Transactions Assessment Scale* (LCPLTAS), was developed and validated [12], with psychometric properties that allow the performance of both healthy older adults and those suffering from different types and stages of dementia to be investigated. It should be noted that individuals with mild cognitive impairment (MCI) are also distinguished via this test. Relevant diagnostic cognitive tests, such the Mini-Mental State Examination (MMSE), were used to predict financial capacity performance, enhancing the validly of the LCPLTAS. In addition, two more instruments, the *Geriatric Depression Scale* (GDS-15) and the *Functional Rating Scale for Symptoms of Dementia* (FRSSD) were used as predictor variables in order to associate LCPLTAS scores with the other scales.

1.2. The Effects of Psychocognitive Resources on Financial Capacity

Research on psychocognitive performance, based on empirical evidence, has established a number of relationships among latent factors related to some mental deficits.

Patients with amnestic mild cognitive impairment (aMCI) are found to be inferior performers in financial capacity tasks compared to healthy individuals [13,14], and the anticipated decline over time in MCI converters is significantly greater than that of the MCI non-converters or healthy control cases [13]. However, there are circumstances where additional factors can concomitantly affect financial capacity, such as comorbid depression. Research has shown that decline in financial capacity in Alzheimer's Disease, Parkinson's Disease, and vascular dementia is observed, specifically when depression is identified during neuropsychological assessment [15,16]. Moreover, studies have supported declining and impaired financial capacity in aMCI individuals [13,14], while some empirical evidence for financial capacity in aMCI with concurrent depressive symptomatology (aMCI-D) has been provided [17].

Regarding methodological issues, all relevant research has been promoted via traditional approaches with linear statistical modeling, the limitations of which are already well known [18]. The present endeavor, fostering the meta-theoretical framework of complexity theory and nonlinear dynamics, aimed to test the nonlinear hypothesis in psychocognitive performance by applying catastrophe theory and implementing financial capacity, GDS, and FRSSD.

Any neuropsychological process is characterized by an inherent complexity. The involved latent constructs, such as financial capacity, are also complex, involving a variety of mental functions, which are operationalized by specific ability tests (e.g., arithmetic, counting coins/currency, paying bills) and judgment decision-making skills [14,15]. All involved mental resources (such as working memory and logical thinking) interact with each other in time via a dynamical process where, in addition to the positively contributing components, counteracting variables and inhibitory factors operate as moderators leading to deteriorated outcomes. Based on the evidence, depression is a moderator factor, which, when combined with additional deficits, leads to a worse performance. It is reasonable to consider that depression not only in AD, but also in other neurocognitive disorders, is a moderator factor for financial capacity [19,20]. This is a hypothesis though that hasn't received systematic investigation providing a coherent and interpretable model that describes the phenomenon.

To this end, the present article proposes a novel approach in exploring medical data in this area by fostering complexity theory and nonlinear dynamics. It is a fundamental theoretical consideration that the latent constructs involved in neuropsychological processes are dynamically interacting and the emergent behavior is described by the notion of complex adaptive systems (CASs) [21]. In a CAS, the behavior is deemed as inherently nonlinear, and changes can often be discontinuous and unpredictable. Complexity science has already gained considerable attention in social sciences [22,23], behavioral sciences [24–28], and life and medical sciences [29–32]. The present endeavor employs catastrophe theory for modeling financial capacity as the state variable dependent measure, while the Geriatric Depression Scale (GDS-15) and the Functional Rating Scale for Symptoms of Dementia (FRSSD) are the predictor variables. Elements of catastrophe theory are presented in the following section.

1.3. Catastrophe Theory

Catastrophe theory as a mathematical theory was founded on the works of Thom [33] and Arnold [34] and concerns the classification of the equilibrium behavior of dynamical systems in the neighborhood of singularities. It proves that, at these critical points, the system can be locally modeled by seven elementary catastrophes, from which cusp catastrophe is the most known and applicable [35]. Catastrophe theory presupposes a dissipating or potential-minimizing system, and the cusp model is expressed by the first derivative of a potential function, U, with respect to the outcome, y, by Equation (1):

$$\frac{\partial U(y,a,b)}{\partial y} = y^3 - by - ay \tag{1}$$

By setting $\partial U(y,a,b)/\partial y = 0$, the resulting equilibrium function is represented by the three-dimensional surface as a function of the two control parameters (α and b).

The development of stochastic catastrophe theory, which is based on the initial work of Cobb [36], allows for testing the relevant models with empirical data. Catastrophe theory is an area of complex dynamical systems and has shown high applicability in behavioral science. The notion of a potential-optimization process is compatible with a neuropsychological system, since it could be considered as pursuing the optimization of some function, e.g., related to adaptation or cognitive dissonance. The description of the cusp model is made via the response surface (Figure 1), where its fundamental features can be observed, such as bimodality, hysteresis, inaccessibility area, divergence, bifurcation, and sudden jumps [37]. The above phenomenology is interpreted via the underlying self-organization processes [38] and is theoretically connected to other areas of nonlinear sciences, such as Prigogine's non-equilibrium [39,40] and Haken's synergetics [41,42].

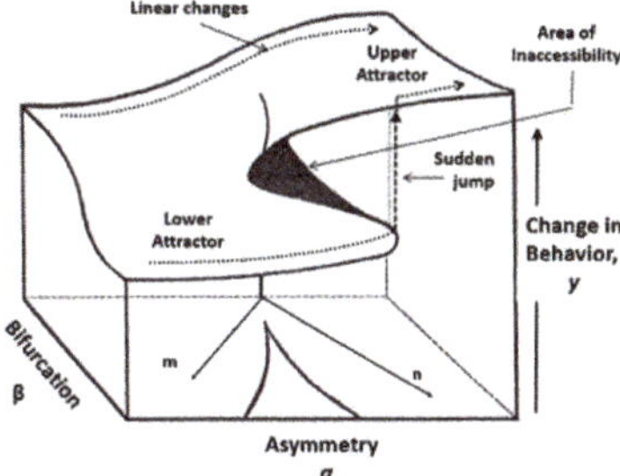

Figure 1. Cusp catastrophe surface.

The interpretation of the cusp model is facilitated by the three-dimensional response surface (see Figure 1), which demonstrates the geometry of behavior. At the back region of the surface, where the bifurcation, *b*, has low values, the surface is smooth and a linear relationship between the state variable (dependent measure) and the asymmetry, *a*, holds. In the forward-facing part, the surface folds and the two regions, i.e., the upper and the lower parts, appear, representing the two behavioral modes that in the language of CDS are called *attractors*. At this region, the probability density function of the empirical data becomes bimodal, whereas, in the area between the two modes, the behavior is unlikely to occur, and it is called the inaccessibility area. Thus, in this region, changes can only occur as jumps or transitions between the two behavioral attractors. Mathematically, these changes are called *discontinuities* and the splitting of the system into two states or different modes of behaviors consists of a bifurcation [43]. Looking from the front of the surface, a sigma-like feature, the *hysteresis effect*, can be observed. These are dynamic effects occurring when the bifurcation variable, *b*, goes beyond a critical value. It is pertinent to emphasize that bifurcation characterizes only nonlinear systems and is considered as a fingerprint of complexity [38].

2. Materials and Methods

2.1. Rationale and Research Hypotheses

In this research and for the relevant diagnoses, well-known neuropsychological assessment tools, such as the GDS, FRSSD, MMSE, and LCPLTAS, were implemented. The instruments operationalize specific psychological resources and are used concomitantly to consolidate conclusions and to help make decisions. The validation of the LCPLTAS [12] was supported with its functional relationships with the rest of the instruments, while the statistical methodology was based on linear modeling. Considering the epistemological and methodological limitations of the traditional linear approaches [18], this research was initiated to examine the applicability of catastrophe theory in this area and to provide insights about the theoretical and practical implications. This neuropsychological endeavor is apparently inductive in nature, since it is a new application in the field, encompassing a supplementary analysis of available data. However, it is also theory-driven, because it is based on the theory of complex adaptive systems (CASs), which is used here to reexamine the outcomes of dynamical processes, such as the neuropsychological processes taking place in assessment procedures. The cognitive factors involved in financial problem solving do not act as parts of a mechanical system, where the outcome can be expressed as a linear function of the contributing mechanisms [18,44,45]. As parts of a CAS system, these components act with no predetermined scenario, but execute their tasks via an iterative dynamical process. The potential nonlinearity can lead to changes encompassing sudden shifts, discontinuities, or transitions, which can be captured by catastrophe theory models. In the present research framework, among the psychological resources involved, *depression* is known as a moderator factor of financial capacity, competing against the positively acting resources. As an inhibitory agent, depression is a potential factor for inducing nonlinear effects.

To this end, the research hypotheses posited in this endeavor concern the potential role of depression in financial capacity, along with testing the applicability of catastrophe theory in neuropsychology, and are stated as follows:

(1) The effect of the GDS and FRSSD on the LCPLTAS can be described via a cusp catastrophe model.
(2) The GDS is the main candidate for acting as a bifurcation factor.
(3) Both the FRSSD and GDS could contribute to both the asymmetry and the bifurcation factors.

2.2. Participants and Measures

The participants were 418 Greek adults (68.2% women), whose age ranged from 45 to 98 years (*mean* = 72.55, *SD* = 8.08, *median* = 72.0). The mean years of education was 8.61 years (*SD* = 4.41, *median* = 6.0). A total of 34.4% were healthy control individuals, while the rest were diagnosed with varying degrees of AD and cognitive impairment. This sample composition ensures large variances in the measured construct and facilitates the variable-centered analyses. The neuropsychological assessments were carried out at the Memory Clinic of Papanikolaou General Hospital and elderly daycare centers during 2012–2016. Written informed consent from each participant was obtained and the study was approved by the Ethics Committee of the Aristotle University of Thessaloniki (protocol code 2.27/3/2013) [12], while the research was performed according to the Declaration of Helsinki.

Financial capacity was assessed with the *Legal Capacity for Property Law Transactions Assessment Scale* (LCPLTAS) short form [12]. The LCPLTAS consists of seven main domains: (1) basic monetary skills, (2) cash transactions, (3) bank statement management, (4) bill payment, (5) financial conceptual knowledge, (6) financial decision making, and (7) knowledge of personal assets [12]. The depressive symptomatology was measured by the *Geriatric Depression Scale* (GDS-15) [46], and the functionality evaluation was provided by the *Functional Rating Scale for Symptoms of Dementia* (FRSSD), which measures activities of daily living (ADLs) [47]. The above three instruments, along with the MMSE scale, are commonly used by psychiatrists, neuropsychologists, and neurologists in Greece [12,48–51]. The reported scores were available for all participants in this sample, since they were important part of assessment protocols in medical settings. Note that in the data set there were no missing values.

2.3. Method

Cusp analysis was carried out via a modeling procedure based on the probability function, *pdf*, of the dependent measure (Equation (2)):

$$pdf(y) = \xi \exp\left[-\frac{1}{4}y^4 + \frac{1}{2}by^2 + ay\right] \tag{2}$$

As the optimization method, the maximum likelihood [52] was used, while the *pdf* was obtained from empirical data. The analysis was performed in R via the cusp package [53]. The cuspfit algorithm utilizes numerical procedures for parameter estimates by minimizing a negative loglikelihood function, on which the model-fit evaluation is based, along with the indices: AIC (Akaike's information criteria), corrected AIC, and BIC (Bayesian information criteria) and the statistically significant coefficients of the model. Moreover, a comparison of the cusp with the linear and logistic alternative model is provided [53]. The literature offers other modeling procedures as well, such as the GEMCAT methodology [54] and a method implementing Equation (2) and least squares as the optimization method [55]. The details of these methods could be found in a lucid review elsewhere [56].

In the cusp analysis, the financial capacity was the dependent measure (LCPLTAS), known as the state variable, while the depressive symptomatology measured by the *Geriatric Depression Scale* (GDS-15) and the *Functional Rating Scale for Symptoms of Dementia* (FRSSD) were the two control variables. Results from a power analysis [57] (power levels of 80%, a medium effect size, two-tailed test with alpha = 0.05, required sample size of 75) showed that the available sample (N = 418) is adequate for testing the multivariate effects under study.

Initially, the model was conceived with the FRSSD as the asymmetry factor and the *Geriatric Depression Scale* (GDS-15) as the bifurcation. The conceptual and mathematical model, however, considers that asymmetry and bifurcation factors that represent antagonistic processes can be operationalized by a combination of the proposed controls, and, consequently, linear functions of the FRSSD and GDS scales were tested as contributing factors to both the asymmetry and bifurcation. The alternative cusp catastrophe models utilize rotated axes [58] and are analogous to the conflict cusp model that has been proposed for Piaget's conservation task [59]. This cusp model implements (FRSSD − GDS) and (FRSSD + GDS) as asymmetry and bifurcation, respectively.

3. Results

Table 1 presents the descriptive statistics (means, standard deviations, minimum and maximum values) for the variables under study. For financial capacity, the LCPLTAS and its short version sLCPLTAS were used.

Table 1. Descriptive statistics.

	Mean	Std. Deviation	Minimum	Maximum
LCPLTAS	160.605	63.550	0.000	212.000
sLCPLTAS	108.708	43.850	0.000	144.000
MMSE	24.892	6.538	0.000	30.000
GDS	2.725	3.561	0.000	21.000
FRSSD	4.641	6.454	0.000	32.000
Age	72.555	8.061	45.000	98.000

Table 2 depicts the correlation matrix for the above variables. Both tables include age and measures of the MMSE, which, however, were not used in the present analysis. Note that both the GDS ($r = -0.220$, $p < 0.001$) and FRSSD ($r = -0.792$, $p < 0.001$) are negatively correlated with financial capacity.

Table 2. Pearson's correlations.

Variable	LCPLTAS	sLCPLTAS	FRSSD	GDS	MMSE	Age
1. LCPLTAS	1					
2. sLCPLTAS	0.998 ***	1				
3. FRSSD	−0.792 ***	−0.789 ***	1			
4. GDS	−0.220 ***	−0.223 ***	0.281 ***	1		
5. MMSE	0.944 ***	0.942 ***	−0.824 ***	−0.201 ***	1	
6. Age	−0.288 ***	−0.289 ***	0.246 ***	−0.018	−0.291 ***	1

* $p < 0.05$, ** $p < 0.01$, *** $p < 0.001$.

Subsequently, cusp catastrophe analysis was carried out, testing a model with the FRSSD and GDS as control variables (*Cusp 1*) and a model with the linear combination of them (*Cusp 2*). Tables 3 and 4 show the slopes, standard errors, Z-tests, and model fit statistics for the cusp and the alternative models.

Table 3. The cusp model estimated by maximum likelihood method: slopes, standard errors, Z-tests, and model fit statistics for cusp and the alternative models. Financial capacity as a function of FRSSD (asymmetry) and Geriatric Depression Scale (bifurcation variable).

Model			b	seb	Z-Value
Cusp 1					
a(Intercept)			1.0628	0.1248	8.52 ***
a[FRSSD]	Functional Rating Scale for Symptoms of Dementia		−1.4557	0.1468	−9.91 ***
b(Intercept)			−1.5417	0.2165	−7.12 **
b[GDS]	Depression Scale		−0.3493	0.0912	−3.83 ***
w(Intercept)			0.8830	0.0355	24.87 ***
w(FC)	Financial Capacity		1.2059	0.02921	41.28 ***
Models' fit statistics (chi-square test of linear vs. cusp model: $\chi^2 = 247.0$, $df = 2$, $p < 0.001$)					
Model	**Pseudo-R^2**	**Npar**	**AIC**	**AICc**	**BIC**
Linear model	0.61	4	781.203	781.300	797.345
Logistic model	0.61	5	744.210	744.351	764.388
Cusp model	0.63	6	538.190	538.392	562.403
Note: *** $p < 0.001$, ** $p < 0.01$, * $p < 0.05$, † $p < 0.05$ (one-tailed); ns = non-significant.					

Table 4. The cusp model estimated by maximum likelihood method: slopes, standard errors, Z-tests, and model fit statistics for cusp and the alternative models. Financial capacity as a function of (FRSSD − GDS) as asymmetry and (FRSSD + GDS) as bifurcation variable.

Model			b	seb	Z-Value
Cusp 2					
a(Intercept)			−0.1606	0.1096	−1.46 ns
a[FRSSD − GDS]	Functional Rating Scale for Symptoms of Dementia		1.3450	0.2181	6.19 ***
b(Intercept)			0.98163	0.3388	2.90 **
b[GDS + FRSSD]	Depression Scale		1.2804	0.1886	6.79 ***
w(Intercept)			0.02605	0.0528	0.50 ns
w(FC)	Financial Capacity		1.02722	0.0472	21.75 ***
Models' fit statistics (chi-square test of linear vs. cusp model: $\chi^2 = 147.6$, $df = 2$, $p < 0.001$)					
Model	Pseudo-R^2	Npar	AIC	AICc	BIC
Linear model	0.39	4	414.767	415.043	426.809
Logistic model	0.47	5	395.039	395.456	410.093
Cusp model	0.63	6	271.213	271.801	289.277

Note: *** $p < 0.001$, ** $p < 0.01$, * $p < 0.05$, † $p < 0.05$ (one-tailed); ns = non-significant.

3.1. Cusp 1

In Cusp 1 (Table 3), the FRSSD acts as the asymmetry factor ($b = -1.4557$, $p < 0.001$) and the GDS acts as the bifurcation factor ($b = -0.3490$, $p < 0.001$). The chi-square test of the linear vs. cusp model gives $\chi^2 = 247.0$, $df = 2$, and $p < 0.001$, and the model fit statistics in terms of AIC, AICc, and BIC favor the cusp catastrophe model. The values for the cusp model (AIC = 538.190, AICc = 538.392, and BIC = 562.404) are minimum compared to the linear (AIC = 781.203, AICc = 781.300, and BIC = 797.345) and logistic (AIC = 744.210, AICc = 744.351, and BIC = 764.388) models, respectively. The values of pseudo R^2 are close, but this index is not reliable, and it is not interpreted as the usual percentage of variance explained.

Figure 2 is a visual display of the lower part of the cusp surface, where the shaded region is the bifurcation area. If at least 10% of the observations fall within this area, it is considered as evidence supporting the cusp model [60]. The size of the dots in Figure 2 is a function of the observed bivariate density of the bifurcation factor's values at that point's location, and the color is evocative of their position relative to the distance between the two parts of the surface (two attractors), i.e., the observations that are darker in color indicate that they are on or closer to the upper attractor and the observations that are lighter in color are on or closer to the lower attractor. Finally, Figure 3, which depicts the three-dimensional cusp surface as a function of the two control variables, provides an additional visual support for the cusp model, showing that the observations are located at the upper and the lower surface, but not within the area of inaccessibility.

3.2. Cusp 2

This cusp catastrophe model utilizes rotated axes [58], analogous to the conflict cusp model [59] using the axes m and n depicted in Figure 1. It implements a combination of the initially proposed controls, specifically their difference (FRSSD − GDS) and their sum (FRSSD + GDS), as asymmetry and bifurcation factors, respectively.

In Cusp 2 (Table 4), (FRSSD − GDS) acts as the asymmetry factor ($b = 1.3450$, $p < 0.001$) and (FRSSD + GDS) as the bifurcation factor ($b = 1.2804$, $p < 0.001$). The chi-square test of the linear vs. cusp model gives $\chi^2 = 147.6$, $df = 2$, and $p < 0.001$, and the model fit statistics in terms of AIC, AICc, and BIC favor the cusp catastrophe model. The values for the cusp model (AIC = 271.213, AICc = 271.801, and BIC = 289.277) are minimum compared to the linear (AIC = 414.767, AICc = 415.043, and BIC = 426.809) and logistic (AIC = 395.039, AICc = 395.456, and BIC = 410.093) alternatives. The values of pseudo R^2 are 0.63, 0.39, and

0.47 for the cusp, linear, and logistic models, respectively; however, they are not counted in the assessment criteria.

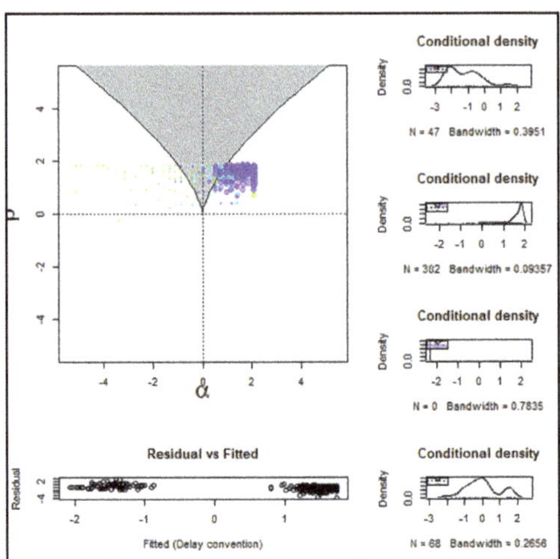

Figure 2. A visual display of the lower part of the cusp response surface of financial capacity using maximum likelihood estimation. FRSSD is the asymmetry factor and depressive symptomatology (GDS-15) is the bifurcation factor.

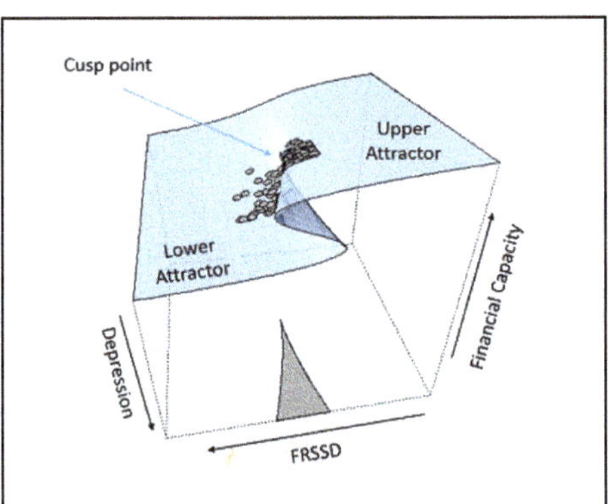

Figure 3. Three-dimensional cusp response surface financial capacity using maximum likelihood estimation. FRSSD is the asymmetry factor and depressive symptomatology (GDS-15) is the bifurcation factor. The gray dots represent observed values from empirical data.

Figure 4, as the visual display of the lower part of the cusp surface, shows that most points are located within the shaded region, the bifurcation area, and in both attractors, the upper and the lower. Figure 5, which depicts the three-dimensional cusp surface as a function of the two control variables, clearly reveals the bifurcation structure with the two diverging slops that are joined at the cusp point and are spreading in each attractor area, while no observations are located in the area of inaccessibility.

3.3. Model Interpretation

For *Cusp 1*, which implements the FRSSD and GDS as control variables (Figure 3), the interpretation of the model suggests that, at low values of depression, changes in the state variable (the financial capacity) occur in a smooth and linear manner. In this region, the linear relationship between the state variable and the asymmetry factor, the FRSSD, holds. At higher values of depression, that is, as approaching the forward-facing part of the surface, where surface folds and two behavioral attractors appear, the changes occur merely as transitions between the two attractors. In this region, people with the same control-factor values can be found at the lower and/or at higher attractor regions. This introduces unpredictability in the system and implies that changes in behavior occur as sudden jumps between two modes.

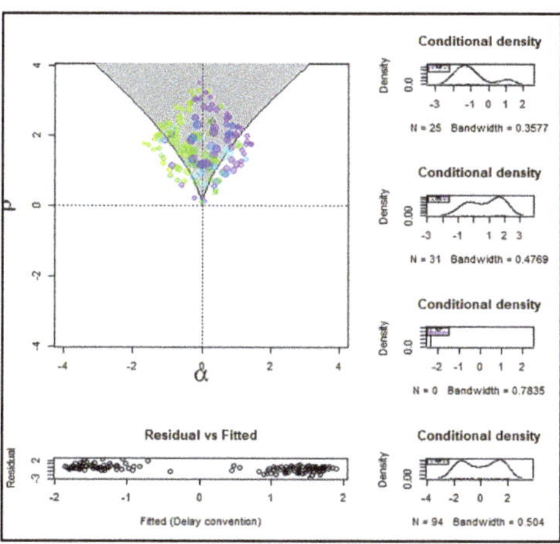

Figure 4. A visual display of the lower part of the cusp response surface of financial capacity using maximum likelihood estimation. (FRSSD − GDS) and (FRSSD + GDS) act as asymmetry and bifurcation factors, respectively.

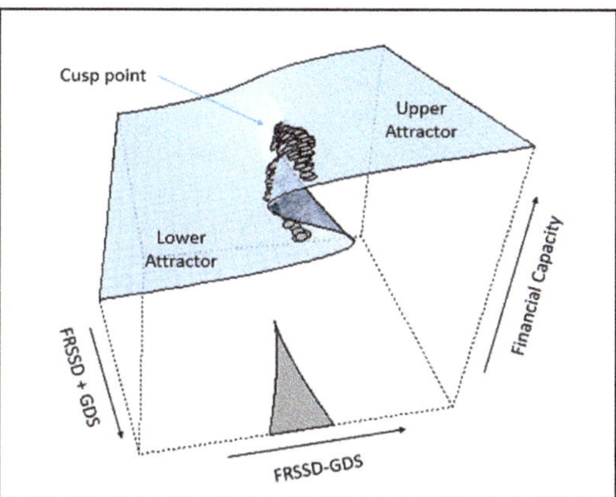

Figure 5. Three-dimensional cusp response surface financial capacity using maximum likelihood estimation. (FRSSD − GDS) and (FRSSD + GDS) act as asymmetry and bifurcation factors, respectively. The gray dots represent observed values from empirical data.

Cusp 2 has an analogous interpretation. Note that both the FRSSD and GDS are negatively associated with the financial capacity. When their difference is large, the negative effect is smaller, and as it increases the outcome increases as well. These changes are expected to be smooth and linear compared to the effect of their sum (FRSSD + GDS). When the net moderating effect of the combined high FRSSD and high GDS becomes unexpectedly increased, in that scale, a threshold value is likely to exist, beyond which abrupt changes occur, inducing bifurcation effects. The present analysis supports the above-described roles by providing empirical evidence and establishes catastrophe phenomena in this type of neuropsychological data. It is imperative to repeat that bifurcations and hysteresis effects are complex phenomena, due to the dynamics of the systems and due to self-organization mechanisms.

4. Discussion

The present catastrophe theory model, to our knowledge, is the first application reported in the domain of neuropsychology, and it has twofold implications. The first is epistemological and regards the underlying theory, while the second concerns measurement issues, classifications, and decision-making. The identification of bifurcation effects challenges the epistemological assumptions that adhere to the linear and mechanistic views of a neuropsychology system. Given that bifurcations only characterize CASs [38], their detection indicates that the underlying system is ontologically a complex adaptive system, and it should be investigated as such, i.e., the linear modeling is inadequate and epistemologically incompatible to describe and interpret the system's behavior [18]. The cusp catastrophe model designated discontinuous changes in a neurocognitive system under gradual changes in the two independent variables, the control factors, namely the asymmetry and the bifurcation. The discontinuous changes occur as transitions between two attractors, which for a neurocognitive system might represent qualitatively distinct modes of behavior, such as a high or low/suboptimal level of performance.

Note that the present cusp analysis was applied to cross-sectional data, but the interpretation of the model also needs to be extended for the dynamical path of the single case. The individual's mind involved in a cognitive task, ontologically acting as a CAS, follows a trajectory driven by self-organization mechanisms and the outcome emerges via a dynamical iterative process [21,61,62]. Bifurcations potently occur in those systems and the interpretation of the present model suggests that, in the course of such a dynamical process, even small random fluctuations in the parameters can induce sudden, unexpected transitions from a state of high performance to a state of failure.

It is pertinent to mention here that bifurcations can be observed and captured analogously when a single case ($N = 1$) is analyzed. Catastrophe phenomena might be relevant and worth examining when dynamical processes are investigated via time series, where nonlinear methods and tools should be employed. Complexity theory offers a theoretical framework and a rich array of methodological tools to support research designs and data analysis. Even though the present investigation used a large sample and cross-sectional data to infer nonlinearity, the effective methodological approach to study CASs is time series analysis [63,64]. This framework has been fruitfully applied in many process approaches [65–68], where bifurcation phenomena are theoretically anticipated and are worth examining. In those cases, catastrophes of this kind represent changes: cognitive, attitudinal, shifts to coherence, or therapeutic changes. Relevant also is the notion of *ergodicity* in a time series of the analysis [69]. Sudden shifts, transitions, and discontinuities denote a *non-ergodic process*, and the present cusp catastrophe structure supports this idea in neuropsychological data.

What has been learned for the AD and aMCI research, is that depression is not merely a linear moderator of mental operators, but it also reacts with other neurocognitive resources and prompts nonlinear effects. To further stimulate a discussion that will bridge the mathematical/methodological domain with the theoretical premises of neuropsychology, it would be pertinent to think and reflect on the role of other coexisting conditions (e.g., diabetes, heart disease) or other factors of biological and/or psychological origin. The common methodological thought suggests that, in addition to the present choices, additional variables could be included in the cusp model specification and tested with empirical data. The effect of additional candidates is an open issue for further research. However, there are some more interesting aspects to reflect on.

A pertinent epistemological remark is that a bifurcation effect should be perceived as a process, where the relevant variables being tested, as factors contributing to the underlying mechanism. Under the CAS perspective, this self-organization mechanism concerns the evolution of an interaction system that possibly includes both biological and psychological factors within a *mutual causality* connective state. The representation of such a system is explicitly the ontology of networks, which is in line with complexity theory assumptions. The proper methodology for this is the *network analysis*, where the latest advances offer a better way to approach and understand those systems [70–72] compared to the traditional methods. In addition, the network ontology explains the possibility of linear and nonlinear changes, and thus catastrophe theory is on the scene.

Another interesting remark that is explicated by catastrophe theory is that in the vicinity singularities, e.g., the bifurcations and discontinuities, the behavior could be described merely by a small number of variables. In the present context, depression is one of them. The levels of depression (GDS), even though might be affected by the dynamic interplay of other factors (biological, medical, and/or psychological), contribute to operationalization of the ensuing bifurcation mechanism, in conjunction with the functional symptoms of dementia (FRSSD).

The existing cusp structure in the data and the operating critical points beyond which nonlinear effects occur, directly concern the measurement issues and the relevant theory. The determination of such thresholds is an open issue and of paramount importance in the actual utilization of the *Legal Capacity for Property Law Transactions Assessment Scale* (LCPLTAS) and financial decision-making. In addition, given that the actual bifurcation process is induced by a composite variable, the determination of the critical point is a challenge. The issue is important because it concerns the measurement processes, diagnosis, and further prevention and treatment.

There are of course limitations in this study, originating from its exploratory character, and since it is the first report with neuropsychological data, the findings should be replicated and extended with other data sets. Cusp analysis could also be tried in other neurocognitive assessments, such as for Parkinson's disease, and in other neurocognitive assessment tools, such as the MMSE or HoNOS and GAF, to extend the model to different socio-medical inquiries. The present report sets a framework for the application of catastrophe theory with neurocognitive resources in AD research and opens new avenues for investigations.

Last, but not least, the message that the present findings convey is mainly epistemological and concerns the adoption of the meta-theoretical framework of CASs, the paradigm shift that is gaining ground in interdisciplinary research.

Author Contributions: Conceptualization, D.S.; methodology, D.S. and J.V.; software, D.S. and J.V.; validation, D.S., J.V., M.T. and V.G.; formal analysis, D.S.; investigation, D.S. and J.V.; resources, D.S., V.G., J.V. and M.T.; data curation, V.G. and M.T.; writing—original draft preparation, D.S. and J.V.; writing—review and editing, D.S., V.G., J.V. and M.T.; visualization, D.S.; supervision, D.S. and M.T.; project administration, V.G. and M.T. All authors have read and agreed to the published version of the manuscript.

Funding: This research received no external funding.

Institutional Review Board Statement: The study was approved by the Ethics Committee of the Aristotle University of Thessaloniki (protocol code 2.27/3/2013), while the research was performed according to the declaration of Helsinki.

Informed Consent Statement: Written informed consent from each participant was obtained.

Data Availability Statement: The data presented in this study are available upon request from the corresponding author.

Conflicts of Interest: The authors declare no conflict of interest.

References

1. American Psychological Association. Judicial Determination of Capacity of Older Adults in Guardianship Proceedings. A Handbook for Judges. 2008. Available online: https://www.apa.org/pi/aging/resources/guides/judges-diminished.pdf (accessed on 13 July 2022).
2. Demakis, G.J. (Ed.) *Civil Capacities in Clinical Neuropsychology: Research Findings and Practical Applications*; Oxford University Press: New York, NY, USA, 2012.
3. Marson, D.C. Loss of Financial Competency in Dementia: Conceptual and Empirical Approaches. *Aging Neuropsychol. Cogn.* **2001**, *8*, 164–181. [CrossRef]
4. Marson, D.C.; Zebley, L. The Other Side of the Retirement Years: Cognitive Decline, Dementia, and Loss of Financial Capacity. *J. Retire. Plan* **2001**, *4*, 30–39.
5. Marson, D.C.; Hebert, K.; Solomon, A.C. Assessing Civil Competencies in Older Adults with Dementia: Consent Capacity, Financial Capacity, and Testamentary Capacity. In *Forensic Neuropsychology: A Scientific Approach*; Larrabee, G.J., Ed.; Oxford University Press: New York, NY, USA, 2012; pp. 401–437.
6. Marson, D.C.; Triebel, K.L.; Knight, A. Financial Capacity. In *Civil Capacities in Clinical Neuropsychology: Research Findings and Practical Applications*; Demakis, G.J., Ed.; Oxford University Press: New York, NY, USA, 2012; pp. 39–68.
7. Marson, D.C.; Sawrie, S.M.; Snyder, S.; McInturff, B.; Stalvey, T.; Boothe, A.; Aldridge, T.; Chatterjee, A.; Harrell, L.E. Assessing Financial Capacity in Patients with Alzheimer Disease. *Arch. Neurol.* **2000**, *57*, 877–884. [CrossRef] [PubMed]
8. Sousa, L.B.; Simões, M.R.; Firmino, H.; Peisah, C. Financial and Testamentary Capacity Evaluations: Procedures and Assessment Instruments underneath a Functional Approach. *Int. Psychogeriatr.* **2014**, *26*, 217–228. [CrossRef]
9. Folia, V.; Kosmidis, M.H. Assessment of Memory Skills in Illiterates: Strategy Differences or Test Artifact? *Clin. Neuropsychol.* **2003**, *17*, 143–152. [CrossRef]
10. Kosmidis, M.H.; Vlahou, C.H.; Panagiotaki, P.; Kiosseoglou, G. The Verbal Fluency Task in the Greek Population: Normative Data, and Clustering and Switching Strategies. *J. Int. Neuropsychol. Soc.* **2004**, *10*, 164–172. [CrossRef]
11. Gerstenecker, A.; Eakin, A.; Triebel, K.; Martin, R.; Swenson-Dravis, D.; Petersen, R.C.; Marson, D.C. Age and Education Corrected Older Adult Normative Data for a Short Form Version of the Financial Capacity Instrument. *Psychol. Assess.* **2016**, *28*, 737–749. [CrossRef]
12. Giannouli, V.; Stamovlasis, D.; Tsolaki, M. Exploring the Role of Cognitive Factors in a New Instrument for Elders' Financial Capacity Assessment. *J. Alzheimer's Dis.* **2018**, *62*, 1579–1594. [CrossRef]
13. Triebel, K.L.; Martin, R.; Griffith, H.R.; Marceaux, J.; Okonkwo, O.C.; Harrell, L.; Clark, D.; Brockington, J.; Bartolucci, A.; Marson, D.C. Declining Financial Capacity in Mild Cognitive Impairment: A 1-Year Longitudinal Study. *Neurology* **2009**, *73*, 928–934. [CrossRef]
14. Giannouli, V.; Tsolaki, M. Unraveling Ariadne's Thread into the Labyrinth of AMCI. *Alzheimer Dis. Assoc. Disord.* **2021**, *35*, 363–365. [CrossRef] [PubMed]
15. Giannouli, V.; Tsolaki, M. Depression and Financial Capacity Assessment in Parkinson's Disease with Dementia: Overlooking an Important Factor? *Psychiatriki* **2019**, *30*, 66–70. [CrossRef] [PubMed]
16. Giannouli, V.; Tsolaki, M. Vascular Dementia, Depression, and Financial Capacity Assessment. *Alzheimer Dis. Assoc. Disord.* **2021**, *35*, 84–87. [CrossRef] [PubMed]
17. Giannouli, V.; Stamovlasis, D.; Tsolaki, M. Longitudinal Study of Depression on Amnestic Mild Cognitive Impairment and Financial Capacity. *Clin. Gerontol.* **2022**, *45*, 708–714. [CrossRef] [PubMed]
18. Stamovlasis, D. Methodological and Epistemological Issues on Linear Regression Applied to Psychometric Variables in Problem Solving: Rethinking Variance. *Chem. Educ. Res. Pract.* **2010**, *11*, 59–68. [CrossRef]
19. Giannouli, V.; Tsolaki, M. Is Depression or Apathy Playing a Key Role in Predicting Financial Capacity in Parkinson's Disease with Dementia and Frontotemporal Dementia? *Brain Sci.* **2021**, *11*, 785. [CrossRef]
20. Giannouli, V.; Tsolaki, M. Mild Alzheimer Disease, Financial Capacity, and the Role of Depression. *Alzheimer Dis. Assoc. Disord.* **2021**, *35*, 360–362. [CrossRef]
21. Prigogine, I.; Stengers, I. *Order Out of Chaos: Man's New Dialogue with Nature*; Bantam: New York, NY, USA, 1984.
22. Glass, L. Dynamical Disease: Challenges for Nonlinear Dynamics and Medicine. *Chaos Interdiscip. J. Nonlinear Sci.* **2015**, *25*, 097603. [CrossRef]
23. Guastello, S.J. *Managing Emergent Phenomena: Nonlinear Dynamics in Work Organizations*; Psychology Press: Mahwah, NJ, USA, 2002; ISBN 1135671958.
24. Burge, S.K.; Katerndahl, D.A.; Wood, R.C.; Becho, J.; Ferrer, R.L.; Talamantes, M. Using Complexity Science to Examine Three Dynamic Patterns of Intimate Partner Violence. *Fam. Syst. Health* **2016**, *34*, 4–14. [CrossRef]
25. Antoniou, F.; AlKhadim, G.; Stamovlasis, D.; Vasiou, A. The Regulatory Properties of Anger under Different Goal Orientations: The Effects of Normative and Outcome Goals. *BMC Psychol.* **2022**, *10*, 106. [CrossRef]
26. Navarro, J.; Rueff-Lopes, R.; Rico, R. New Nonlinear and Dynamic Avenues for the Study of Work and Organizational Psychology: An Introduction to the Special Issue. *Eur. J. Work. Organ. Psychol.* **2020**, *29*, 477–482. [CrossRef]
27. Sideridis, G.D.; Stamovlasis, D.; Antoniou, F. Reading Achievement, Mastery, and Performance Goal Structures Among Students with Learning Disabilities: A Nonlinear Perspective. *J. Learn. Disabil.* **2016**, *49*, 631–643. [CrossRef]

28. Stamovlasis, D.; Vaiopoulou, J. The Role of Dysfunctional Myths in a Decision-Making Process under Bounded Rationality: A Complex Dynamical Systems Perspective. *Nonlinear Dyn. Psychol. Life Sci.* **2017**, *21*, 267–288.
29. Martin, C.; Sturmberg, J. Universal Health (UHC) and Primary Health Care (PHC)—A Complex Dynamic Endeavor. *J. Eval. Clin. Pract.* **2022**, *28*, 332–334. [CrossRef] [PubMed]
30. Ma, Y.; Sun, S.; Peng, C.-K. Applications of Dynamical Complexity Theory in Traditional Chinese Medicine. *Front. Med.* **2014**, *8*, 279–284. [CrossRef]
31. Caffrey, L.; Wolfe, C.; McKevitt, C. Embedding Research in Health Systems: Lessons from Complexity Theory. *Health Res. Policy Syst.* **2016**, *14*, 54. [CrossRef] [PubMed]
32. Brochet, T.; Lapuyade-Lahorgue, J.; Huat, A.; Thureau, S.; Pasquier, D.; Gardin, I.; Modzelewski, R.; Gibon, D.; Thariat, J.; Grégoire, V.; et al. A Quantitative Comparison between Shannon and Tsallis–Havrda–Charvat Entropies Applied to Cancer Outcome Prediction. *Entropy* **2022**, *24*, 436. [CrossRef] [PubMed]
33. Thom, R. *Structural Stability and Morphogenesis*; Benjamin-Cummings Press: Reading, MA, USA, 1975; ISBN 9780429493027.
34. Arnol'd, V.I. *Geometrical Methods in the Theory of Ordinary Differential Equations*; Springer: Berlin/Heidelberg, Germany, 1988.
35. Castrigiano, D.P.L.; Hayes, S.A. *Catastrophe Theory*; Westview Press: Boulder, CO, USA, 2004; ISBN 9780813341255.
36. Cobb, L. Stochastic Catastrophe Models and Multimodal Distributions. *Behav. Sci.* **1978**, *23*, 360–374. [CrossRef]
37. Gilmore, R. *Catastrophe Theory for Scientists and Engineers*; Dover Publications: New York, NY, USA, 1993.
38. Nicolis, G.; Nicolis, C. *Foundations of Complex Systems*; World Scientific Publishing Co.: Singapore, 2007; ISBN 9789812775658.
39. Nicolis, G.; Prigogine, I. *Self-Organization in Nonequilibrium Systems*; Wiley: New York, NY, USA, 1977.
40. Prigogine, I. *Introduction to Thermodynamics of Irreversible Processes*; Wiley: New York, NY, USA, 1961.
41. Haken, H. *Information and Self Organization: A Macroscopic Approach to Complex Systems*; Springer: Berlin/Heidelberg, Germany, 1988.
42. Haken, H. *Synergetics: An Introduction: Nonequilibrium Phase Transition and Self-Organization in Physics, Chemistry, and Biology*; Springer: Berlin/Heidelberg, Germany, 1983.
43. Poston, T.; Stewart, I. *Catastrophe Theory and Its Applications*; Dover Publications: New York, NY, USA, 1978.
44. Stamovlasis, D. Nonlinear Dynamics and Neo-Piagetian Theories in Problem Solving: Perspectives on a New Epistemology and Theory Development. *Nonlinear Dyn. Psychol. Life Sci.* **2011**, *15*, 145–173.
45. Vaiopoulou, J.; Tsikalas, T.; Stamovlasis, D.; Papageorgiou, G. Nonlinear Dynamic Effects of Convergent and Divergent Thinking in Conceptual Change Process: Empirical Evidence from Primary Education. *Nonlinear Dyn. Psychol. Life Sci.* **2021**, *25*, 335–355.
46. Fountoulakis, K.N.; Tsolaki, M.; Iacovides, A.; Yesavage, J.A.; O'Hara, R.; Kazis, A.; Ierodiakonou, C. The Validation of the Short Form of the Geriatric Depression Scale (GDS) in Greece. *Aging Clin. Exp. Res.* **1999**, *11*, 367–372. [CrossRef]
47. Hutton, J.T. Alzheimer's Disease. In *Conn's Current Therapy*; Rakel, R.E., Ed.; W. B. Saunders: Philadelphia, PA, USA, 1990; pp. 778–781.
48. Giannouli, V.; Tsolaki, M. Financial Incapacity of Patients with Mild Alzheimer's Disease: What Neurologists Need to Know about Where the Impairment Lies. *Neurol. Int.* **2022**, *14*, 90–98. [CrossRef] [PubMed]
49. Lazarou, I.; Moraitou, D.; Papatheodorou, M.; Vavouras, I.; Lokantidou, C.; Agogiatou, C.; Gialaoutzis, M.; Nikolopoulos, S.; Stavropoulos, T.G.; Kompatsiaris, I.; et al. Adaptation and Validation of the Memory Alteration Test (M@T) in Greek Middle-Aged, Older, and Older-Old Population with Subjective Cognitive Decline and Mild Cognitive Impairment. *J. Alzheimer's Dis.* **2021**, *84*, 1219–1232. [CrossRef] [PubMed]
50. Tsolaki, M.; Fountoulakis, K.; Nakopoulou, E.; Kazis, A.; Mohs, R.C. Alzheimer's Disease Assessment Scale: The Validation of the Scale in Greece in Elderly Demented Patients and Normal Subjects. *Dement. Geriatr. Cogn. Disord.* **1997**, *8*, 273–280. [CrossRef] [PubMed]
51. Papadopoulos, F.C.; Petridou, E.; Argyropoulou, S.; Kontaxakis, V.; Dessypris, N.; Anastasiou, A.; Katsiardani, K.P.; Trichopoulos, D.; Lyketsos, C. Prevalence and Correlates of Depression in Late Life: A Population Based Study from a Rural Greek Town. *Int. J. Geriatr. Psychiatry* **2005**, *20*, 350–357. [CrossRef]
52. Cobb, L. *An Introduction to Cusp Surface Analysis*; Corrales Software Development: Carbondale, CO, USA, 1998.
53. Grasman, R.; van der Maas, H.L.J.; Wagenmakers, E.-J. Fitting the Cusp Catastrophe in R: A Cusp Package Primer. *J. Stat. Softw.* **2009**, *32*, 1–27. [CrossRef]
54. Oliva, T.A.; Desarbo, W.S.; Day, D.L.; Jedidi, K. GEMCAT: A General Multivariate Methodology for Estimating Catastrophe Models. *Behav. Sci.* **1987**, *32*, 121–137. [CrossRef]
55. Guastello, S.J. Discontinuities and Catastrophes with Polynomial Regression. In *Nonlinear Dynamics Systems Analysis for the Behavioral Sciences Using Real Data*; Guastello, S.J., Gregson, R.A.M., Eds.; CRC Press: Boca Raton FL, USA, 2011; pp. 252–280.
56. Stamovlasis, D. Catastrophe Theory: Methodology, Epistemology, and Applications in Learning Science. In *Complex Dynamical Systems in Education: Concepts, Methods and Applications*; Koopmans, M., Stamovlasis, D., Eds.; Springer: Cham, Switzerland, 2016; pp. 141–175; ISBN 9783319275772.
57. Chen, D.-G.; Chen, X.; Lin, F.; Tang, W.; Lio, Y.; Guo, Y. Cusp Catastrophe Polynomial Model: Power and Sample Size Estimation. *Open J. Stat.* **2014**, *4*, 803–813. [CrossRef]
58. Zeeman, E.C. Catastrophe Theory. *Sci. Am.* **1976**, *234*, 65–83. [CrossRef]
59. van der Maas, H.L.J.; Molenaar, P.C.M. Stagewise Cognitive Development: An Application of Catastrophe Theory. *Psychol. Rev.* **1992**, *99*, 395–417. [CrossRef]

60. van der Maas, H.L.J.; Kolstein, R.; van der Pligt, J. Sudden Transitions in Attitudes. *Sociol. Methods Res.* **2003**, *32*, 125–152. [CrossRef]
61. Haken, H.; Portugali, J. Information and Selforganization: A Unifying Approach and Applications. *Entropy* **2016**, *18*, 197. [CrossRef]
62. Haken, H.; Portugali, J. Information and Self-Organization II: Steady State and Phase Transition. *Entropy* **2021**, *23*, 707. [CrossRef] [PubMed]
63. Koopmans, M. Ergodicity and the Merits of the Single Case. In *Complex Dynamical Systems in Education: Concepts, Methods and Applications*; Koopmans, M., Stamovlasis, D., Eds.; Springer International Publishing: Cham, Switzerland, 2016; pp. 119–139; ISBN 9783319275772.
64. Koopmans, M. *Using Time Series to Analyze Long-Range Fractal Patterns*; SAGE Publications: Los Angeles, CA, USA, 2021.
65. Tschacher, W.; Haken, H. *The Process of Psychotherapy: Causation and Chance*; Springer: Cham, Switzerland, 2019; ISBN 9783030127473.
66. Guastello, S.J.; Peressini, A.F. Development of a Synchronization Coefficient for Biosocial Interactions in Groups and Teams. *Small Group Res.* **2017**, *48*, 3–33. [CrossRef]
67. Avdi, E.; Paraskevopoulos, E.; Lagogianni, C.; Kartsidis, P.; Plaskasovitis, F. Studying Physiological Synchrony in Couple Therapy through Partial Directed Coherence: Associations with the Therapeutic Alliance and Meaning Construction. *Entropy* **2022**, *24*, 517. [CrossRef] [PubMed]
68. Orsucci, F.; Petrosino, R.; Paoloni, G.; Canestri, L.; Conte, E.; Reda, M.A.; Fulcheri, M. Prosody and Synchronization in Cognitive Neuroscience. *EPJ Nonlinear Biomed. Phys.* **2013**, *1*, 6. [CrossRef]
69. Molenaar, P.C.M. A Manifesto on Psychology as Idiographic Science: Bringing the Person Back into Scientific Psychology, This Time Forever. *Meas. Interdiscip. Res. Perspect.* **2004**, *2*, 201–218. [CrossRef]
70. Epskamp, S.; Rhemtulla, M.; Borsboom, D. Generalized Network Psychometrics: Combining Network and Latent Variable Models. *Psychometrika* **2017**, *82*, 904–927. [CrossRef]
71. Epskamp, S.; Maris, G.; Waldorp, L.J.; Borsboom, D. Network Psychometrics. In *The Wiley Handbook of Psychometric Testing: A Multidisciplinary Reference on Survey, Scale and Test Development*; Irwing, P., Booth, T., Hughes, D.J., Eds.; Wiley: New York, NY, USA, 2018; pp. 953–986.
72. Cramer, A.O.J.; van Borkulo, C.D.; Giltay, E.J.; van der Maas, H.L.J.; Kendler, K.S.; Scheffer, M.; Borsboom, D. Major Depression as a Complex Dynamic System. *PLoS ONE* **2016**, *11*, e0167490. [CrossRef]

Article

Dynamics of Remote Communication: Movement Coordination in Video-Mediated and Face-to-Face Conversations

Julian Zubek [1,*], Ewa Nagórska [1], Joanna Komorowska-Mach [1,2], Katarzyna Skowrońska [1], Konrad Zieliński [1] and Joanna Rączaszek-Leonardi [1]

[1] Human Interactivity and Language Lab, Faculty of Psychology, University of Warsaw, 00-927 Warsaw, Poland; ewa.nagorska@psych.uw.edu.pl (E.N.); j.komorowska-mach@uw.edu.pl (J.K.-M.); katarzyna.skowronska@student.uw.edu.pl (K.S.); konrad.zielinski@psych.uw.edu.pl (K.Z.); raczasze@psych.uw.edu.pl (J.R.-L.)

[2] Faculty of Philosophy, University of Warsaw, 00-927 Warsaw, Poland

* Correspondence: j.zubek@uw.edu.pl

Abstract: The present pandemic forced our daily interactions to move into the virtual world. People had to adapt to new communication media that afford different ways of interaction. Remote communication decreases the availability and salience of some cues but also may enable and highlight others. Importantly, basic movement dynamics, which are crucial for any interaction as they are responsible for the informational and affective coupling, are affected. It is therefore essential to discover exactly how these dynamics change. In this exploratory study of six interacting dyads we use traditional variability measures and cross recurrence quantification analysis to compare the movement coordination dynamics in quasi-natural dialogues in four situations: (1) remote video-mediated conversations with a self-view mirror image present, (2) remote video-mediated conversations without a self-view, (3) face-to-face conversations with a self-view, and (4) face-to-face conversations without a self-view. We discovered that in remote interactions movements pertaining to communicative gestures were exaggerated, while the stability of interpersonal coordination was greatly decreased. The presence of the self-view image made the gestures less exaggerated, but did not affect the coordination. The dynamical analyses are helpful in understanding the interaction processes and may be useful in explaining phenomena connected with video-mediated communication, such as "Zoom fatigue".

Keywords: remote communication; movement coordination; recurrence quantification analysis

Citation: Zubek, J.; Nagórska, E.; Komorowska-Mach, J.; Skowrońska, K.; Zieliński, K.; Rączaszek-Leonardi, J. Dynamics of Remote Communication: Movement Coordination in Video-Mediated and Face-to-Face Conversations. *Entropy* **2022**, *24*, 559. https://doi.org/10.3390/e24040559

Academic Editors: Franco Orsucci and Wolfgang Tschacher

Received: 1 February 2022
Accepted: 11 April 2022
Published: 15 April 2022

1. Introduction

When two people engage in a dialogue, they do much more than just exchanging strings of words. According to Fusaroli et al. [1], dialogue participants coordinate on multiple levels, establishing a functional organization fit to a particular situation. Essentially, they form a coupled system within which meanings are co-created, and interaction dynamics are essential to this process [2]. The ability to coordinate movements during interaction is already present in infancy [3] and constitutes the most basic form of bonding with others [4]. Movement coordination allows the establishment of informational and affective coupling [5,6]. This has consequences for various processes of social cognition. As demonstrated by numerous empirical studies, spontaneous movement coordination of people engaged in natural conversations can predict rapport [7], affiliation [8], empathic accuracy [9], joint-action task performance [10,11] or psychotherapy outcomes [12]. The connections between movement coordination and social interaction may go in both directions: particular patterns of movement coordination may be constitutive factors for the interaction or they can be merely indicators of a successful interaction taking place [13]. In any case, by analyzing interpersonal movement coordination, we can infer much regarding the quality of an interaction.

In the present pandemic, many social interactions have moved online. Remote video calls are used as an alternative to face-to-face conversations, both in professional and casual contexts. Video-mediated interactions indeed allow the use of visual cues (gestures, face expressions, body posture) and the establishment of some form of functional movement coordination between participants, which is not possible in audio-only interactions. Studies comparing video-mediated communication to audio-only communication report benefits such as increased effectiveness of group problem-solving, shorter discussion time, and increased emotional bonding [14,15]. However, the experience with video-mediated interactions is not always smooth. In some cases, people were more satisfied with audio-only interactions than with video-mediated interactions, and audio-only interactions seemed more efficient [16–18]. Recently, there have been discussions regarding "Zoom fatigue", a form of exhaustion experienced by participants of video conference meetings [19–22]. The possible causes of this phenomenon include both a lack of proper social cues (i.e., eye contact, body language), leading to increased cognitive effort, and information overload (i.e., self-image visible, multiple faces visible on the screen), leading to additional stress [23].

A deeper understanding of video-mediated communication can be gained by studying the process of interaction itself [16,24]. Different media provide characteristic constraints and afford specific communicative actions with different degrees of synchronicity. This shapes the ongoing interaction process and, consequently, interaction outcomes. In the case of video-mediated interactions, disrupted social cues and visual information overload may affect the capabilities of nonverbal communication, leading to different coordination dynamics than in face-to-face interactions. We suspect that altered coordination capabilities in online communication may influence informational and affective couplings between participants, may be a possible cause of decreased satisfaction with an interaction, as stated in the recent literature, and may also cause decreased effectiveness of communication as compared to face-to-face interactions.

1.1. Dynamics of Video-Mediated Interactions

Patterns of social interaction dynamics are emergent properties shaped by multiple interrelated factors [13,25]. In the case of natural conversation, any change in a participant's impression of their interlocutor influences the way the participant responds, which in turn influences the interlocutor. This ongoing feedback loop, constituting patterns of interaction dynamics, may work differently in mediated interactions. A communication medium—such as a video-conferencing setup—is one of the factors that may significantly constrain interaction dynamics. In the language of dynamical systems, if a medium offers fewer possibilities for interaction than the number of available options in unmediated communication, the number of degrees of freedom of the system is reduced. On the one hand, when the preferred interaction means are taken away, it may disrupt the interaction. On the other hand, when the redundant modes of communication are reduced, it may present a case of functional reduction in degrees of freedom facilitating the interaction. Either way, the patterns of interaction dynamics are changed.

Constraints imposed by the communication medium can be traced through the analysis of interactions between a person and the medium. In this case, the ecological psychology notion of affordance is helpful [26]. Affordances are opportunities for action and perception offered by the environment to an active subject. They are not simply objective properties of the external objects (shape, size), but meaningful relations in which complementarity between the subject and its environment manifests (graspability, possibility to sit upon). In the social realm, affordances are created and used dynamically by each interactant "on the fly" [27]. Introducing a video-based communication medium creates new possibilities for actions and forms of interaction, while precluding others. The landscape of affordances available for the individuals and the dyad changes, which changes their behavior and cocreated meanings [28].

Affordances of video-mediated interaction are significantly changed by the presence of video latency—a mean delay between the moment the movement is made, and the moment it is visible on another user's screen. Another aspect is jitter—variability of the delay, caused by the different length of time each data packet takes to arrive. If the jitter is large, movement in the video is not smooth. Video latency during a high quality video call may be 150 ms with 40 ms jitter [29], but these values may vary depending on the network traffic, connection bandwidth and hardware configuration/quality. Since latency works in both directions, the effective time between a communicative action and the perceived response may double. Additionally, glitches in the form of video freezing or distorted images are common during video calls. These factors modify the affordances of interaction participants, for instance, by limiting the possibility of reacting quickly to each other thus constraining their patterns of coordination. It is known that people are able to perceive delays of 200 ms [30,31], which suggests that even relatively small video latency may affect coordination in a video-mediated interaction. Boland et al. [32] studied turn-taking during face-to-face and Zoom conversations and discovered that delays introduced by the latter significantly disrupted the rhythm of conversation, increasing the average turn transition time from 135 ms to 487 ms. Such altered coordination patterns may have further consequences for communication. The length of the gap between turns may provide information on the valence of the upcoming response, with preferred responses coming quicker and taking simpler forms [33]. A gap as short as 300 ms may be sufficient to project that a straightforward acceptance is less probable [34]. Because of the prolonged gaps due to the video latency, speakers may erroneously expect more dispreferred reactions than in face to face communication. Additionally, according to the studies on telephone communication, the longer the delays are, the more interlocutors are perceived as less attentive, less friendly, less extraverted and less conscientious [35].

Another aspect that differentiates video-mediated and face-to-face interactions is the way the image of the conversation partner is presented to the interaction participant. In natural face-to-face conversations, people typically face each other, moving their glances between the face, body and hands of the interaction partner [36], which provides them with specific means to fluently structure the interaction (see, e.g., Rączaszek-Leonardi and Nomikou [37]). In contrast, in a typical video-mediated interaction (for instance, using a laptop computer with a built-in camera), the captured field of vision is much narrower, limiting visual cues concerning whole body movement and hand gestures. This may severely limit nonverbal communication, as hand gestures play an important role in supplementing speech with additional content, disambiguating expressions or organizing turn-taking [38–40]. It is possible to compensate for this through the use of other modalities such as head gestures, which are captured well in video-conferencing settings. Head gestures are considered to be important for coordinating interaction, providing confirmatory feedback for the speaker [41] and signaling turn claims [42]. In many cultures head nodding and head shaking are associated with affirmative and negative responses, respectively (Refs. [43–45], but with exceptions [46]). Being able to convey approval through head gestures during conversation would be an important factor contributing to the perceived naturalness of an interaction. Additionally, the need to fit within the field of view of the camera may limit the overall movement and induce a feeling of being physically trapped [21]. In face-to-face meetings, people can shift their position and stretch, but during video communication their mobility is limited to a narrow space. This reduced mobility may undermine cognitive performance [47], further disrupting communicative abilities.

Moreover, in many video conferencing programs, there is a setting in which a self-image of the participant is displayed along with the image of their interaction partner. This may be potentially disturbing in several ways. It may change the basic gaze dynamics, which was claimed to serve as a "glue" for interaction [48], and introduce effects on individuals' behavior similar to the presence of a mirror. Research in social psychology shows that seeing the self-image in a mirror can heighten self-focused attention, which in the case of longer exposition can have negative psychological consequences, including decreased mood or even depression ([19,49–51], but see [52,53]). In the interactive context, self-focused attention was reported to decrease prosocial behavior in some contexts [54], although it is possible to find conditions in which it enhances prosocial behavior [55]. Finally, seeing oneself in a mirror provides visual feedback—an additional affordance that might be used for more precise control of one's appearance and expression. Little is known about the consequences of the visible self-image for coordination with a conversation partner.

Our investigation complements existing studies on the naturalness of online interactions through the introduction of the movement coordination perspective and the dynamical systems methodology, which goes beyond individual cognitive processes by focusing on coupling. We show how the movements of individuals are constrained in video-mediated interactions, and what patterns of interpersonal coordination emerge.

1.2. Current Study and Hypotheses

The goal of our study was to explore movement coordination dynamics shaped by the affordances altered by video-mediated means of communication. We identified factors such as: restricted mobility in front of the camera, video latency and jitter, and the optional presence of one's own mirror image. All these components potentially constrain movement of the individual, modify adopted nonverbal communication strategies, and, finally, reshape interactive patterns of interpersonal coordination. To disentangle the influences of the video medium and the mirror image, we adopted an experimental design in which casual, friendly conversations of the same dyads were recorded in four conditions: (I) video-mediated remote conversation with the mirror image displayed, (II) video-mediated remote conversation without the mirror image, (III) face-to-face conversation with the mirror image, and (IV) face-to-face conversation without the mirror image. We expected the differences to be manifested at the individual level and at the dyadic coordination level. At the individual level:

Hypothesis 1. *Overall movement will be more restricted in remote interactions, because of the need to stay visible (in the field of view of the camera) and to see the interlocutor.*

Hypothesis 2. *Intentional communicative gestures will be exaggerated (in comparison to the overall movement) in remote interactions to compensate for potential disruptions.*

Hypothesis 3. *The availability of the self mirror image in remote interactions will allow participants to calibrate their expressions, making the movement more natural and less exaggerated. No such effect is expected for face-to-face interactions, where natural instantaneous feedback is available through the partner's reactions.*

Regarding interpersonal movement coordination, we expected that:

Hypothesis 4. *Coordination will be more stable in face-to-face interactions, and episodes of coordination will be longer.*

Hypothesis 5. *Coordination will be less stable with the mirror image present, as it presents an additional distraction (participants captivated by their own movement may be less attentive to their partners).*

To operationalize our hypotheses, we tracked participants' head movement during conversations using OpenPose software [56]. We focused on head movements, as they were important and visible both in face-to-face and remote conversations. According to the existing literature, the dominant head gesture during conversations is nodding, which is associated with vertical motion [40,41]. Head nodding (vertical motion) and head shaking (horizontal motion) are typically distinguished as they are associated with positive and negative responses, respectively [44]. Head nodding was reported to increase the perceived likability and approachability of a person [43]. Following this logic, we decided to differentiate between vertical and horizontal motion in our analyses. After watching the collected video material, we discovered that there were multiple episodes of head nodding in response to the partner, but hardly any head shaking. This was consistent with the friendly character of the conversations, where head nodding is expected to be much more prominent than head shaking [57]. Horizontal head movements in our recordings seemed to result not from head shaking, but mostly from body sways and position adjustments less connected with the conversation dynamics. Thus, at the risk of oversimplification and with the limits of cross-cultural generalization in mind, we interpreted vertical head movement as an indicator of intentional communicative gestures expressing positive reaction to the interlocutor, and horizontal head movement was treated as a control—an indicator of general body movement.

When operationalizing interpersonal coordination, we decided to focus on the congruence of head movement direction within the dyad. Two people moving their heads in the same direction (nodding, tilting, turning, etc.) simultaneously or with a constant delay examplify coordinated behavior. We quantified the coordination using cross-recurrence quantification analysis (cRQA) [58], a nonlinear technique providing measures of coordination stability.

2. Materials and Methods

2.1. Participants and Setup

The examined material consisted of 24 recordings (137 min in total), collected from interactions of two groups of three people: Group A consisted of three men, and Group B consisted of three women (age 22–35). All participants were university students. The study was approved by the ethics committee of the Faculty of Psychology, University of Warsaw. Participants gave their consent to record their conversations and use them for research purposes.

Participants were students in the same program. Their level of acquaintance was assessed through a short interview. Participants from Group A were attending online courses together and had a chance to get to know each other while doing a group project together. Participants from Group B were engaged in research within the same research group and spent some time socializing before participating in the study. They can be described as colleagues, but there were no close friends within either group. All conversations were held in English, which was the second language for all participants. All participants had previous experience using videoconferencing software and were used to this form of communication.

Within each group, everyone was paired up, therefore creating six dyads (three per group) in total. Each dyad engaged in two conversations: one conducted remotely and one face-to-face, and each of these conversations was divided into two parts: with the mirror image and without. Each part lasted approximately five minutes. We briefed the participants regarding the purpose of the study, length of the conversations and the differences between experimental conditions. Participants knew that their movement will be tracked and their coordination will be analyzed. They were not informed on the detailed study hypotheses. Participants were instructed to keep the conversations casual and choose the topic freely. Most of the conversations started with a general opening question ("What's up?") and then developed spontaneously. Topics such as university studies, work, vacations, hobbies, etc., emerged. All conversations were friendly in tone, and no controversial topics or heated debates occurred.

Figure 1 presents the general schema of the four experimental conditions.

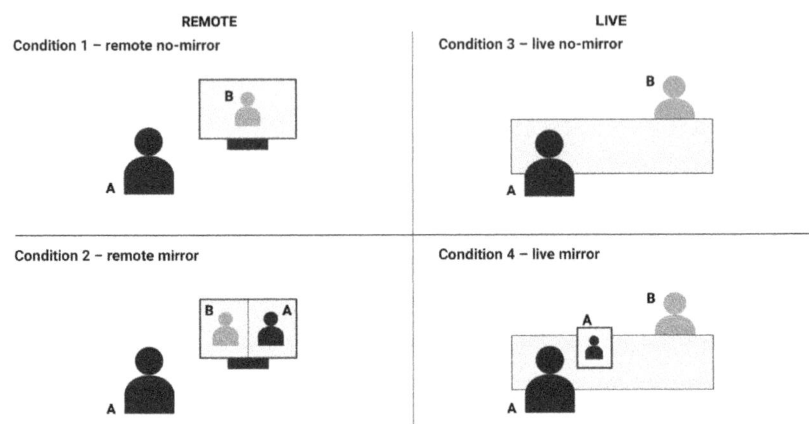

Figure 1. General schema of experimental conditions. In Condition (1), "remote no-mirror", the participant sees their partner on the screen; in Condition (2), "remote mirror", the participant sees their partner and their own mirror image side by side. In Condition (3), the "live no-mirror" participant sits in front of their partner with a dimmed smartphone screen placed in between, and in Condition (4), the "live mirror" participant sits in front of their partner with a smartphone displaying mirror image placed in between.

The remote conversations took place on the Google Meet platform. Two participants engaged in the conversation, and the researcher joined the meeting and recorded the interaction using OBS Studio software for screen recording. The researcher recorded the meeting in a "gallery view" mode, where images of the two interlocutors were placed side by side. Both participants were recorded with lag characteristic for the videoconferencing platform. In the mirror condition (with self-view), the participants saw both the other person and their own face, while in the no-mirror condition (without self-view) they could only see their interlocutor. They conducted a single 10-min conversation starting without self-view and switching self-view after 5 min. Participants used their own laptops with built-in video cameras.

Before the actual recordings of remote conversations, trial recording sessions took place during which participants were able to familiarize themselves with the setup. After the trial sessions, participants were instructed to adjust their setup (position of the camera, lighting) to improve the quality of the recordings.

Face-to-face conversations were recorded via a smartphone camera connected to a laptop (using Droidcam OBS and OBS studio software). We connected two smartphones to the same laptop via a local WiFi network and used OBS studio to combine the two image streams into a single output video file in which images of two interlocutors were placed side-by-side (as in the typical videoconference setup). We placed each smartphone in front of one of the interlocutors, with the front camera filming one's face and upper body. The participants were given a few minutes to sit down and adjust their positions to make them feel comfortable and ensure they fit into the video frame. The mirror condition was reproduced by showing the person's face and upper body position and movements in real time on the smartphone screen. The participants had a single 10-min conversation, in which smartphone screens were dimmed for the first half and were switched on for the second half.

2.2. Movement Tracking

We converted the video recordings to a common video format with 20 FPS. Each video frame contained the images of two participants side by side. We cropped the videos to obtain a separate video file for each participant during each conversation. The minimal resolution of the cropped video was 530 × 304 pixels. All videos were downscaled to this resolution. We processed the videos with OpenPose software [56] to obtain the x-y coordinates of key body parts (see Figure 2). For our analyses we extracted coordinates of points P0 (tip of the nose) and P1 (point in the middle of the torso on the shoulder level). There were missing values due to the algorithm not identifying a keypoint on a particular frame. In the recordings of one male dyad in the remote condition, the numbers of missing values were particularly large (16–55%). We removed these two recordings from the analysis. A small number (<5%) of missing values in other recordings were imputed using linear interpolation. Afterward, we applied a running median filter with a window size of five for each coordinate separately to remove possible outliers.

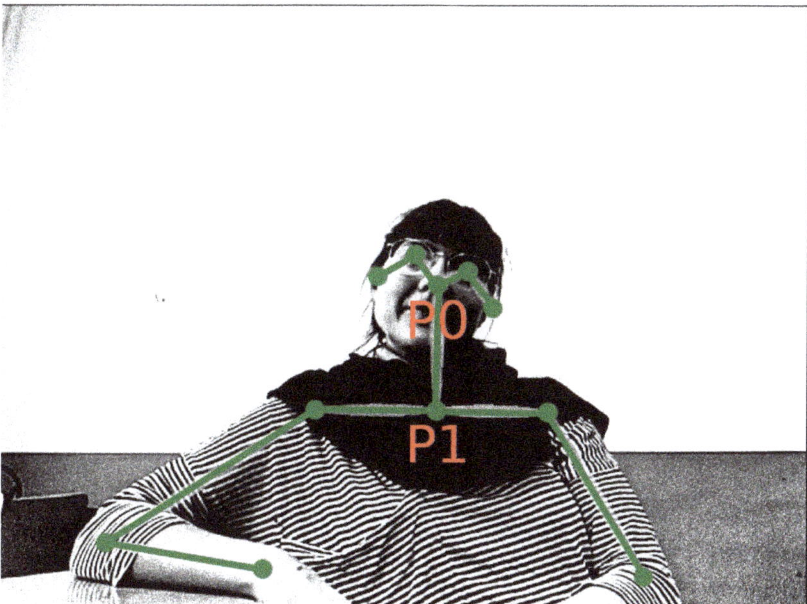

Figure 2. Output from OpenPose program: a video frame with detected key points marked. The two key points used in our analysis are P0 (tip of the nose) and P1 (point in the middle of the torso on the shoulder level).

2.3. *Measures and Data Analysis Techniques*

In our analyses, we focused on the movement of two points: the tip of the nose (point P0), as an indicator of head movement, and the middle of the torso (point P1), as the reference (see Figure 2). To normalize the data, we used the average P0-P1 distance for each person as a natural scale of movement. To operationalize our hypotheses regarding individual movement, we introduced the following measures:

- Horizontal mobility—standard deviation of the horizontal P0 coordinate divided by the average P0-P1 distance. It is interpreted as a general indicator of participant mobility.
- Vertical mobility—standard deviation of the vertical P0 coordinate divided by the average P0-P1 distance. It is interpreted as an indicator of communicative nodding gestures.
- Horizontal-vertical mobility ratio—ratio between horizontal and vertical mobility. It is interpreted as a ratio between overall movement and communicative nodding gestures.

The described measures were calculated separately for each of the two members of the six dyads in each of the four conditions, which should result in 48 data points. Since we excluded two recordings of the particular dyad in the remote condition (see Section 2.2), the final number of analyzed data points was 44.

To analyze the properties of interpersonal coordination, we focused on the direction of frame-to-frame movement of P0 point. For each frame we calculated a 2D vector, representing the shift in position from the previous frame. All vectors were normalized to have unit length. A low-pass Butterworth filter was used to smoothen the data. Then we calculated the interpersonal coordination statistics using the methodology inspired by multidimensional cross-recurrence quantification analysis [59]. We constructed separately for x and y coordinates time-delayed embeddings using a delay of 7 frames and embedding dimension 4 (values chosen using minimal mutual information heuristic for delay and false nearest neighbors for dimension [60]). Embeddings for the two coordinates were concatenated, resulting in a final dataset with eight columns. We constructed a recurrence matrix by calculating distances between all pairs of 8-dimensional vectors and thresholding them using a fixed value. All distances below the threshold formed recurrent points. We chose the threshold value for each matrix separately to ensure that the fraction of recurrence points was always 10%. In this way, RQA statistics were normalized across dyads and experimental conditions (This methodology is different from some other studies using RQA (e.g., Rączaszek-Leonardi et al. [11]), where threshold value is fixed across all samples and the fraction of recurrent points (RR) was compared across conditions. In the case of our data, differences in optimal threshold level were too large for this kind of comparison.).

In layman's terms, a cross-recurrence matrix represents the temporal structure of "meetings" of two evolving systems. A recurrent point with coordinates (i, j) means that system A at time point i was in the same state as system B at time point j. In the context of participants of our study, recurrence means that two participants moved in the same direction relative to their cameras. Recurrent points on the main matrix diagonal indicate that participants' movements were synchronized, while recurrent points outside the main diagonal indicate more complex kinds of coordination. We controlled for the fraction of recurrent points—denoting the overall strength of coordination— and quantified characteristic patterns of coordination through the analysis of diagonal and vertical lines formed by recurrent points. We will use the following notation: l—length of diagonal line, $P(l)$—probability of a diagonal line of length l occurring, v—length of vertical line, $P(v)$—probability of a vertical line of length v occurring. Then, popular recurrence quantification measures can be defined as follows:

- Determinism, fraction of recurrent points forming diagonal lines.

$$DET = \frac{\sum_{l=l_{min}}^{N} lP(l)}{\sum_{l=1}^{N} lP(l)}$$

  A large DET means that there are stable episodes of coordination and that coordination is more predictable. In interaction it suggests that partners may anticipate each other's actions and successfully maintain coordination.
- Entropy of the distribution of diagonal line lengths.

$$ENTR = -\sum_{l=l_{min}}^{N} P(l) \ln P(l)$$

  A large ENTR means that the coordination is more complex with more characteristic patterns of coordination. This suggests that the interaction process is more varied.
- Average length of a diagonal line.

$$L = \frac{\sum_{l=l_{min}}^{N} lP(l)}{\sum_{l=l_{min}}^{N} P(l)}$$

  A large L means that the episodes of coordination are longer on average.

- Lmax – maximum length of a diagonal line. A large Lmax means that it is possible to maintain coordination for a longer time.
- Laminarity, fraction of recurrent points forming vertical lines.

$$LAM = \frac{\sum_{v=v_{min}}^{N} vP(v)}{\sum_{v=1}^{N} vP(v)}$$

  Vertical lines form when one participant remains in the same state (moving uniformly or being still) for some time. A large LAM indicates that participants' movement is steadier.
- Trapping time, average length of a vertical line.

$$TT = \frac{\sum_{v=v_{min}}^{N} vP(v)}{\sum_{v=v_{min}}^{N} P(v)}$$

  A large TT means that the episodes of steady movement are longer on average.

We counted only diagonal and vertical lines of length 10 or more ($l_{min} = v_{min} = 10$ corresponds to episodes of coordination or steady movement lasting 0.5 s or more; this value was chosen empirically to ensure sufficient variability of DET and LAM statistics). RQA measures were calculated for each of the 6 dyads across 4 conditions, except for the one dyad for which recordings of remote interactions were excluded from the analysis (see Section 2.2). The final sample consisted of 22 observations.

We performed statistical analysis using mixed-effects linear models adequate for the repeated measures experimental design. All analyses were performed in Julia programming language using the packages DynamicalSystems.jl [61] and MixedModels.jl [62].

3. Results

3.1. Horizontal and Vertical Mobility

We started by comparing participants' mobility along horizontal and vertical dimensions across the experimental conditions (see Figure 3). The differences were quantified using mixed-effects linear models, with model coefficients presented in Table 1.

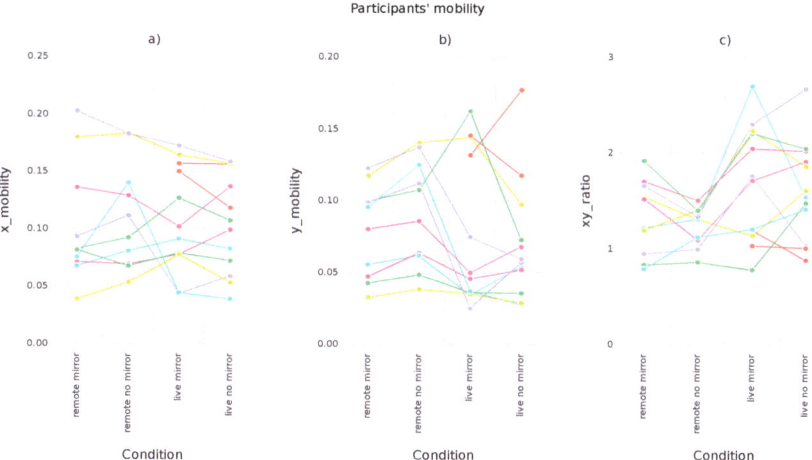

Figure 3. Average participants' mobility in horizontal (**a**) and vertical (**b**) dimensions, and their ratio (**c**) across experimental conditions. Mobility is defined as the standard deviation of the participant position on the video frame. For each dyad, two lines are drawn: one for Participant A, and one for Participant B (same color lines for participants in each dyad).

Horizontal mobility was similar in all conditions, vertical mobility was larger in remote conditions ($p = 0.001$), and the ratio was significantly larger in face-to-face conditions ($p < 0.001$). Additionally, studying the plot (Figure 3b) suggested that vertical mobility might be slightly larger in the "remote no-mirror" condition in comparison to the "remote mirror" condition. To verify this, we applied an additional paired samples Student's t-test which compared the two conditions. We obtained $t = -4.8922$ ($DF = 10$) and $p < 0.001$, which gives support to the hypothesis that the conditions differ.

Interpreting the results in the light of research hypotheses, we had to reject Hypothesis 1, as neither horizontal nor vertical mobility was visibly restricted in remote interactions. Hypothesis 2—stating that in remote interaction, participants exaggerate communicative gestures—was confirmed by the differences in vertical mobility and horizontal-vertical mobility ratio. Larger vertical mobility and a smaller ratio in remote conditions suggest that participants increased their range of nodding movements while restricting other movements. Finally, comparison of vertical mobility between the "remote mirror" and "remote no-mirror" conditions supports Hypothesis 3: the presence of self-image in the mirror condition reduced exaggerated nodding gestures.

Table 1. Coefficients of mixed-effects linear models comparing horizontal and vertical mobility across experimental conditions.

	Est.	**SE**	z	p	σ
		Horizontal mobility			
(Intercept)	0.1046	0.0126	8.27	$<10^{-15}$	0.0398
remote	0.0088	0.0066	1.33	0.1825	
no mirror	0.0015	0.0063	0.24	0.8104	
Residual	0.0210				
		Vertical mobility			
(Intercept)	0.0724	0.0123	5.87	$<10^{-8}$	0.0374
remote	0.0235	0.0074	3.19	0.0014	
no mirror	0.0024	0.0071	0.35	0.7295	
Residual	0.0235				
		Horizontal-vertical mobility ratio			
(Intercept)	1.6989	0.1281	13.27	$<10^{-39}$	0.3627
remote	−0.4709	0.0910	−5.18	$<10^{-6}$	
no mirror	−0.0827	0.0876	−0.94	0.3450	
Residual	0.2906				

3.2. Interpersonal Movement Coordination

Figure 4 presents cRQA statistics for interactions in all four conditions, while Table 2 contains coefficients of mixed-effects linear regression models verifying the strengths of effects for each statistic. As we can see, differences between remote and face-to-face interactions are evident on all measures except TT, which is congruent with Hypothesis 4. We found no visible effect of mirror image presence on movement coordination; there is no support for Hypothesis 5.

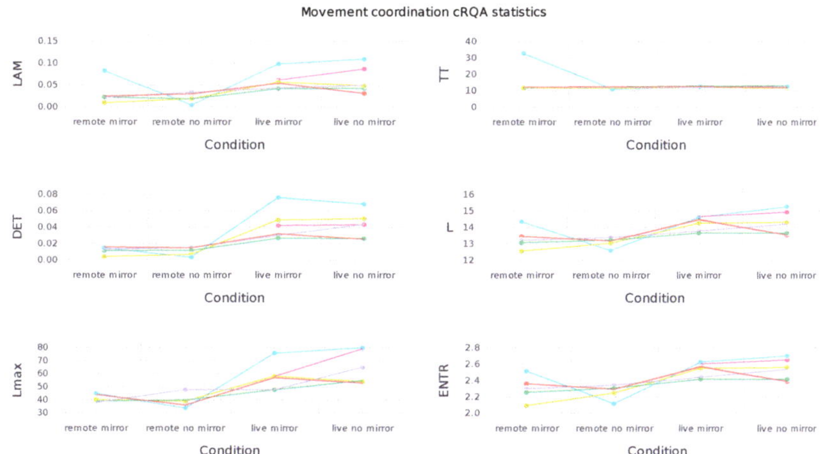

Figure 4. cRQA statistics describing properties of participants' movement coordination across experimental conditions. For each dyad a single line is drawn.

Table 2. Coefficients of mixed-effects linear models comparing various RQA measures across experimental conditions.

	Est.	**SE**	**z**	p	σ
			ENTR		
(Intercept)	2.5486	0.0362	70.41	$<10^{-99}$	0.0000
remote	−0.2672	0.0418	−6.39	$<10^{-9}$	
no mirror	−0.0138	0.0418	−0.33	0.7408	
Residual	0.1024				
			DET		
(Intercept)	0.0429	0.0042	10.23	$<10^{-23}$	0.0043
remote	−0.0323	0.0044	−7.33	$<10^{-12}$	
no mirror	−0.0001	0.0044	−0.03	0.9776	
Residual	0.0108				
			L		
(Intercept)	14.3208	0.1730	82.78	$<10^{-99}$	0.1239
remote	−1.1179	0.1910	−5.85	$<10^{-8}$	
no mirror	−0.0553	0.1910	−0.29	0.7720	
Residual	0.4679				
			Lmax		
(Intercept)	59.5000	2.9122	20.43	$<10^{-92}$	1.5260
remote	−21.5000	3.2848	−6.55	$<10^{-10}$	
no mirror	2.8333	3.2848	0.86	0.3884	
Residual	8.0462				
			LAM		
(Intercept)	0.0593	0.0088	6.71	$<10^{-10}$	0.0120
remote	−0.0315	0.0085	−3.71	0.0002	
no mirror	0.0024	0.0085	0.28	0.7764	
Residual	0.0208				

Table 2. *Cont.*

	Est.	SE	z	p	σ
			TT		
(Intercept)	13.0525	1.5470	8.44	$<10^{-16}$	0.0000
remote	2.1017	1.7863	1.18	0.2394	
no mirror	−0.9629	1.7863	−0.54	0.5899	
Residual	4.3755				

4. Discussion

Our results show that shifting to remote communication changes the dynamics of movement manifested by individuals and on the dyadic level. Our intuitions that participants move differently during remote and live interactions were confirmed by quantitative analyses. During remote interactions, they exaggerated their nodding gestures, which may stem both from the awareness of their lesser visibility by the partner and compensation for the unnaturalness of the situation. This effect was reduced when the self-image was present. One of the possible explanations is that in remote conversations participants lacked some immediate feedback from their interlocutors and were unsure whether their gestures were visible. The self-image might have provided compensatory feedback allowing them to calibrate their expression.

On the dyadic level, we demonstrated that in video-mediated remote conversations interpersonal coordination was less stable (smaller DET), less complex (smaller ENTR) and occurred in shorter episodes (smaller L and Lmax). Our findings suggest that partners interacting remotely do not form a coupled system with the same properties as in natural face-to-face interactions. According to De Jaegher et al. [13], particular dynamics of social interaction enable processes of social cognition. With the altered interaction dynamics, these processes might be disrupted, diminishing mutual understanding between interaction partners.

Contrary to our expectations, the presence of the self-image in the mirror condition had no visible effect on movement coordination. It is possible that the movements we captured do not reflect the changes that might be induced by this presence—such as changes in gaze behavior. In any case, these changes did not result in altered coordination. It is also possible that the effects were too subtle to be detected in the current experimental design, e.g., due to the brevity of five-minute conversations.

The study of movement coordination not only provided objectively measurable determinants of the quality of communication but also allowed us to transfer the analysis from the level of the individual to the level of dyad dynamics. This is in line with the embodied and enacted perspectives on social interactions [2,13] and compatible with Burgoon's "principle of interactivity" [16], suggesting that the process of interaction afforded by a communication medium should be characterized first before investigating interaction outcomes. Our investigation of movement coordination complements individualistic studies pertaining to individual satisfaction and cognitive load during online conversations [19,20].

The interactive perspective might potentially provide an alternative explanation of the "Zoom fatigue" phenomenon. Our results demonstrate that interaction properties deemed to enable social cognition [13] are altered, and the coordination is overall less complex (smaller DET and ENTR) in remote interactions. In that case, what is missing are not so much individual social cues (such as gestures or facial expressions) but rather "interactive cues"—specific properties of the interaction dynamics that allow us to tell an affiliative conversation from a quarrel, the continuation of an ongoing conversation topic from the beginning of a new topic, etc. Lack of this interactional scaffolding might lead to confusion and frustration. Further research could test this hypothesis by combining the two perspectives and checking how the satisfaction reported by the respondents participating in video-mediated interactions is reflected in their coordination. This would confirm whether coordination properties are actually connected with the experienced fatigue. The results could also be compared with previous studies associating movement synchrony with positive outcomes in face-to-face interactions [7,8,63].

Another intriguing perspective that may provide a framework for reflection on the sources of perturbations in video communication is the comparison with audio-only communication. From an information theory perspective, a video call offers a channel of greater capacity—it allows transmitting more information than a phone call. However, despite audio communication being more limited, we observe no "phone fatigue" phenomenon. This may suggest that perhaps extra information provided by video communication is actually more cognitively demanding than helpful. For the "receiver", the nonverbal message might be more difficult to interpret because some contextual social cues facilitating the interpretation are altered in remote interactions (for instance, response times allowing the prediction of positive or negative reactions [34]). Increased channel capacity in the case of video-mediated interactions may also be more demanding from the sender's perspective. For example, being aware that at least some parts of their body are visible to the partners and therefore gestures are an important source of information on interaction, senders feel obliged to use their body language in the same manner as in a normal face-to-face conversation. This makes a difference with audio-only interactions, since the same body language that is appropriate during a phone call is no longer appropriate within video conversation. At the same time, remote communication limits the possibility of the natural use of body language, as demonstrated by our result of more exaggerated nodding gestures, which may lead to an experience of frustration or fatigue. Examination of the impact of these factors in comparison between video and audio-only conversations is another interesting line of further research, especially with an attempt to untangle the experience related to sender and receiver perspectives.

Continuing the information-theoretic considerations, we should also discuss the role of noise in the communication channel or the reliability of a medium. From the user point of view, a tool that offers less functionality but is more predictable is still more effective than a more powerful but unreliable tool [64]. A video call is a channel of greater capacity than an audio-only call, but at the same time, it is more affected by noise due to latency and jitter. Video calls are prone to image and audio lags and disturbances, which even if they are minor and seemingly insignificant, may keep both sender and receiver in a state of constant uncertainty about how much information is lost during the transmission. Shorter episodes of stable coordination in video-mediated interaction discovered in the study may be a sign of low reliability of this medium: whenever participants began to coordinate on a nonverbal level, an unpredictable signal distortion might have destroyed the coordination.

Our small exploratory study does not allow us to formulate any strong recommendations concerning preferred forms of remote communication. Nevertheless, some cautious observations can be formulated. Despite worries that the presence of the self-image makes the conversation less natural, it may have its use as a source of compensatory feedback during interaction. Using this option can thus be recommended. As coordination in remote interaction is overall less stable, some conscious effort could be made to stabilize it. The simplest idea would be to deliberately slow down and avoid fast gestures, which could be misinterpreted due to video lag. Assessing such a strategy would require additional studies.

Limitations

Although our results confirmed that movement coordination is impaired in remote communication, they do not allow us to draw conclusions as to the main factors that contribute to this result. We note that our data are not conclusive on the effect of the presence of the self-image in the mirror condition, which might be one possible source of distraction. We observed no significant differences in movement coordination comparing these two conditions of both live and remote conversations; the ineffectiveness of the variable manipulation may be the underlying reason. The participants, being aware that they were being recorded, might have a lower tendency to focus attention on their image than in a natural environment. Additionally, looking in a mirror while talking to someone across the table is much less natural than seeing one's own image during a video call, which could have had an impact on our results in face-to-face conversations. Furthermore, as our sample was very small, it did not allow us to study interindividual differences in responses to online interactions.

The study can be extended through tracking whole body position during conversations and including hand gestures, body positions, etc., in the analysis. It would be possible to supplement coordination measures with the measure of behavior matching, that is body position mirroring [9]. Specific gestures or expressions could be identified automatically using machine learning techniques [65]. To better render the differences in coordination in remote and live interactions it would be crucial to obtain measures on other "coupling means" in dyadic conversations than the movement itself, such as gaze coordination and vocal dynamics. Related to body movement coordination they would inform about the use of the relevant cues as affordances for interaction and allow for forming a fuller picture of the relevant differences.

5. Conclusions

The differences between video-mediated and face-to-face interactions cannot be explained by either the technical properties of the medium or individual cognitive processes alone. In this study, we tried to apply an interactive perspective to identify key factors shaping our experience of online interactions. In line with this perspective, our study revealed significant differences in patterns of interlocutors' coordination between video-mediated remote and live interactions. We demonstrated that in video communication, the stability of movement coordination is lower, which may have a negative impact on the overall quality of interaction. The presence of the mirror image did not have a detectable effect on coordination; however, it seems that the mirror image helped to control one's expression during remote interactions, making the communicative gestures less exaggerated. Vast differences in coordination patterns indicate that the remote medium radically alters the landscape of affordances for communicative actions. It remains to be seen which affordances result in those differences when they are altered.

Author Contributions: Conceptualization, J.Z., E.N., J.K.-M., K.S., K.Z. and J.R.-L.; data collection J.Z., E.N. and K.S.; methodology, J.Z.; investigation, J.Z.; writing—original draft preparation, J.Z., E.N., J.K.-M., K.S., K.Z., writing—review and editing, J.R.-L. All authors have read and agreed to the published version of the manuscript.

Funding: This work was supported by the Faculty of Psychology, University of Warsaw, from the funds awarded by Ministry of Science and Higher Education in the form of subsidy for the maintenance and development of research potential in 2021 (501-D125-01-1250000, zlec. 5011000615).

Institutional Review Board Statement: The study was approved by the Ethical Committe of the Faculty of Psychology, University of Warsaw.

Informed Consent Statement: Informed consent was obtained from all subjects involved in the study.

Data Availability Statement: The data presented in this study are openly available in OSF at DOI 10.17605/OSF.IO/8YA47.

Acknowledgments: We are grateful to Urszula Kuczma, Fernando Garcia Hipola, Rati Tsiklauri and Yevhenii Rachynskyi for help in data collection.

Conflicts of Interest: The authors declare no conflict of interest. The funders had no role in the design of the study; in the collection, analyses, or interpretation of data; in the writing of the manuscript, or in the decision to publish the results.

References

1. Fusaroli, R.; Rączaszek-Leonardi, J.; Tylén, K. Dialog as Interpersonal Synergy. *New Ideas Psychol.* **2014**, *32*, 147–157. [CrossRef]
2. De Jaegher, H.; Di Paolo, E. Participatory Sense-Making. *Phenomenol. Cogn. Sci.* **2007**, *6*, 485–507. [CrossRef]
3. Trevarthen, C. Communication and Cooperation in Early Infancy: A Description of Primary Intersubjectivity. In *Before Speech: The Beginning of Human Communication*; Bullowa, M., Ed.; Cambridge University Press: London, UK, 1979; pp. 321–347.

4. Oullier, O.; de Guzman, G.C.; Jantzen, K.J.; Lagarde, J.; Kelso, J.S. Social Coordination Dynamics: Measuring Human Bonding. *Soc. Neurosci.* **2008**, *3*, 178–192. [CrossRef] [PubMed]
5. López Pérez, D.; Leonardi, G.; Niedźwiecka, A.; Radkowska, A.; Rączaszek-Leonardi, J.; Tomalski, P. Combining Recurrence Analysis and Automatic Movement Extraction from Video Recordings to Study Behavioral Coupling in Face-to-Face Parent-Child Interactions. *Front. Psychol.* **2017**, *8*, 2228. [CrossRef] [PubMed]
6. Jensen, T.W. Emotion in Languaging: Languaging as Affective, Adaptive, and Flexible Behavior in Social Interaction. *Front. Psychol.* **2014**, *5*, 720. [CrossRef]
7. Bernieri, F.J. Coordinated Movement and Rapport in Teacher-Student Interactions. *J. Nonverbal Behav.* **1988**, *12*, 120–138. [CrossRef]
8. Latif, N.; Barbosa, A.V.; Vatiokiotis-Bateson, E.; Castelhano, M.S.; Munhall, K.G. Movement Coordination during Conversation. *PLoS ONE* **2014**, *9*, e105036. [CrossRef]
9. Fujiwara, K.; Daibo, I. Empathic Accuracy and Interpersonal Coordination: Behavior Matching Can Enhance Accuracy but Interactional Synchrony May Not. *J. Soc. Psychol.* **2021**, *162*, 71–88. [CrossRef]
10. Fusaroli, R.; Bjørndahl, J.S.; Roepstorff, A.; Tylén, K. A Heart for Interaction: Shared Physiological Dynamics and Behavioral Coordination in a Collective, Creative Construction Task. *J. Exp. Psychol. Hum. Percept. Perform.* **2016**, *42*, 1297–1310. [CrossRef]
11. Rączaszek-Leonardi, J.; Krzesicka, J.; Klamann, N.; Ziembowicz, K.; Denkiewicz, M.; Kukiełka, M.; Zubek, J. Cultural Artifacts Transform Embodied Practice: How a Sommelier Card Shapes the Behavior of Dyads Engaged in Wine Tasting. *Front. Psychol.* **2019**, *10*, 2671. [CrossRef]
12. Ramseyer, F.; Tschacher, W. Nonverbal Synchrony in Psychotherapy: Coordinated Body Movement Reflects Relationship Quality and Outcome. *J. Consult. Clin. Psychol.* **2011**, *79*, 284–295. [CrossRef] [PubMed]
13. De Jaegher, H.; Di Paolo, E.; Gallagher, S. Can Social Interaction Constitute Social Cognition? *Trends Cogn. Sci.* **2010**, *14*, 441–447. [CrossRef] [PubMed]
14. Porter, R. Business Meetings: A Comparison of the Effectiveness of Audio and Video Conferencing in Dispersed Teams. Ph.D. Thesis, University of Maine, Orono, ME, USA, 2012.
15. Sherman, L.E.; Michikyan, M.; Greenfield, P.M. The Effects of Text, Audio, Video, and in-Person Communication on Bonding between Friends. *Cyberpsychol. J. Psychosoc. Res. Cyberspace* **2013**, *7*. [CrossRef]
16. Burgoon, J.K.; Bonito, J.A.; Ramirez, A., Jr.; Dunbar, N.E.; Kam, K.; Fischer, J. Testing the Interactivity Principle: Effects of Mediation, Propinquity, and Verbal and Nonverbal Modalities in Interpersonal Interaction. *J. Commun.* **2002**, *52*, 657–677. [CrossRef]
17. Chillcoat, Y.; DeWine, S. Teleconferencing and Interpersonal Communication Perception. *J. Appl. Commun. Res.* **1985**, *13*, 14–32. [CrossRef]
18. Federman, M. On the Media Effects of Immigration and Refugee Board Hearings via Videoconference. *J. Refug. Stud.* **2006**, *19*, 433–452. [CrossRef]
19. Bailenson, J.N. Nonverbal Overload: A Theoretical Argument for the Causes of Zoom Fatigue. *Technol. Mind Behav.* **2021**, *2*. [CrossRef]
20. Nesher Shoshan, H.; Wehrt, W. Understanding "Zoom Fatigue": A Mixed-Method Approach. *Appl. Psychol.* **2021**. [CrossRef]
21. Fauville, G.; Luo, M.; Muller Queiroz, A.C.; Bailenson, J.; Hancock, J. Zoom Exhaustion & Fatigue Scale. *SSRN Electron. J.* **2021**, *4*, 100119. [CrossRef]
22. Fauville, G.; Luo, M.; Muller Queiroz, A.C.; Bailenson, J.N.; Hancock, J. *Nonverbal Mechanisms Predict Zoom Fatigue and Explain Why Women Experience Higher Levels than Men*; SSRN Scholarly Paper ID 3820035; Social Science Research Network: Rochester, NY, USA, 2021. [CrossRef]
23. Riedl, R. On the Stress Potential of Videoconferencing: Definition and Root Causes of Zoom Fatigue. *Electron. Mark.* **2021**. [CrossRef]
24. Burgoon, J.K.; Bonito, J.A.; Bengtsson, B.; Ramirez, A.; Dunbar, N.E.; Miczo, N. Testing the Interactivity Model: Communication Processes, Partner Assessments, and the Quality of Collaborative Work. *J. Manag. Inf. Syst.* **1999**, *16*, 33–56. [CrossRef]
25. Froese, T.; Di Paolo, E.A. Toward Minimally Social Behavior: Social Psychology Meets Evolutionary Robotics. In *Advances in Artificial Life. Darwin Meets von Neumann*; Lecture Notes in Computer Science; Kampis, G., Karsai, I., Szathmáry, E., Eds.; Springer: Berlin/Heidelberg, Germany, 2011; pp. 426–433. [CrossRef]
26. Gibson, J.J. *The Ecological Approach to Visual Perception*; Houghton Mifflin: Boston, MA, USA, 1979.
27. Rączaszek-Leonardi, J.; Nomikou, I.; Rohlfing, K.J. Young Children's Dialogical Actions: The Beginnings of Purposeful Intersubjectivity. *IEEE Trans. Auton. Ment. Dev.* **2013**, *5*, 210–221. [CrossRef]
28. Rietveld, E.; Kiverstein, J. A Rich Landscape of Affordances. *Ecol. Psychol.* **2014**, *26*, 325–352. [CrossRef]
29. Accessing Meeting and Phone Statistics. 2021. Available online: https://support.zoom.us/hc/en-us/articles/202920719-Accessing-meeting-and-phone-statistics (accessed on 20 January 2022).
30. Kohrs, C.; Angenstein, N.; Brechmann, A. Delays in Human-Computer Interaction and Their Effects on Brain Activity. *PLoS ONE* **2016**, *11*, e0146250. [CrossRef] [PubMed]
31. Hirshfield, L.M.; Bobko, P.; Barelka, A.; Hirshfield, S.H.; Farrington, M.T.; Gulbronson, S.; Paverman, D. Using Noninvasive Brain Measurement to Explore the Psychological Effects of Computer Malfunctions on Users during Human-Computer Interactions. *Adv. Hum. Comput. Interact.* **2014**, *2014*, 2. [CrossRef]

32. Boland, J.E.; Fonseca, P.; Mermelstein, I.; Williamson, M. Zoom Disrupts the Rhythm of Conversation. *J. Exp. Psychol. Gen.* **2021**. [CrossRef]
33. Clayman, S.E. Sequence and Solidarity. In *Advances in Group Processes*; Emerald Group Publishing Limited: Bingley, UK, 2002; Volume 19, pp. 229–253. [CrossRef]
34. Kendrick, K.H.; Torreira, F. The Timing and Construction of Preference: A Quantitative Study. *Discourse Process.* **2015**, *52*, 255–289. [CrossRef]
35. Schoenenberg, K.; Raake, A.; Koeppe, J. Why Are You so Slow?—Misattribution of Transmission Delay to Attributes of the Conversation Partner at the Far-End. *Int. J. Hum. Comput. Stud.* **2014**, *72*, 477–487. [CrossRef]
36. Hessels, R.S. How Does Gaze to Faces Support Face-to-Face Interaction? A Review and Perspective. *Psychon. Bull. Rev.* **2020**, *27*, 856–881. [CrossRef]
37. Rączaszek-Leonardi, J.; Nomikou, I. Beyond Mechanistic Interaction: Value-Based Constraints on Meaning in Language. *Front. Psychol.* **2015**, *6*, 1579. [CrossRef]
38. Clark, H.H.; Wilkes-Gibbs, D. Referring as a Collaborative Process. *Cognition* **1986**, *22*, 1–39. [CrossRef]
39. Bavelas, J.B. Gestures as Part of Speech: Methodological Implications. *Res. Lang. Soc. Interact.* **1994**, *27*, 201–221. [CrossRef]
40. Wagner, P.; Malisz, Z.; Kopp, S. Gesture and Speech in Interaction: An Overview. *Speech Commun.* **2014**, *57*, 209–232. [CrossRef]
41. Wlodarczak, M.; Buschmeier, H.; Malisz, Z.; Kopp, S.; Wagner, P. Listener Head Gestures and Verbal Feedback Expressions in a Distraction Task. In Proceedings of the Interdisciplinary Workshop on Feedback Behaviors in Dialog, INTERSPEECH2012 Satellite Workshop, Portland, OR, USA, 9–13 September 2012.
42. Duncan, S. Some Signals and Rules for Taking Speaking Turns in Conversations. *J. Personal. Soc. Psychol.* **1972**, *23*, 283–292. [CrossRef]
43. Osugi, T.; Kawahara, J.I. Effects of Head Nodding and Shaking Motions on Perceptions of Likeability and Approachability. *Perception* **2018**, *47*, 16–29. [CrossRef]
44. Moretti, S.; Greco, A. Truth Is in the Head. A Nod and Shake Compatibility Effect. *Acta Psychol.* **2018**, *185*, 203–218. [CrossRef]
45. Moretti, S.; Greco, A. Nodding and Shaking of the Head as Simulated Approach and Avoidance Responses. *Acta Psychol.* **2020**, *203*, 102988. [CrossRef]
46. Andonova, E.; Taylor, H.A. Nodding in Dis/Agreement: A Tale of Two Cultures. *Cogn. Process.* **2012**, *13*, 79–82. [CrossRef]
47. Oppezzo, M.; Schwartz, D.L. Give Your Ideas Some Legs: The Positive Effect of Walking on Creative Thinking. *J. Exp. Psychol. Learn. Mem. Cogn.* **2014**, *40*, 1142–1152. [CrossRef]
48. Nomikou, I.; Leonardi, G.; Rohlfing, K.J.; Rączaszek-Leonardi, J. Constructing Interaction: The Development of Gaze Dynamics. *Infant Child Dev.* **2016**, *25*, 277–295. [CrossRef]
49. Ingram, R.E.; Cruet, D.; Johnson, B.R.; Wisnicki, K.S. Self-Focused Attention, Gender, Gender Role, and Vulnerability to Negative Affect. *J. Personal. Soc. Psychol.* **1988**, *55*, 967–978. [CrossRef]
50. Fejfar, M.; Hoyle, R. Effect of Private Self-Awareness on Negative Affect and Self-Referent Attribution: A Quantitative Review. *Personal. Soc. Psychol. Rev.* **2000**, *4*, 132–142. [CrossRef]
51. Gonzales, A.L.; Hancock, J.T. Mirror, Mirror on My Facebook Wall: Effects of Exposure to Facebook on Self-Esteem. *Cyberpsychol. Behav. Soc. Netw.* **2011**, *14*, 79–83. [CrossRef] [PubMed]
52. Horn, R.; Behrend, T. Video Killed the Interview Star: Does Picture-in-Picture Affect Interview Performance? *Pers. Assess. Decis.* **2017**, *3*, 5. [CrossRef]
53. Kuhn, K.M. The Constant Mirror: Self-view and Attitudes to Virtual Meetings. *Comput. Hum. Behav.* **2022**, *128*, 107110. [CrossRef]
54. Weltzien, S.; Marsh, L.E.; Hood, B. Thinking of Me: Self-focus Reduces Sharing and Helping in Seven- to Eight-Year-Olds. *PLoS ONE* **2018**, *13*, e0189752. [CrossRef]
55. Gibbons, F.X.; Wicklund, R.A. Self-Focused Attention and Helping Behavior. *J. Personal. Soc. Psychol.* **1982**, *43*, 462–474. [CrossRef]
56. Cao, Z.; Hidalgo, G.; Simon, T.; Wei, S.E.; Sheikh, Y. OpenPose: Realtime Multi-Person 2D Pose Estimation Using Part Affinity Fields. *IEEE Trans. Pattern Anal. Mach. Intell.* **2021**, *43*, 172–186. [CrossRef]
57. Fusaro, M.; Vallotton, C.D.; Harris, P.L. Beside the Point: Mothers' Head Nodding and Shaking Gestures during Parent–Child Play. *Infant Behav. Dev.* **2014**, *37*, 235–247. [CrossRef]
58. Marwan, N.; Carmen Romano, M.; Thiel, M.; Kurths, J. Recurrence Plots for the Analysis of Complex Systems. *Phys. Rep.* **2007**, *438*, 237–329. [CrossRef]
59. Wallot, S. Multidimensional Cross-Recurrence Quantification Analysis (MdCRQA)—A Method for Quantifying Correlation between Multivariate Time-Series. *Multivar. Behav. Res.* **2019**, *54*, 173–191. [CrossRef] [PubMed]
60. Kennel, M.B.; Brown, R.; Abarbanel, H.D.I. Determining Embedding Dimension for Phase-Space Reconstruction Using a Geometrical Construction. *Phys. Rev. A* **1992**, *45*, 3403–3411. [CrossRef] [PubMed]
61. Datseris, G. DynamicalSystems.Jl: A Julia Software Library for Chaos and Nonlinear Dynamics. *J. Open Source Softw.* **2018**, *3*, 598. [CrossRef]
62. Bates, D.; Alday, P.; Kleinschmidt, D.; José Bayoán Santiago Calderón, P.; Zhan, L.; Noack, A.; Arslan, A.; Bouchet-Valat, M.; Kelman, T.; Baldassari, A.; et al. JuliaStats/MixedModels.Jl: V4.6.0. Zenodo. 2022. Available online: https://zenodo.org/record/5825693#.YlZkn9NBxPY (accessed on 20 January 2022).
63. Tschacher, W.; Rees, G.M.; Ramseyer, F. Nonverbal Synchrony and Affect in Dyadic Interactions. *Front. Psychol.* **2014**, *5*, 1323. [CrossRef] [PubMed]

64. Weber, F.; Haering, C.; Thomaschke, R. Improving the Human–Computer Dialogue With Increased Temporal Predictability. *Hum. Factors* **2013**, *55*, 881–892. [CrossRef] [PubMed]
65. Beugher, S.D.; Brône, G.; Goedemé, T. A Semi-Automatic Annotation Tool for Unobtrusive Gesture Analysis. *Lang. Resour. Eval.* **2018**, *52*, 433–460. [CrossRef]

MDPI

Article

Changes in the Complexity of Limb Movements during the First Year of Life across Different Tasks

Zuzanna Laudańska [1,2,*], David López Pérez [1,*], Alicja Radkowska [1], Karolina Babis [1], Anna Malinowska-Korczak [1], Sebastian Wallot [3] and Przemysław Tomalski [1,*]

1 Institute of Psychology, Polish Academy of Sciences, Jaracza 1, 00-378 Warsaw, Poland; a.radkowska@psych.pan.pl (A.R.); kbabis@psych.pan.pl (K.B.); a.malinowska-korczak@psych.pan.pl (A.M.-K.)
2 Graduate School for Social Research, Polish Academy of Sciences, Nowy Świat 72, 00-330 Warsaw, Poland
3 Institute of Psychology, Leuphana University of Lüneburg, Universitätsallee 1, 21335 Lüneburg, Germany; sebastian.wallot@leuphana.de
* Correspondence: zlaudanska@psych.pan.pl (Z.L.); d.lopez@psych.pan.pl (D.L.P.); ptomalski@psych.pan.pl (P.T.)

Abstract: Infants' limb movements evolve from disorganized to more selectively coordinated during the first year of life as they learn to navigate and interact with an ever-changing environment more efficiently. However, how these coordination patterns change during the first year of life and across different contexts is unknown. Here, we used wearable motion trackers to study the developmental changes in the complexity of limb movements (arms and legs) at 4, 6, 9 and 12 months of age in two different tasks: rhythmic rattle-shaking and free play. We applied Multidimensional Recurrence Quantification Analysis (MdRQA) to capture the nonlinear changes in infants' limb complexity. We show that the MdRQA parameters (entropy, recurrence rate and mean line) are task-dependent only at 9 and 12 months of age, with higher values in rattle-shaking than free play. Since rattle-shaking elicits more stable and repetitive limb movements than the free exploration of multiple objects, we interpret our data as reflecting an increase in infants' motor control that allows for stable body positioning and easier execution of limb movements. Infants' motor system becomes more stable and flexible with age, allowing for flexible adaptation of behaviors to task demands.

Keywords: complexity; motor development; multidimensional recurrence quantification analysis; infants; limb movements

Citation: Laudańska, Z.; López Pérez, D.; Radkowska, A.; Babis, K.; Malinowska-Korczak, A.; Wallot, S.; Tomalski, P. Changes in the Complexity of Limb Movements during the First Year of Life across Different Tasks. *Entropy* **2022**, *24*, 552. https://doi.org/10.3390/e24040552

Academic Editors: Franco Orsucci and Wolfgang Tschacher

Received: 9 February 2022
Accepted: 12 April 2022
Published: 15 April 2022

1. Introduction

One of the fascinating phenomena in human development is how quickly infants learn new motor skills. Infants' movements advance from being disorganized to having a more recognizable adult-like pattern in the first years of life [1]. The development of motor behavior involves learning through practice as infants improve their skills over time and learn to optimize their actions to the demands of any specific task. However, how motor coordination patterns emerge in development and across different actions is unknown.

Initially, reflexes and general movements are controlled at the spinal and brain stem levels during the neonatal period. Later, motor control at the subcortical level of the central nervous system emerges and matures mainly throughout the first year of life, followed by the activation of the cortical level of motor control [2]. The increase in motor control allows for body positioning and stability, which also facilitates the execution of limb movements [3,4]. Initially, the pattern of spontaneous movements seems to involve all the limbs simultaneously, and it refines to a more selective inter-limb coordination with age [5,6]. The dissociation between arms and legs mainly emerges in the second half of the first year [7], facilitating object manipulation and playing with toys [8]. Moreover, the leg activity becomes more stable with age, while the inverse pattern is observed in the arms [9].

Additionally, the increase in postural control allows for using upper limbs for purposes other than the stabilization of body position. Infants aged 6 and 7 months present trunk control mostly in the thoracic region [10], and the acquisition of trunk control in the lumbar region between 4 and 6 months of age has a positive impact on the quality of reaching behavior [11]. Full trunk control is presented by infants from 8 to 9 months of age [10]. The emerging postural control is also characterized by increasing complexity, where the upper limbs become more involved in skilled manual reaching and less in stabilizing the body posture [12]. During the first three or four months after birth, infants' head and trunk control are poor, and they mainly lie down if not supported. Around 6 months of age, infants begin to gain sufficient stability to sit independently, allowing them to move their arms more freely. Later, around 8–9 months of age, most infants learn the first ways of locomotion, such as crawling. Finally, towards the end of their first year, infants stand freely and walk around, opening new possibilities to explore the environment.

Motor control development always occurs in a rapidly changing environment consisting of constant constraints (e.g., gravity) and variable and constantly changing elements, such as objects or people. Thus, to understand the development of the complexity of limb movements, we need to consider that they are embedded in a given context and constrained by situational demands [13]. On the one hand, particular contexts may encourage highly structured and repetitive patterns of limb movements—for example, rhythmic activities such as drumming or rattle shaking. Infants' movements during drumming become faster and more regular with age [14], and the rhythmic synchronization is usually not limited to arm movements but diffuses throughout the body [15]. This increase in the regularity of movements may result in a developmental decrease in the complexity of limb movements. On the other hand, the lack of structure in unconstrained free play may be related to a developmental increase in the complexity of limb movements as older infants can selectively use hands in varied ways to manipulate objects while using legs to stabilize their position or move around. Thus, the context and task demands are also important when evaluating the complexity of limb movements.

The rapid evolution of wearable devices has opened new avenues for recording and analyzing infant movement, which might help to understand the changes in the complexity of infants' spontaneous movements during different activities in greater detail. Advanced wearable sensors—Inertial Motion Units (IMUs)—combine information from accelerometers, gyroscopes and magnetometers, resulting in a more precise estimation of the position and orientation of body parts. Given the portability, mobility, small size and low weight of this wearable technology, it is becoming widely used in infant studies (e.g., [16–21]). Although wearable sensors may cause some discomfort in clinical pediatric populations (as suggested in [22]), studies in typically developing infants have reported that wearables do not affect infant movement (e.g., [23]). An alternative method is using markerless algorithms to detect movements from videos (e.g., [24–28]). However, this approach is challenging in multi-person set-ups with older infants that move around freely since obtaining a clear view of them at all times remains difficult and the resulting occlusions may significantly limit the accuracy of tracking ([26]). Therefore, the IMUs can currently be considered a gold standard for quantifying infants' 3D kinematics in multi-person and unconstrained settings.

In this study, we use wearables to investigate the developmental changes in the complexity of limb movements in two tasks that differ in the level of structuring—more constrained rattle-shaking and free play with a larger set of toys. Parent–infant dyads were invited to the lab four times: when infants were around 4, 6, 9 and 12 months age, as these ages reflect significant changes in motor control and gross motor development. As Abney et al. [9] demonstrated, infant development can be studied as a complex system with the analytical tools derived from nonlinear dynamics. Studies on motor development have traditionally focused on quantifying changes in individual limb movements (i.e., reaching hand) or in pairs (either hands or legs). Since the pattern of spontaneous movements initially involves all the limbs shifting simultaneously and it refines with age, in this paper, we focus on the changes in the movement complexity of all limbs together. To achieve this, we use the Multidimensional Recurrence Quantification Analysis (MdRQA, [29]). Many methods of inferring complexity measures from a time series allow for the inclusion of a maximum one (e.g., fractals, recurrence quantification analysis, entropy measures) or two (e.g., cross-recurrence quantification analysis) time series and cannot be used to determine potential higher-level interactions in the movement of all limbs together. MdRQA, in contrast to other methods, is a dynamical systems method that allows for quantifying the dynamics of a multidimensional system at different levels of description by combining information from multiple variables (n > 2) and can be used to infer the shared dynamics of multiple time series [29]. Those shared dynamics are later summarized in a series of parameters that provide information about the complexity of the time series (see description in Section 2.5). Here, we combine wearable motion trackers and MdRQA to study the developmental trajectories of the complexity of infants' limb movements in two play contexts: rattle-shaking and free play. To our knowledge, the coordination between all four limbs has not been previously studied simultaneously in a longitudinal design and across tasks that differ in their level of constraints. We hypothesize that the trajectories of the complexity of limb movements will differ between the tasks, with the age-related decrease in complexity in the rattles task and the increase in complexity in the free play task.

2. Materials and Methods

2.1. Participants

Participants were 26 mother–infant dyads from an ongoing longitudinal study. Participants were invited to the lab when infants were around 4 (T1), 6 (T2), 9 (T3) and 12 (T4) months old. Four infants contributed data at all four time points, whereas nineteen infants missed one visit (mostly due to COVID-19 related restrictions). Therefore, 12 infants contributed data at T1, T2 and T3; 7 at T2, T3 and T4; and 3 at T1, T3 and T4 (see Table 1 for an overview of sample characteristics). Participants were from predominantly middle-class families living in a city with >1.5 million inhabitants. The majority of the mothers had completed higher education: 22 held a master's degree, 2 held a bachelor's and 2 completed high school. For their participation, infants received a diploma and a small gift (a baby book). The study received clearance from the local institution's ethics committee.

Table 1. Sample Characteristics.

Time Point	N	Mean Age in Months (SD)	Min Age in Months	Max Age in Months
T1	19	4.41 (0.30)	4.00	5.20
T2	21	6.57 (0.36)	6.00	7.20
T3	26	9.14 (0.41)	8.60	10.20
T3	17	12.14 (0.46)	11.60	13.10

2.2. Equipment

Infants' and caregivers' movements were recorded at 60 Hz using wearable motion trackers (MTw Awinda, Xsens Technologies B.V., Enschede, The Netherlands): an Awinda station receiver (Xsens Technologies B.V.) and MT Manager Software (Xsens Technologies B.V.). Overall, 12 sensors were used (on infant's arms, legs, head and torso, see Figure 1, and on caregiver's arms, head and torso), but in this paper, we report data only from 4 sensors placed on infant's arms and legs.

Figure 1. Placement of infant's motion trackers: legs, hands, torso, head. Signed permission of the caregiver was acquired for the publication of the image.

2.3. Procedure

Interactions were recorded in a laboratory room on a carpeted play area. Upon the family's arrival, an experimenter explained the study protocol and obtained parental consent. Once the infant was familiarized with the laboratory, the wearable motion trackers attached to the elastic bands were put on the infant's and caregiver's bodies. Then, a set of parent–child interaction tasks with different sets of age-appropriate toys took place. The sets for infants aged 4 and 6 months were slightly different from those for infants aged 9 and 12 months to maintain their interest in a given task (see Figure 2). There were 6–7 different tasks during each meeting, but here, we report data comparing two of them—free play and rattle-shaking. In a rattle-shaking task, which lasted approx. 5 min, the dyads were given two maracas rattles and two rattles of different types (the barbell rattles for younger infants and teddybear rattles for older ones). In a free play task, which lasted approx. 10 min, the younger infants were offered a large, standard set of toys that included baby books, teethers, rattles, rubber blocks and plush toys. The set for older infants included block sorter, cars, stackable cups, rubber blocks, puppets, rattles, plush toys and a wooden box with a drawer and a ball. Caregivers were instructed to play with their infants using each set of toys in their preferred way, as they usually do at home.

Figure 2. Photos of the toys used in the free play at T1 and T2 (**a**) and T3 and T4 (**b**) and the rattle-shaking task at T1 and T2 (**c**) and T3 and T4 (**d**). Signed permission of the caregiver was acquired for the publication of the images.

2.4. Data Pre-Processing

IMU data from sensors placed on both wrists and ankles of an infant were processed in Matlab (Mathworks, Inc., Natick, MA, USA) using in-house scripts. First, missing values were identified and interpolated using the *interp1* function with a Spline interpolation that applies a cubic interpolation of the values at neighboring grid points. Then, we collapsed the three-dimensional acceleration data obtained from the IMUs to a one-dimensional overall acceleration time series by calculating the magnitude of acceleration for each three-dimensional data point. Next, data were smoothed using the *medfilt1* function that applies a third-order median filter to remove one-point outliers by replacing each value with the median of three neighboring entries (see Figure 3a for an example of the sensor time series). Finally, to avoid the possibility that data from any limb with higher variance may bias the outcome of the complexity analysis and because we were interested in the sequential properties of the data, each individual time series was z-scored before further analysis.

IMUs record gyroscopic and magnetometer data, providing more detailed orientation information. Combining this information with accelerometer data, one can create quaternions [30], an alternate way to describe orientation or rotations on limb movements. Supplementary data using quaternions are included to test the robustness of the IMUs data (see Supplementary Information 1.1–1.3).

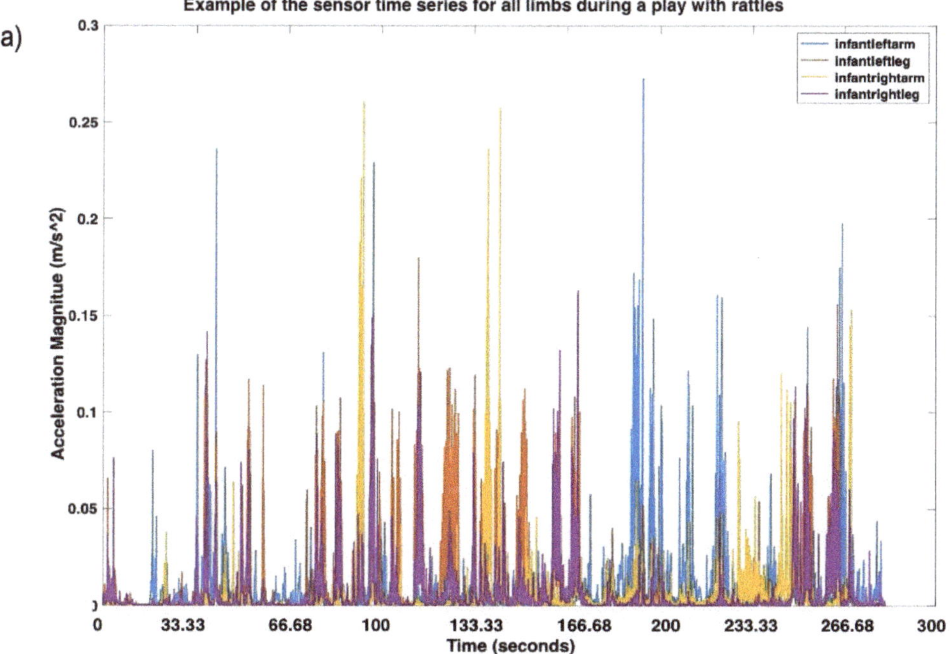

Figure 3. *Cont.*

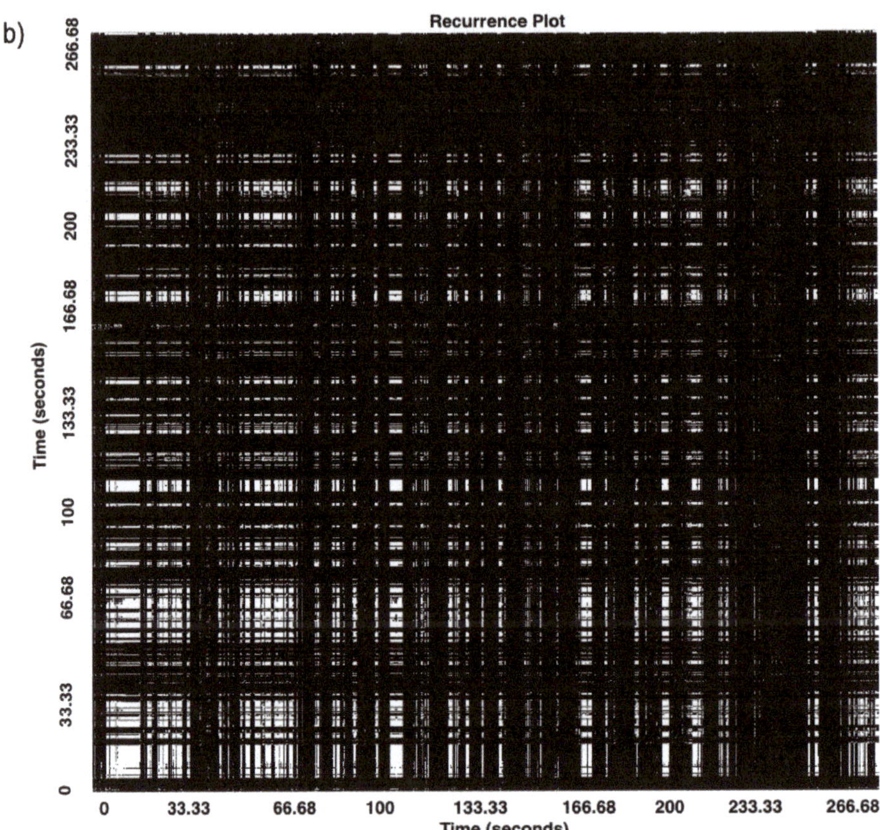

Figure 3. Examples of the sensor time series for all limbs during a play with rattles (**a**) and its correspondent recurrence plot (**b**). Recurrences in the plot are marked by a white dot, while non-recurrences are marked by a black dot.

2.5. Complexity Analysis

We used MdRQA [29] to quantify the simultaneous coupling of four limbs' time series. MdRQA is a multivariate extension of Recurrence Quantification Analysis that captures recurring patterns within a multidimensional time series. This is achieved by calculating the distances between all coordinate pairs of data points (e.g., using Euclidean distance norm) in a multidimensional time series and by thresholding this distance matrix, where distances below the threshold are treated as recurrent, and distances exceeding the threshold are treated as nonrecurrent [31]. That thresholded matrix is called the recurrence plot, where values are coded as 1 or 0 depending on whether the values are recurrent or not for each of the values within the all time series (see Figure 3b for an example). From the final recurrence plot, we extracted three main measures:

- Entropy (Ent): it is the Shannon entropy of the distribution of the diagonal lines on the recurrence plot, capturing repeating movement patterns;
- Recurrence Rate (RR): it is the density of recurrence points in a recurrence plot, and it corresponds to the probability that a specific state will recur;
- Mean Line (ML): it is the average length of repeating patterns in the system, which can be understood as a measure of overall system's stability.

These three measures allow for describing different yet supplementary aspects of the system's behavior, such as stability and adaptability. When infants acquire a new motor skill, their repertoire becomes more complex allowing for increased adaptability to situational demands (e.g., [32]). Furthermore, when infants master these new skills, their motor coordination patterns become more stable over time. In this context, entropy acts as a measure of the complexity or flexibility of limb movements. An unconstrained movement signal will carry low entropy since the probability of finding recurrent patterns would be lower than in a constrained situation with interaction-dominant dynamics, which postulates that the system's structure is emergent and context-dependent. In contrast, component-dominant dynamics proposes that all components (in this case, infants' limbs) contribute to the system dynamics in a stable and independent way [33–35]. When infants are playing in an unconstrained situation (free play task in our study), they adjust their movements to the needs of the task at hand, i.e., perform various types of movement (e.g., reaching, banging, touching) with different types of objects. Consequently, their movements are less regular and form more random patterns. In this case, there will be low variability in the length of the recurrent states, leading to lower entropy. In the constrained situation (rattle-shaking task), infants move their arms in a rhythmic way to produce the sound, and the rattles placed in their hands may reduce the number of degrees of freedom of movement. Therefore, the rattle-shaking task decreases in complexity as infants attempt to perform periodic/semi-periodic movements, introducing higher variability in the patterns of recurrences and increasing the overall entropy. On the other hand, the recurrence rate and mean line are measures of the stability of the limb movements. In a constrained situation, such as rattle-shaking, the more the infants' movements would follow interaction-dominant dynamics (i.e., the infants learn with age how to move the rattles synchronously), the more recurrence rate and mean line would increase.

Three critical parameters need to be set to calculate the recurrence plots (see [36]). First, we estimated the delay of embedding using the *mdDelay* function, which estimates the delay in a multidimensional time series using the average mutual information method. Second, we estimated the embedding dimension using the *mdFnn* function, which applies a false nearest neighbor estimation. We obtained an average value of 1 for the delay and 14 for the embedding dimension, which is consistent with the typical values recommended for biological signals [37]. Finally, we adapted the radius for each infant individually. To this end, we fixed the recurrence rate sufficiently low (i.e., RR = 5% [38] and used the embedding dimension and delay previously computed. We carried this out for the first visit data of each infant and fixed these parameters for the consequent visits to estimate the changes in complexity over time.

Control analyses were performed using the same approach but with shuffling the movement data in a random order within each time series. This allows us to compare the results from the entropy and mean line and prove that temporal dynamics did not arise randomly (e.g., [39]).

2.6. Statistical Analysis

To assess the repeated-measures effects of age (4) and task (2), we ran the General Estimating Equations (GEEs) with a Bonferroni correction for pairwise comparisons. GEEs are particularly adequate for longitudinal data because they take into account the dependency and ordering of the data within subjects in repeated-measures designs. Furthermore, in the GEE analysis, even if a subject is missing one or more of the repeated measurements, the remaining data of that subject are used in the analysis (e.g., [40,41]). Data analysis was conducted in IBM SPSS Statistics 26. Figure 4 was created using R [42] and RStudio [43] and ggplot2 package [44].

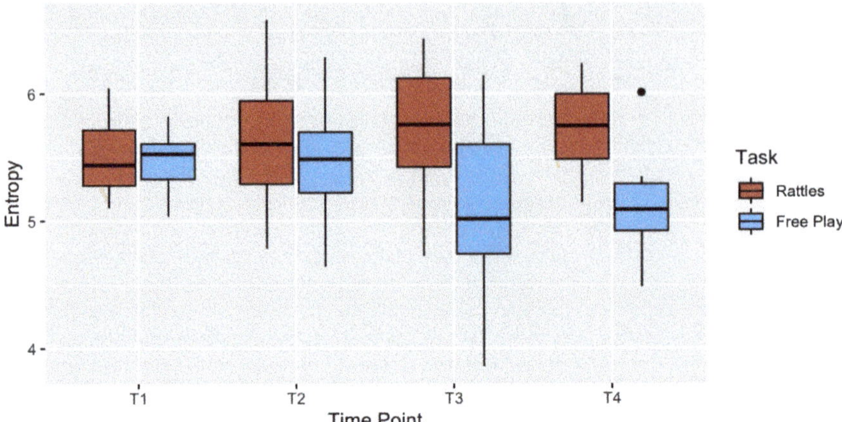

Figure 4. Boxplots showing entropy in each time point in rattle-shaking (red) and free play (blue). Horizontal lines represent median value, boxes are drawn from the first quartile to the third quartile, whiskers indicate min and max value and the dot indicates an outlier.

3. Results

3.1. Complexity Measures

3.1.1. Entropy

The GEE with age (4) and task (2) as within-subjects factors showed a significant difference in entropy level between rattle-shaking and free play (Wald $\chi^2(1) = 36.888$, $p < 0.001$; see Figure 4). There was no effect of time point (Wald χ^2 (3) = 3.365, $p = 0.339$), but the interaction between task and time point was significant (Wald χ^2 (3) = 26.634, $p < 0.001$). Post hoc pairwise comparisons revealed that there were no task-related differences at T1 and T2. The entropy was higher in rattle-shaking than free play at T3 ($p < 0.001$) and T4 ($p = 0.010$). Within free play, entropy was also higher at T2 than at T3 ($p = 0.037$). See Table 2 for descriptive data.

Table 2. Entropy (Ent), Recurrence Rate (RR) and Mean Line (ML) values at each time point and each task.

		T1			T2			T3			T4		
		Mean (SD)	Min	Max	Mean (SD)	Min	Max	Mean (SD)	Min	Max	Mean (SD)	Min	Max
Rattles	*Ent*	5.51 (0.30)	5.10	6.04	5.62 (0.44)	4.78	6.58	5.72 (0.45)	4.72	6.43	5.73 (0.37)	5.14	6.24
	RR	5.03 (0.05)	4.93	5.09	7.28 (5.13)	2.07	19.14	9.17 (7.19)	0.69	27.55	7.78 (5.01)	0.95	15.60
	ML	19.48 (6.61)	5.07	35.20	23.06 (8.59)	11.96	50.95	23.55 (9.02)	1.66	41.79	23.79 (5.78)	15.13	32.19
Free Play	*Ent*	5.46 (0.20)	5.04	5.82	5.48 (0.42)	4.64	6.29	5.10 (0.59)	3.86	6.14	5.08 (0.42)	4.48	6.01
	RR	5.05 (0.04)	4.98	5.09	5.47 (3.84)	0.14	14.54	4.51 (4.79)	0.02	16.26	2.99 (3.52)	0.17	13.28
	ML	18.96 (2.75)	14.17	23.79	21.04 (7.36)	10.96	39.05	16.15 (6.28)	7.44	30.03	15.39 (4.53)	9.92	26.78

3.1.2. Recurrence Rate

There was a significant difference in the recurrence rate between rattle-shaking and free play (Wald χ^2 (1) = 11.281, $p = 0.001$). There was no effect of time-point (Wald χ^2 (3) = 4.353, $p = 0.226$), but the interaction between task and time-point was significant (Wald χ^2 (3) =18.660, p < 0.001) as the recurrence rate in rattle-shaking at T3 was significantly higher than in free play at T3 (p = 0.001) and T4 (p = 0.029).

3.1.3. Mean Line

There was a significant difference in the mean line between rattle-shaking and free play (Wald χ^2 (1) = 8.919, p = 0.003). The interaction effect between task and time-point was also significant (Wald χ^2 (3) =17.739, p < 0.001) as the mean line in free play at T3 was lower than in rattle-shaking at T3 (p < 0.001). There was no effect of time-point (Wald χ^2 (3) = 3.618, p < 0.306).

3.2. Control Analysis

To check whether the effects did not arise randomly, we compared observed and shuffled versions using paired t-tests at each time point. At each time point, the observed versions were significantly different from those shuffled for each measure. At T1: entropy t(34) = 76.675, p < 0.001; recurrence rate t(34) = 162.090, p < 0.001; mean line t(34) = 21.334, p < 0.001. At T2: entropy t(45) = 79.094, p < 0.001; recurrence rate t(45) = 9.841, p < 0.001; mean line: t(45) = 19.053, p < 0.001. At T3: entropy: t(48) = 53.352, p < 0.001; recurrence rate t(48) = 7.580, p < 0.001; mean line t(48) =16.382, p < 0.001. At T4: entropy t(22) = 43.767, p < 0.001; recurrence rate t(22) = 5.331, p < 0.001; mean line t(22) =14.276, p < 0.001.

4. Discussion

In this paper, we showed that the complexity of limb movements changes across infancy. In a longitudinal study, we recorded infants' limb movements at around 4, 6, 9 and 12 months of age in two tasks that differed structurally—more constrained and repetitive rattle-shaking and free play with a larger set of toys. To investigate the changes in the complexity of all four limbs, we applied the Multidimensional Recurrence Quantification Analysis (MdRQA, [29]). We showed that the complexity measures (entropy, recurrence rate and mean line) are modulated by task at 9 and 12 months but not at 4 or 6 months of age. We interpret this finding as reflecting an increase in infants' motor control that allows for stable body positioning and easier execution of limb movements. Increased motor control is related to an overall increase in the motor system's complexity as the infant can adjust movements specifically to the task. In our case, higher entropy in the rattle-shaking task may reflect the capacity to flexibly adapt behaviors to environmental demands. Furthermore, a longer mean line and a higher recurrence rate suggest that an infant's motor system is more stable during rattle-shaking and has a more confined attractor state.

Our results provide further insight into the early developmental organization of motor actions. The global pattern of inter-limb coordination varies with changing contexts because the behaviors are adapted and selected to fit a given task [1]. The motor action system continues to specialize across infancy to respond to particular environmental pressures [45]. In our case, each task qualitatively required different acts—rhythmic body movements to produce the rattling sound or various reaching and holding acts to explore different objects—and infants learned how to adjust their behaviors to the specific context with age. This suggests that limb movement organization becomes context-specific by the end of the first year of life. This is in line with recent studies showing that less experienced infants generate multiple inconsistent coordination patterns, while more experienced infants tailor their coordination patterns to body–environment relations and flexibly switch solutions (e.g., [32,46,47]).

This study is an important step in understanding changes in the complexity of limb movements in infancy. We showed that the MdRQA measures are sensitive to changes in the dynamics of limb movements between tasks and that the observed patterns do not form randomly, as was shown in comparisons with the shuffled time series. This result is in line with previous studies suggesting that infants' development can be studied as a complex system with the tools from nonlinear dynamics [9]. Moreover, MdRQA goes one step further than traditional methods as it allows estimating the complex dynamics of multiple effectors (n > 2) and, therefore, characterizing the complexity of the developmental organization of motor actions in more detail. Nevertheless, MdRQA can be further extended to assess the coupling between multidimensional time series [48]. Therefore, methods such as MdRQA open new possibilities to understand the role of limb movement for other domains of development (e.g., vocal production or visual attention) or studying coupling and leader–follower relationships during parent–infant interactions (e.g., parent limb movements vs. infant limb movements or parent vocalizations vs. infant limb movements).

Several limitations arise from this study. First, there were some missing values in the sensor data in 10.1% of cases. However, control analysis with excluded cases with over 15% of missing data showed the same final pattern of the results (see Supplementary Information 2.1–2.3). Second, we used only accelerometer data in this study, while IMUs offer more possibilities (magnetometer and gyroscope data). To establish whether our results are limited to accelerometric data only, we conducted a supplementary analysis using quaternion data and showed a similar pattern of results with respect to task modulation and age-related changes (see Supplementary Information 1.1–1.3), but further studies should consider the possibility of expanding this work and explore not only changes in acceleration but also rotational movements. Third, in this study, we compared tasks that differed in overall duration (5 min in rattle-shaking vs. 10 min in free play). Variable length of analyzed time series are commonly used in studies using RQA-based approaches (see, for example, [24,49]) since capturing the overall dynamics of the phenomenon is more important than task duration (and comparison of observed data with shuffled time series allows checking whether the effects were not random). Fourth, although we observed different age-related trajectories in the complexity of limb movements over the first year of life, there is high variability in the way infants develop. Therefore, future longitudinal studies with more time-points are necessary to more accurately depict the patterns of variability and the shape of the developmental trajectories of inter-limb coordination. This is especially important since stable execution of gross motor skills is usually preceded by many transitions when the skills vacillate between occurrence and absence [50], which could reflect phase transition periods when the entire motor system undergoes reorganization. Thus, nonlinear methods combined with a more dense sampling of behavior across development could shed more light on the developmental trajectories of movement coordination and capture both phase transitions and periods of stability. Fifth, data were collected in a laboratory room, and therefore, future studies could explore the possibilities of continuous measurement of limb coordination across different contexts "in the wild". The wearable motion trackers can be worn for the entire day or multiple days without the presence of an experimenter and record densely sampled data during infants' everyday experiences [17,21]. A dense sampling of infants' daily experiences would help understand how caregivers scaffold infants' actions and create "social affordances" [51] and understand the influence of social influences in context-dependent changes in infants' inter-limb coordination. Moreover, it could also help to identify atypical patterns of motor development. Lower complexity of movements might represent more repetitive motor behaviors, which are diagnostic symptoms of several neurodevelopmental disorders, such as autism spectrum disorder [52] or developmental delay [53]. Finally, future studies should investigate whether a similar pattern of results could be observed using other ways of movement tracking, such as marker-less video-based algorithms (see [24–28]), to make sure that wearing sensors does not affect infant movement.

5. Conclusions

Our study explored the developmental changes in the complexity of limb movements in infancy using a multidimensional nonlinear approach (MdRQA). We showed that infants' movements become more complex with age and that the age-related changes in complexity are context-dependent. We interpret these changes in the complexity of the motor system as an increase in motor control that allows the infant to adjust movements specifically to the task. These findings may have important implications for the study of atypical patterns of motor development.

Supplementary Materials: The following are available online at https://www.mdpi.com/article/10.3390/e24040552/s1.

Author Contributions: Conceptualization: Z.L., D.L.P., A.R., K.B., A.M.-K. and P.T.; Investigation: Z.L., A.R., K.B. and A.M.-K.; Data curation: D.L.P., Z.L. and S.W.; Formal analysis: Z.L. and D.L.P.; Supervision: P.T.; Funding acquisition: P.T.; Visualization: Z.L., D.L.P. and A.M.-K.; Writing—original draft: Z.L. and D.L.P.; Writing—review and editing: Z.L., D.L.P., P.T., S.W., A.R., A.M.-K. and K.B. All authors have read and agreed to the published version of the manuscript.

Funding: This research and the APC were funded by the Polish National Science Centre grant number 2018/30/E/HS6/00214 to PT. ZL received organizational support from the Graduate School for Social Research, Polish Academy of Sciences, and additional funding from the Polish National Agency for Academic Exchange, NAWA, "International Scholarship Exchange of PhD Candidates and Academic Staff," PROM, project (contract No. PPI/PRO/2019/1/00043/U/00001). PROM is financed by the European Social Fund within Operational Program Knowledge Education Development as a non-competition project and is implemented as part of the Action specified in the application for project funding No. POWR.03.03.00-00-PN13/18. S.W. acknowledges funding from the German Science Foundation (DFG—grant number: 442405852).

Institutional Review Board Statement: The study was conducted according to the guidelines of the Declaration of Helsinki and approved by the Research Ethics Committee at the Institute of Psychology, Polish Academy of Sciences.

Informed Consent Statement: Prior to each testing session, all caregivers gave written informed consent.

Data Availability Statement: The data that support the findings will be available upon request from the corresponding authors following an embargo from the date of publication to allow for finalization of the ongoing longitudinal project. The computer code used in this study is openly available in GitHub: https://github.com/Mirandeitor/entropyPaper, accessed on 10 February 2022.

Conflicts of Interest: The authors declare no conflict of interest.

References

1. Thelen, E.; Smith, L.B. *A Dynamic Systems Approach to the Development of Cognition and Action*; The MIT Press: Cambridge, MA, USA, 1994.
2. Kobesova, A.; Kolar, P. Developmental kinesiology: Three levels of motor control in the assessment and treatment of the motor system. *J. Bodyw. Mov. Ther.* **2014**, *18*, 23–33. [CrossRef] [PubMed]
3. Dusing, S.C.; Harbourne, R.T. Variability in Postural Control During Infancy: Implications for Development, Assessment, and Intervention. *Phys. Ther.* **2010**, *90*, 1838–1849. [CrossRef]
4. Westcott, S.L.; Lowes, L.P.; Richardson, P.K. Evaluation of postural stability in children: Current theories and assessment tools. *Phys. Ther.* **1997**, *77*, 629–645. [CrossRef] [PubMed]
5. Piek, J.P.; Gasson, N. Spontaneous kicking in fullterm and preterm infants: Are there leg asymmetries? *Hum. Mov. Sci.* **1999**, *18*, 377–395. [CrossRef]
6. Piek, J.P.; Gasson, N.; Barrett, N.; Case, I. Limb and gender differences in the development of coordination in early infancy. *Hum. Mov. Sci.* **2002**, *21*, 621–639. [CrossRef]
7. Kanemaru, N.; Watanabe, H.; Taga, G. Increasing selectivity of interlimb coordination during spontaneous movements in 2- to 4-month-old infants. *Exp. Brain Res.* **2012**, *218*, 49–61. [CrossRef]
8. Watanabe, H.; Taga, G. Flexibility in infant actions during arm- and leg-based learning in a mobile paradigm. *Infant Behav. Dev.* **2009**, *32*, 79–90. [CrossRef]
9. Abney, D.H.; Warlaumont, A.S.; Haussman, A.; Ross, J.M.; Wallot, S. Using nonlinear methods to quantify changes in infant limb movements and vocalizations. *Front. Psychol.* **2014**, *5*, 771. [CrossRef]
10. Greco, A.L.R.; da Costa, C.S.N.; Tudella, E. Identifying the level of trunk control of healthy term infants aged from 6 to 9 months. *Infant Behav. Dev.* **2018**, *50*, 207–212. [CrossRef]
11. Rachwani, J.; Santamaria, V.; Saavedra, S.L.; Wood, S.; Porter, F.; Woollacott, M.H. Segmental trunk control acquisition and reaching in typically developing infants. *Exp. Brain Res.* **2013**, *228*, 131–139. [CrossRef]
12. Hadders-Algra, M. Development of Postural Control During the First 18 Months of Life. *Neural Plast.* **2005**, *12*, 99–108. [CrossRef] [PubMed]

13. Adolph, K.E.; Hoch, J.E. Motor Development: Embodied, Embedded, Enculturated, and Enabling. *Annu. Rev. Psychol.* **2019**, *70*, 141–164. [CrossRef] [PubMed]
14. Rocha, S.; Southgate, V.; Mareschal, D. Infant Spontaneous Motor Tempo. *Dev. Sci.* **2021**, *24*, e13032. [CrossRef] [PubMed]
15. Hoehl, S.; Fairhurst, M.; Schirmer, A. Interactional synchrony: Signals, mechanisms and benefits. *Soc. Cogn. Affect. Neurosci.* **2021**, *16*, 5–18. [CrossRef]
16. Trujillo-Priego, I.A.; Smith, B.A. Kinematic characteristics of infant leg movements produced across a full day. *J. Rehabil. Assist. Technol. Eng.* **2017**, *4*, 205566831771746. [CrossRef]
17. Deng, W.; Trujillo-Priego, I.A.; Smith, B.A. How Many Days Are Necessary to Represent an Infant's Typical Daily Leg Movement Behavior Using Wearable Sensors? *Phys. Ther.* **2019**, *99*, 730–738. [CrossRef]
18. Patel, P.; Shi, Y.; Hajiaghajani, F.; Biswas, S.; Lee, M.H. A novel two-body sensor system to study spontaneous movements in infants during caregiver physical contact. *Infant Behav. Dev.* **2019**, *57*, 101383. [CrossRef]
19. Zhou, J.; Schaefer, S.Y.; Smith, B.A. Quantifying Caregiver Movement when Measuring Infant Movement across a Full Day: A Case Report. *Sensors* **2019**, *19*, 2886. [CrossRef]
20. Airaksinen, M.; Räsänen, O.; Ilén, E.; Häyrinen, T.; Kivi, A.; Marchi, V.; Gallen, A.; Blom, S.; Varhe, A.; Kaartinen, N.; et al. Automatic Posture and Movement Tracking of Infants with Wearable Movement Sensors. *Sci. Rep.* **2020**, *10*, 169, [CrossRef]
21. Franchak, J.M.; Scott, V.; Luo, C. A Contactless Method for Measuring Full-Day, Naturalistic Motor Behavior Using Wearable Inertial Sensors. *Front. Psychol.* **2021**, *12*, 701343. [CrossRef]
22. Khan, M.H.; Helsper, J.; Boukhers, Z.; Grzegorzek, M. Automatic recognition of movement patterns in the vojta-therapy using RGB-D data. In Proceedings of the 2016 IEEE International Conference on Image Processing (ICIP), Phoenix, AZ, USA, 25–28 September 2016; pp. 1235–1239. [CrossRef]
23. Jiang, C.; Lane, C.J.; Perkins, E.; Schiesel, D.; Smith, B.A. Determining if wearable sensors affect infant leg movement frequency. *Dev. Neurorehabilit.* **2018**, *21*, 133–136. [CrossRef] [PubMed]
24. López Pérez, D.; Leonardi, G.; Niedźwiecka, A.; Radkowska, A.; Rączaszek-Leonardi, J.; Tomalski, P. Combining Recurrence Analysis and Automatic Movement Extraction from Video Recordings to Study Behavioral Coupling in Face-to-Face Parent-Child Interactions. *Front. Psychol.* **2017**, *8*, 2228. [CrossRef] [PubMed]
25. López Pérez, D.; Laudańska, Z.; Radkowska, A.; Babis, K.; Kozioł, A.; Tomalski, P. Do we need expensive equipment to quantify infants' movement? A cross-validation study between computer vision methods and sensor data. In Proceedings of the 2021 IEEE International Conference on Development and Learning (ICDL), Beijing, China, 23–26 August 2021; pp. 1–6. [CrossRef]
26. Khan, M.H.; Schneider, M.; Farid, M.S.; Grzegorzek, M. Detection of Infantile Movement Disorders in Video Data Using Deformable Part-Based Model. *Sensors* **2018**, *18*, 3202. [CrossRef] [PubMed]
27. Khan, M.H.; Zöller, M.; Farid, M.S.; Grzegorzek, M. Marker-Based Movement Analysis of Human Body Parts in Therapeutic Procedure. *Sensors* **2020**, *20*, 3312. [CrossRef] [PubMed]
28. Baccinelli, W.; Bulgheroni, M.; Simonetti, V.; Fulceri, F.; Caruso, A.; Gila, L.; Scattoni, M.L. Movidea: A Software Package for Automatic Video Analysis of Movements in Infants at Risk for Neurodevelopmental Disorders. *Brain Sci.* **2020**, *10*, 203. [CrossRef] [PubMed]
29. Wallot, S.; Roepstorff, A.; Mønster, D. Multidimensional Recurrence Quantification Analysis (MdRQA) for the Analysis of Multidimensional Time-Series: A Software Implementation in MATLAB and Its Application to Group-Level Data in Joint Action. *Front. Psychol.* **2016**, *7*, 1835. [CrossRef]
30. Roetenberg, D.; Luinge, H.J.; Baten, C.T.; Veltink, P.H. Compensation of magnetic disturbances improves inertial and magnetic sensing of human body segment orientation. *IEEE Trans. Neural Syst. Rehabil. Eng.* **2005**, *13*, 395–405. [CrossRef]
31. Gordon, I.; Wallot, S.; Berson, Y. Group-level physiological synchrony and individual-level anxiety predict positive affective behaviors during a group decision-making task. *J. Psychophysiol.* **2021**, *9*, 58. [CrossRef]
32. Ossmy, O.; Adolph, K.E. Real-Time Assembly of Coordination Patterns in Human Infants. *Curr. Biol.* **2020**, *30*, 4553–4562.e4. [CrossRef]
33. Kello, C.T.; Beltz, B.C.; Holden, J.G.; Van Orden, G.C. The Emergent Coordination of Cognitive Function. *J. Exp. Psychol. Gen.* **2007**, *136*, 551–568. [CrossRef]
34. Ihlen, E.A.; Vereijken, B. Interaction-dominant dynamics in human cognition: Beyond 1/f αfluctuation. *J. Exp. Psychol. Gen.* **2010**, *139*, 436–463. [CrossRef] [PubMed]
35. Wijnants, M.L.; Hasselman, F.; Cox, R.F.; Bosman, A.M.; van Orden, G. An interaction-dominant perspective on reading fluency and dyslexia. *Ann. Dyslexia* **2012**, *62*, 100–119. [CrossRef] [PubMed]
36. Wallot, S.; Mønster, D. Calculation of Average Mutual Information (AMI) and false-nearest neighbors (FNN) for the estimation of embedding parameters of multidimensional time series in matlab. *Front. Psychol.* **2018**, *9*, 1679. [CrossRef] [PubMed]
37. Marwan, N.; Webber, C.L. *Recurrence Quantification Analysis. Understanding Complex Systems*; Springer: New York, NY, USA, 2015. [CrossRef]
38. Marwan, N.; Wessel, N.; Meyerfeldt, U.; Schirdewan, A.; Kurths, J. Recurrence-plot-based measures of complexity and their application to heart-rate-variability data. *Phys. Rev. E Stat. Phys. Plasmas Fluids Relat. Interdiscip. Top.* **2002**, *66*, 026702. [CrossRef]
39. Richardson, D.C.; Dale, R. Looking to understand: The coupling between speakers' and listeners' eye movements and its relationship to discourse comprehension. *Cogn. Sci.* **2005**, *29*, 1045–1060. [CrossRef]

40. Twisk, J.; de Vente, W. Attrition in longitudinal studies: How to deal with missing data. *J. Clin. Epidemiol.* **2002**, *55*, 329–337. [CrossRef]
41. Diggle, P.; Diggle, D.; Allgemeine Tierzucht, F.; Press, O.U.; Diggle, P.; Heagerty, P.; Liang, K.; Zeger, S.; Zeger, B. *Analysis of Longitudinal Data*; Oxford Statistical Science Series; OUP: Oxford, UK, 2002.
42. R Core Team. *R: A Language and Environment for Statistical Computing*; R Foundation for Statistical Computing: Vienna, Austria, 2020.
43. RStudio Team. *RStudio: Integrated Development Environment for R*; RStudio, PBC.: Boston, MA, USA, 2020.
44. Wickham, H. *ggplot2: Elegant Graphics for Data Analysis*; Springer: New York, NY, USA, 2016.
45. Goldfield, E.C. *Emergent forms: Origins and Early Development of Human Action and Perception*; Oxford University Press: Oxford, UK, 1995.
46. Soska, K.C.; Galeon, M.A.; Adolph, K.E. On the other hand: Overflow movements of infants' hands and legs during unimanual object exploration. *Dev. Psychobiol.* **2012**, *54*, 372–382. [CrossRef]
47. D'Souza, H.; Cowie, D.; Karmiloff-Smith, A.; Bremner, A.J. Specialization of the motor system in infancy: From broad tuning to selectively specialized purposeful actions. *Dev. Sci.* **2017**, *20*, e12409. [CrossRef]
48. Wallot, S. Multidimensional Cross-Recurrence Quantification Analysis (MdCRQA)—A Method for Quantifying Correlation between Multivariate Time-Series. *Multivar. Behav. Res.* **2019**, *54*, 173–191. [CrossRef]
49. De Jonge-Hoekstra, L.; Van der Steen, S.; Van Geert, P.; Cox, R.F.A. Asymmetric Dynamic Attunement of Speech and Gestures in the Construction of Children's Understanding. *Front. Psychol.* **2016**, *7*, 473. [CrossRef]
50. Adolph, K.E.; Robinson, S.R.; Young, J.W.; Gill-Alvarez, F. What is the shape of developmental change? *Psychol. Rev.* **2008**, *115*, 527–543. [CrossRef] [PubMed]
51. Borghi, A.M.; Binkofski, F. *Words as Social Tools: An Embodied View on Abstract Concepts*; SpringerBriefs in Psychology; Springer: New York, NY, USA, 2014. [CrossRef]
52. Wilson, R.B.; Vangala, S.; Elashoff, D.; Safari, T.; Smith, B.A. Using wearable sensor technology to measure motion complexity in infants at high familial risk for autism spectrum disorder. *Sensors* **2021**, *21*, 616. [CrossRef] [PubMed]
53. Abrishami, M.S.; Nocera, L.; Mert, M.; Trujillo-Priego, I.A.; Purushotham, S.; Shahabi, C.; Smith, B.A. Identification of developmental delay in infants using wearable sensors: Full-day leg movement statistical feature analysis. *IEEE J. Transl. Eng. Health Med.* **2019**, *7*, 2800207. [CrossRef] [PubMed]

 MDPI

Article

A Dynamic Autocatalytic Network Model of Therapeutic Change

Kirthana Ganesh *,† and Liane Gabora *,†

Fipke Centre for Innovative Research, Department of Psychology, University of British Columbia, 3247 University Way, Kelowna, BC V1V 1V7, Canada
* Correspondence: kirthana.ganesh@ubc.ca (K.G.); liane.gabora@ubc.ca (L.G.)
† These authors contributed equally to this work.

Abstract: Psychotherapy involves the modification of a client's worldview to reduce distress and enhance well-being. We take a human dynamical systems approach to modeling this process, using Reflexively Autocatalytic foodset-derived (RAF) networks. RAFs have been used to model the self-organization of adaptive networks associated with the origin and early evolution of both biological life, as well as the evolution and development of the kind of cognitive structure necessary for cultural evolution. The RAF approach is applicable in these seemingly disparate cases because it provides a theoretical framework for formally describing under what conditions systems composed of elements that interact and 'catalyze' the formation of new elements collectively become integrated wholes. In our application, the elements are mental representations, and the whole is a conceptual network. The initial components—referred to as *foodset items*—are mental representations that are innate, or were acquired through social learning or individual learning (of *pre-existing* information). The new elements—referred to as *foodset-derived items*—are mental representations that result from creative thought (resulting in *new* information). In clinical psychology, a client's distress may be due to, or exacerbated by, one or more beliefs that diminish self-esteem. Such beliefs may be formed and sustained through distorted thinking, and the tendency to interpret ambiguous events as confirmation of these beliefs. We view psychotherapy as a creative collaborative process between therapist and client, in which the output is not an artwork or invention but a more well-adapted worldview and approach to life on the part of the client. In this paper, we model a hypothetical albeit representative example of the formation and dissolution of such beliefs over the course of a therapist–client interaction using RAF networks. We show how the therapist is able to elicit this worldview from the client and create a conceptualization of the client's concerns. We then formally demonstrate four distinct ways in which the therapist is able to facilitate change in the client's worldview: (1) challenging the client's negative interpretations of events, (2) providing direct evidence that runs contrary to and counteracts the client's distressing beliefs, (3) using self-disclosure to provide examples of strategies one can use to diffuse a negative conclusion, and (4) reinforcing the client's attempts to assimilate such strategies into their own ways of thinking. We then discuss the implications of such an approach to expanding our knowledge of the development of mental health concerns and the trajectory of the therapeutic change.

Keywords: autocatalytic network; creativity; conceptual network; psychotherapy; therapeutic change; uncertainty; worldview

Citation: Ganesh, K.; Gabora, L. A Dynamical Autocatalytic Network Model of Therapeutic Change. *Entropy* **2022**, *24*, 547. https://doi.org/10.3390/e24040547

Academic Editors: Franco Orsucci and Wolfgang Tschacher

Received: 1 February 2022
Accepted: 22 March 2022
Published: 13 April 2022

1. Introduction

While the efficacy of psychotherapy as a form of treatment has been clearly established [1], there is uncertainty about *why* it works [2]. Statistical approaches model the psychotherapeutic process using moderator and mediator variables [3,4], but this does not go far toward explaining each mind's unique, self-organizing network of associations, how this structure took shape, and how it responds to psychotherapy. We have only a hazy understanding of how specific elements of the psychotherapy process contribute to therapeutic change [5]. This paper aims to take a step forward toward a more precise understanding of what happens in psychotherapy and what makes it effective using a dynamical systems framework to model the interconnected, self-organizing nature of an individual's worldview, and its dynamical change over time. In so doing, we aim to both strengthen the theoretical bases of psychotherapy, and sharpen our capacity to improve it in practice.

We view psychotherapy as a creative collaborative process between therapist and client, in which the output is not an artwork or invention, but an outlook and approach to life on the part of the client. A client's outlook and behavior flows from the web of knowledge and experience that collectively constitute a way of seeing the world and being in the world, i.e., a *worldview*. What makes network approaches to cognition particularly promising is that this web of knowledge and experience can be described as a network consisting of loosely connected clusters (i.e., the network has an intermediate degree of modularity) that can be characterized using tools from network science [6–12]. Thus, the network approach offers a novel way of understanding mental health concerns and their treatment [13]. Psychotherapy (whether it be behavioral, cognitive, etc.) attempts to 'destabilize' a distressing or pathological mental state and shift the individual toward a healthier mental state [13–15]. What distinguishes the 'autocatalytic' approach to cognition taken here from other network-based models of cognition is its capacity to generate new elements (such as mental representations or schemas) out of interactions between existing ones. This makes it useful in the context of describing psychotherapeutic change, though the approach has been applied more broadly to other forms of cognitive change as well [16–18]. Since therapy entails change in the structure and dynamics of this network, network science seems a natural place to start in modeling the therapeutic process.

We note that it is not simply the case that positive interpretations (i.e., narratives that make the individual feel good) are adaptive while negative interpretations (i.e., narratives that make the individual feel bad) are maladaptive. It is often necessary that negative situations be acknowledged as such to spur action and find solutions; however, one worldview may predispose the individual to overcoming challenges and finding opportunities, while another leads to unnecessary distress and feelings of helplessness. Thus, the therapist strives to help the client to 'unravel' their worldview just enough to 'reweave' it into one that is, for that client, adaptive.

One's society and culture provides stories, narratives, scripts, and schemas, as well as larger conceptual frameworks (such as science or religion) that offer prescriptions for integrating them into a worldview; however, since no one else is privy to an individual's entire repertoire of personal experiences and intimate observations, the worldview one weaves is ultimately unique. Much as, for any given set of dots there are multiple ways of connecting them, for any given set of experiences or mental representations, there may be many ways of integrating them into a worldview. However, this sentence is not to be interpreted as implying that mental representations are, indeed, anything like 'dots'; they clearly have a context-dependent inner structure. A worldview may selectively include, or exclude, certain experiences and positively (or negatively) valenced items, (or weight them more strongly). Some worldviews may be more adaptive than others, i.e., more conducive to thriving, personal growth, and the well-being of the individual and their social and environmental milieu.

We model the psychotherapeutic process using a certain kind of network referred to as a Reflexively Autocatalytic foodset-derived (RAF) network. Though the term 'autocatalytic networks' reflects their initial application to the origin of life [19,20], RAFs provide a general mathematical setting for studying networks that arose out of earlier work in graph theory [21]. The term *reflexively* is used in its mathematical sense, meaning that each part is related to the whole. The term *autocatalytic* will be defined more precisely shortly, but for now it refers to the fact that the whole can be reconstituted through interactions amongst its parts. The term *foodset* refers to the elements that are initially present, as opposed to those that are the products of interactions amongst them. As in other network science approaches, the nodes of the network represent units of information such as words, concepts, memories, or mental representations of concrete or abstract knowledge, and connections between nodes (by way of free association, shared features, or co-occurrences) are represented as edges (For example, 'chair' and 'wood' are nodes, and the relationship between them, i.e., wood can be used to create a chair, is represented as an edge).[1]

What differentiates RAFs from other approaches in network science is that the nodes are not just passive transmitters of activation; they actively galvanize, or 'catalyze' the synthesis of novel ('foodset-derived') nodes from existing ones (the 'foodset'). The generalized RAF setting is conducive to the development of efficient (i.e., polynomial time), algorithms for questions that are computationally intractable (i.e., NP-hard. [30]). These features make RAFs uniquely suited to model how new structure grows out of earlier structure, i.e., generative network growth [30]. Such generativity may result in phase transitions to a network that is self-sustaining and self-organizing [31–33], as well as potentially able to evolve, i.e., exhibit cumulative, adaptive, open-ended change [34,35]. For this reason, RAFs have been used to model the origins of evolutionary processes, both biological—the origin of life (OOL) [36,37]—and cultural—the origin of culture (OOC), or more specifically, the kind of cognitive structure capable of generating cumulative, adaptive, open-ended innovation [17,38–41]. In a OOL context, RAFs were used to develop the hypothesis that life began as, not as a single self-replicating molecule, but as a set of molecules that, through catalyzed reactions, collectively reconstituted the whole [20]. Autocatalytic network theory has successfully demonstrated—mathematically or using simulations [36,42], and with real biochemical systems [37]—how self-maintaining structures that evolve and replicate can emerge from nonliving molecules. Because RAF nodes modify network structure, the RAF framework is consistent with the goal of understanding not just how networks are structured but also how they dynamically restructure themselves in response to internal and external pressures.

When autocatalytic models are applied in a cognitive context as they are here, they model not just network structure, but how the network reconfigures itself on the fly in response to changing needs and experiences. The observation that, similar to living organisms, cognitive networks are self-sustaining, self-organizing, and self-reproducing [43–46] suggests that cognitive networks constitute a second level of autocatalytic structure. By *cognitive network,* we refer to an individual's web of concepts, language terms, and their associations, as well as knowledge and memories, and how they are structured. The self-sustaining nature of a cognitive network is evident in the tendency to reduce cognitive dissonance, resolve inconsistencies, and preserve existing schemas in the face of new information. Although the contents of a cognitive network change over time, it maintains integrity as a relatively coherent whole. Its spontaneously self-organizing nature is evident in the capacity to combine remote associates [47] (such as combining CHOCOLATE and BUNNY to invent CHOCOLATE BUNNY).[2] The cognitive autocatalytic network replicates in a piecemeal manner through social learning and story-telling. Psychotherapeutic change facilitates the piecemeal replication of adaptive perspectives and habits, as well as the reorganization of relationships between elements of the client's worldview, and the RAF approach is well-suited to model this.

We begin with an introduction to the psychotherapeutic process. We then introduce RAF networks, and elaborate how they are used in this paper. Next, we present A RAF network model of therapeutic change facilitated by the therapist. The paper closes with implications of the model for fostering a concrete understanding of psychotherapeutic techniques, and suggestions for extending and testing it. A list of abbreviations, and a glossary of terms and their definitions can be found in the Appendix A.

[1] As explained elsewhere [16], the 'autocatalytic' approach taken here is consistent with distributed models of mental representations in memory [22], and with quantum models of their interactions [23–29].

[2] This proposed cognitive level of autocatalytic structure is not merely an extension of organismal needs; indeed, the biological and cognitive/cultural levels of endogenous control can be at odds (e.g., a scientist immersed in solving a problem may neglect offspring, or forget to eat).

2. The Therapeutic Process

Psychotherapy, or 'talk therapy' is rooted in formal Western medicine since the late 1800s and practices to alleviate human distress through conversation, known as the 'moral cure' has existed formally and informally for centuries [48]. Despite the fact that many effective forms of psychotherapy have been developed, there is uncertainty regarding the mechanisms of therapeutic change [2,5]. Different constructs across therapies show overlap, leading to difficulty with defining their roles and relative importance in the therapeutic process [49]. Thus, a paradigm shift towards dimensional, systemic, and interactional approaches to understanding mental illness and psychotherapy is warranted [50–52]. Multi-modal, multi-perspective research methods that enable us to capture the process of therapy in real-time are the future of psychotherapy research [53]. A complex systems approach can thus help re-conceptualize mental health concerns and treatment to more accurately represent the dynamic interactions involved [13], and help us understand therapeutic change [54].

The collaborative nature of psychotherapy is dyadic, and each member of the dyad influences the other through verbal, nonverbal, and physiological synchrony [55–58]. However, the client and therapist rarely have independently corroborative estimates of important process variables such as the therapeutic alliance [59], and any conceptualization of therapeutic change should make room for both perceptions. RAF networks can accommodate both perspectives within a single framework. The model presented in this paper focuses more on the change in the client's mind, but the approach has the potential to be expanded to include the therapist, and thereby capture the bidirectional exchange more comprehensively.

Clients generally enter therapy to alleviate distress and increase well-being. Sometimes the decision is prompted by a specific problem, a troubling experience or belief, or something that is difficult to accept. A client may report symptoms of depression, such as sadness, hopelessness, and decreased motivation due to, for example, difficulties with interpersonal relationships. As such, the client's approach towards such relationships, whether it is the thoughts, emotions, or behaviors involved, are currently insufficient/ineffective in helping them achieve their goals. Therapy may bring about modification of their cognitive network, by enabling them to find a new perspective on a problem or a troubling experience or belief, or come to terms with something they could not accept, thus integrating it into their worldview.

There are significant parallels between the creative process involved in, say, inventing something new, and the process of problem-solving in psychotherapy. Creativity flourishes in situations that involve a tension between uncertainty and constraints [60], or what has been referred to as *enabling constraints* [17,61]. We posit that the forging of a new, healthy, integrated conception of the world and one's place in it is a creative process, and by cultivating a client-tailored therapeutic interaction, the therapist acts as the midwife of this process.[3]

3. Rationale for the Approach

Similar to other cognitive network approaches, RAF networks are hierarchical yet decentralized, and they can be analyzed with respect to density, connectedness, and size. They also draw upon the conventional cognitive science notion of *spreading activation* through nodes of a concept graph, and techniques such as shortest path distance and clustering analysis. However, the RAF approach differs from other network approaches used psychology and cognitive science in a number of respects:

[3] Many describe the generative aspect of cognition as a 'birthing' of new ideas or attitudes, and this word directly captures the therapist's role in facilitating this process.

3.1. 'Reactions' and 'Catalysis'

A RAF consists of not just nodes connected by edges, but also *reactions*, or interactions that trigger, or 'catalyze' new nodes. Thus, RAF nodes are not merely passive recipients of spreading activation; they actively redirect it. For example, seeing a Superman movie might spark a child who believes she is powerless to draw herself as a superhero. In this example, the 'reactants' are the mental representation of herself as powerless, as well as her drawing skills. The 'catalyst' is the Superman movie, and the 'product' is the drawing of herself as a superhero.

In chemistry, and in applications of RAFs to the origin of life, the term 'reaction' refers to an interaction between molecules. For consistency, in cognitive applications of RAFs, the term *reaction* is used to refer to an interaction between *mental representations* (MRs) in a cognitive network. It may involve *representational redescription* (RR): the re-coding of information in working memory by modifying, restructuring, elaborating, and/or performing mental operations upon it, or possibly in the absence of an external cue [62]. RR can also involve a shift of perspective, and it can result in a flash of insight, or a newly perceived application for an old idea.[4] The issue of which concepts participate in a given reaction is discussed and mathematically modeled in [40].

RAFs also have two kinds of edges: reaction edges and catalysis edges. *Reaction edges* are similar to the edges in conventional network science approaches. (They can be thought of as the 'anatomy' of the network). *Catalysis edges* are more dynamic. (They can be thought of as the 'physiology' of the network). MRs are *catalytic* because they not only participate in certain reactions, but also facilitate—or catalyze—other reactions.[5] In chemistry, a catalyst speeds up a reaction that would otherwise occur very slowly if at all. By endowing cognitive network models with the capacity for catalysis we can model how one idea or environmental stimulus, triggers a mental operation (such as concept combination, or RR) that would otherwise occur very slowly or not at all. For example, realization of a novel or creative outcome (such as the drawing of a superhero version of oneself) may not have occurred without the galvanizing or 'catalyzing' impact of the experience of watching a Superman movie.

As in chemistry, the cognitive equivalent of a 'catalyzed reaction' may trigger another reaction, and so forth, resulting in a *reaction sequence*. In cognitive models, this reaction sequence is a stream of thought, which may ultimately have arisen from a problem, question, or cognitive dissonance. For example, the ultimate source of the cognitive reaction sequence culminating in the creation of a superhero character may be the desire never to feel powerless.

The rationale for treating mental representations (MRs) as catalysts comes, in part, from the literature on concepts, which provides extensive evidence that when concepts act as contexts for each other, their meanings change [66,67]. For example, an ISLAND has the property 'surrounded by water', but (hopefully) not a KITCHEN ISLAND. KITCHEN momentarily reconfigures the cognitive network, altering the perceived meaning of ISLAND. Such alterations in meaning are often nontrivial, and defy classical logic [68]; however, quantum models of concept interactions provided a means of formalizing the process by which a context (such as the goal of creating a spot to cut food) spontaneously bridges remote associates (such as KITCHEN and ISLAND) [23–25]. Although cognitive RAF models are influenced by how context is modeled in these quantum models of concepts, it is not committed to any formal approach to modeling context. Context is considered to be anything in the external environment, or anything from long-term memory that influences how a MR is instantiated in working memory. The extent to which one MR modifies the meaning of another is referred to as its *reactivity*.

[4] Creative insights often arise subconsciously from just beyond the bounds of working memory [63].

[5] The use of the word 'catalyze' in a cognitive context extends beyond autocatalytic models of cognition [64,65], though these other approaches are purely descriptive.

In sum, the RAF approach incorporates not just cognitive change due to adjustments in association strengths, but also cognitive change due to the prompting or 'catalysis' of new nodes. The resulting network is dynamic both in terms of structure (e.g., new nodes can be generated), and information flow (e.g., newly generated nodes can result in new information pathways).

3.2. Foodset versus Foodset-Derived

Another key feature of RAF models is the distinction between *foodset* items, which came into existence *outside* the network in question, and *foodset-derived* items, which come about through 'catalyzed reactions' *within* the network in question. An individual's 'mental foodset', or simply, *foodset* includes memories of direct experiences, i.e., that came about by way of the senses, including any knowledge that has come about through individual learning (of pre-existing information) by way of direct experience in the world, or through social learning processes such as imitation or classroom learning. The foodset may also include innate responses, such as the fear of heights and corresponding inclination to back away from a precipice. Together, these innate responses, direct experiences, and socially transmitted knowledge constitute the *raw materials* from which the individual's cognitive network is built. Thus, the worldview is "grounded in perception" because it grows out from this foodset.

Much as bricks and bags of mortar do not constitute a house, the foodset does not constitute a mental model of the world, a worldview. The set of *foodset-derived* items consists of mental contents that were generated by that individual from scratch (and constitute new information) using foodset elements, or perhaps other foodset-derived elements, as ingredients. The generation of foodset-derived items occurs by way of mental operations such as problem solving, insight, deduction, induction, and abduction. Since the elements of the worldview described by foodset-derived items are not grounded in perception, they can be viewed as 'useful fictions'. Thus, if the therapist responds to powerlessness in a certain way, and the client learns and (later) copies that response, that way of responding is an item in the client's foodset; however, if the therapist acts as a midwife for the client's expression of emotions associated with powerlessness during a therapy session, the artwork is a foodset-derived item[6]. The approach thereby distinguishes between conceptual shifts originating within the mind of a given individual, and those that originated by others, and were learnt or assimilated by that individual. What foodset items all have in common is that they are raw materials the individual has at his/her disposal to work with in the generation of new MRs, and this generation of new MRs is a key component of to the conceptual change that occurs during psychotherapy.

In cognitive networks, the distinction between foodset and foodset-derived provides a natural means of grounding abstract concepts in direct experiences; foodset-derived elements emerge through 'reactions,' that can be traced back to foodset items. This enables us to identify the necessary precursor ideas ideas for the emergence of new understandings, and the mental operations a given individual carried out to generate a particular idea. This capacity to model the reconfiguration of a cognitive network makes RAFs ideal for the study of change that occurs in psychotherapy.

3.3. Generational Cognitive/Cultural Change

Because of the distinction between foodset and foodset-derived MRs it is possible to tag new insights with their point of origin (i.e., keep track of whose mind did each idea arose in), and track cumulative change step by step within and across individuals. We posit that a mind can be described in terms of nested and overlapping RAFs, and these RAFs are what evolve through culture. Thus, each human lifetime constitutes a small segment of our collective cultural evolutionary lineage (see [34,38,39]). Each generation builds

[6] This distinction is not as black and white as portrayed here, but for simplicity, we do not address that subtlety for now.

on the accomplishments of the previous generation, such that items that were foodset-derived for one generation become elements of the foodset for the next, and this kind of cumulative cultural evolution has also been modeled, both computationally [69–72], and mathematically using RAFs [17,34,38–40]. For example, an early hominid invented the first tool by realizing that repeatedly striking one stone with another can produce a stone that is sharp, and the mental script of how to make this tool is described as a foodset-derived item in that individual's mind. This mental script was shared with peers, who in turn transmitted it to others, and in their minds it was a foodset item. As a more psychological example, 'flattery makes friends' constituted a foodset-derived item in the mind of the first person to have this thought. He or she may have shared this notion with others, and in their mind it is a foodset item, but one of them may build on it by realizing that imitation can be flattering, and therefore a route to friendship, in which case this new version is a foodset-derived idea (i.e., 'imitation is the sincerest form of flattery'). Thus, our worldviews consist largely of information that has already been preprocessed into scripts, schemas, stories, and narratives by previous generations, and such 'chunks' constrain the shape of one's worldview.

3.4. Potential to Scale Up

In this initial application of RAF networks to the therapeutic process, the examples used are fairly simple; however, a significant strength of the approach is that RAFs can scale up. The RAF approach can be used to analyze and detect phase transitions in extremely complex networks (such as the phase transition from no-RAF to RAF in Kauffman's [20] binary polymer model) that have proven intractable using other analytic approaches [37,73].

4. Reflextively Autocatalytic Foodset-Derived Networks (RAFs)

Let us now define the term *Reflexively Autocatalytic and foodset-derived network* (RAF) more precisely [30–32,35,74]. The term *reflexive* is used in its mathematical sense to mean that each component is related (directly or indirectly) to the whole. As mentioned in Section 1, the term *autocatalytic* refers to the fact that the whole can be reconstituted through interactions amongst its components. A network qualifies as A RAF network if it meets the following two criteria:

(1) It is *reflexively autocatalytic*: each reaction $r \in \mathcal{R}'$ is catalyzed by at least one element type that is either produced by $\mathcal{R}'$ or is present in the foodset F. This is sometimes referred to as *closure*.
(2) It is *F-generated*: all reactants in $\mathcal{R}'$ can be generated from the foodset F by using a series of reactions only from $\mathcal{R}'$ itself.

Thus, an RAF is a non-empty subset $\mathcal{R}' \subseteq \mathcal{R}$ of reactions that meets these two criteria: it is reflexively autocatalytic, and F-generated.

The term *catalytic reaction system* refers to a network consisting of components that can catalyze the generation of other components, and a catalytic reaction system can consist of one or more RAFs. The largest RAF, which subsumes all other RAFs, is referred to as the *maxRAF*. All other RAFs are referred to as *subRAFs*. A RAF that cannot be broken down into smaller RAFs is referred to as an irreducible RAFs, or *irrRAF*. It is not necessarily the case that a catalytic reaction system contains a RAF, but if it does contain one or more RAFs, it has a unique maxRAF. To put this more formally, if the network contains a RAF, then the collection of all its RAFs forms a partially ordered set (i.e., a poset) under set inclusion, with the maxRAF as its unique maximal element. RAFs can evolve, as demonstrated both mathematically and in simulation studies, through selective proliferation and drift acting on possible subRAFs of the maxRAF [32,75].

The catalytic reaction system is a tuple $\mathcal{Q} = (X, \mathcal{R}, C, F)$ consisting of a set X of types, a set $\mathcal{R}$ of reactions, a catalysis set C indicating which molecule types catalyze which reactions, and a subset F of X referred to as the foodset. A subset $\mathcal{R}'$ of the full reaction set $\mathcal{R}$ of a catalytic reaction system $\mathcal{Q}$ forms a RAF if is both *collectively autocatalytic* (by the first criterion, because each of its reactions is catalyzed by some component in the system), and *self-sustaining* (because of the F-generated criterion).

RAFs can enlarge and combine. The union of any two (or more) subRAFs forms a RAF (which explains why there is a unique maximal RAF). These two subRAFs may be disjoint, or they may have some reactions in common. A subRAF $\mathcal{R}'$ can also expand by combining with a 'co-RAF', where a *co-RAF* is any nonempty set of reactions that is not A RAF but, when combined with $\mathcal{R}'$, forms A RAF. RAF expansion can also be extrinsically driven. For example, it can be due to social learning of a new story or skill, i.e., a change in the foodset. External stimuli may even trigger a 'reaction'; for example, the instruction to 'think creatively' may 'catalyze' the generation of new ideas. In a therapeutic context, this could take the form of of a question or suggestion, such as to try seeing a particular interpersonal situation from the other person's perspective.

RAFs emerge in a system of interacting components when their complexity passes a critical threshold [20,33]. In applications of RAF networks to model the origin of life, the components are polymers: molecules made up of repeated units called monomers. In applications of RAF networks to model cognitive networks, the components are MRs. The RAF framework provides a means of analyzing the emergence of complex networks, identifying how phase transitions might occur, and at what parameter values. The phase transition from no RAF to A RAF has been analyzed (mathematically and through simulations), and applied to biochemical [31–33,36,42], cognitive [38–40], and ecological [76], systems.

During childhood, the individual assimilates experiences, stories, narratives, scripts, and schemas, and gradually weaves them into a network of understandings, and this process has been analyzed using the RAF framework [16]. Eventually these pieces of knowledge condense into a maxRAF, which grows and changes through childhood and beyond. Once the maxRAF encompasses the majority of these fragments of knowledge they are mutually accessible. At this point, the child no longer requires a cue or reminder in the environment to access something from memory because the maxRAF provides a route from any one idea to any other. The maxRAF enables the individual to make plans and predictions, generate metaphors, and adapt old techniques or ideas to new circumstances; however, while an integrated maxRAF network helps the individual think creatively and effectively negotiate the environment, it may be conducive to distorted thinking, and other biases that are *emotionally* dysfunctional, and result in mental health concerns.

An individual's worldviews could be said to be self-contained in that there exists a maxRAF—meta-RAF of sorts—that encompasses the majority of the individual's subRAFs. We have modeled not just how this maxRAF forms over the course of child development [16], but how the capacity for such a maxRAF evolved over the course of human history [34,38–40]. The worldviews of different individuals are interconnected in that and subRAFs of one individual are mirrored in subRAFs of another, and indeed, RAF structure can 'flow' and extend across individuals [17,18].

Cognitive RAFs

Whern RAFs are used to model cognition, all MRs in a given individual i are denoted X_i, and a specific MR $x = x_i$ in X_i is denoted by writing $x \in X_i$. MRs are either *foodset MRs*, or *foodset-derived MRs*. The *foodset* of individual i, denoted F_i, encompasses MRs that are either innately present, or that are the result of direct experience in the world, whether it be by way of social learning, or by way of natural or artificial stimuli. Thus, F_i has multiple components:

- $\mathbb{S}_i$ denotes the set of MRs arising through direct experience that have been encoded in individual i's memory. It includes:
  - MRs obtained through social learning from the communication of an MR x_j by another individual j, denoted $\mathbb{S}_i[x_j]$.
  - MRs obtained through individual learning, denoted $\mathbb{S}_i[\ell]$.
- Any *innate knowledge* with which individual i is born, denoted I_i.

F_i includes information obtained through social interaction with *someone else* who acquired this knowledge as a result of their own creative or analytical thought processes. (For example, if individual i learns from individual j that it is ok to say no, this is an instance of social learning, and "it's ok to say no" becomes a member of F_i. In contrast, if individual i realizes on their own that it is ok to say no, then "it's ok to say no" is not a member of F_i.) F_i includes everything in individual i's long-term memory that did not result from individual i engaging in RR. F_i also includes pre-existing information obtained by i through individual learning (which, as stated earlier, involves learning from the environment by non-social means), so long as this information retains the form in which it was originally perceived (and does not undergo redescription or restructuring through abstract thought). The crucial distinction between foodset and non-foodset items is not whether another person was involved, nor whether the MR was originally obtained through abstract thought (by *someone*), but whether the abstract thought process originated in the mind of the individual i in question.

Foodset-derived elements are denoted $\neg F_i$. Thus, $\neg F_i$ refers to mental contents that are *not* part of F_i (i.e. $\neg F_i$ consists of all the products $b \in B$ of all reactions $r \in R_i$). In particular, $\neg F_i$ includes the products of any reactions derived from F_i and encoded in individual i's memory. Its contents come about through mental operations *by the individual in question* on the foodset; in other words, foodset-derived items are the direct product of RR. Thus, $\neg F_i$ includes everything in long-term memory that *was* the result of one's own thought processes. $\neg F_i$ may include a MR in which social learning played a role, so long as the most recent modification to this MR was a catalytic event (i.e., it involved RR).

A single instance of RR in individual i is referred to as a *reaction*, and denoted $r \in \mathcal{R}_i$. RR is often applied recursively, such that the output of one thought serves as the input to the next. The set of reactions that can be catalyzed by a given MR x in individual i is denoted $C_i[x]$. The entire set of MRs either *undergoing* or *resulting from* r is denoted A or B, respectively, and a member of the set of MRs undergoing or resulting from reaction r is denoted $a \in A$ or $b \in B$. Thus, for example, if a client has the idea of expressing her grief at the passing of her father by painting a scene in which the clouds evoke her deceased father, the concepts FATHER and CLOUD are reactants in A, and the resulting concept FATHER-CLOUD is a product in B. This conceptual shift, treated as a 'reaction', is 'catalyzed' by the client's desire to process the death of her father. It is in this way that the RAF approach tags novelty (in this case, the painting) with its point of origin (by showing in which mind in a cultural lineage it was a foodset-derived item).

The set of *all* possible reactions in individual i is denoted $\mathcal{R}_i$. The mental contents of the mind, including all MRs and all RR events, is denoted $X_i \oplus \mathcal{R}_i$. Recall that the set of all MRs in individual i, including both the food set and the food set-derived items, is denoted X_i. $\mathcal{R}_i$ and C_i are not prescribed up front; because C_i includes remindings and associations on the basis of one or more shared features, different kinds of interactions are possible between any given pair of MRs. Nonetheless, it makes sense mathematically to refer to $\mathcal{R}_i$ and C_i as sets.

5. Model

We now apply RAF theory to the modification of a client's worldview in psychotherapy. To make this more concrete, we explain our model using a hypothetical interaction between a fictional yet representative therapist named Thera, and a client, named Clive. We show how the therapist elicits adaptive change in a dysfunctional belief in the client's worldview.

5.1. Intake Form and Thera's Emerging Mental Model of Clive

Thera learns from an intake form that Clive is a young man with no strong friendships, who has experienced debilitating social anxiety for years. His decision to start therapy was prompted by a recent incident in which his wife called him a 'moron' during an intense disagreement. This, in conjunction with several other earlier incidents, have forced him to conclude that he is 'stupid'.

A portion of Thera's mental model of her client after reading this report is shown in the first panel (panel (a)) of Figure 1. For the relationship between Thera and Clive to develop, they need to establish some form of psychological contact [77]. Thera welcomes Clive to the room and sits down. As she introduces herself, her body posture is relaxed. She provides several forms of non-verbal encouragement, such as smiling, nodding, supportive interjections, and eye contact. This makes him feel like someone worthy of the attention of another, which lifts his self-confidence, and allows him to speak more comfortably.

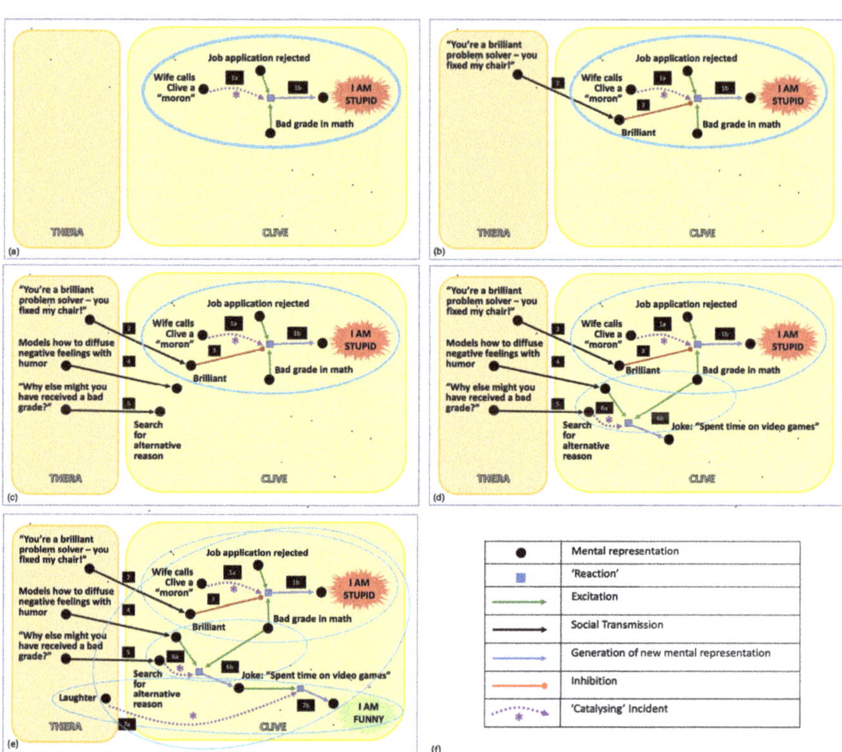

Figure 1. (**a**) RAF model of how client's worldview is altered over the course of a psychotherapy session. Initially, Thera's conception of Clive consists solely of what she read on his intake form. Following a 'catalyzing incident' in which his wife called him a "moron", he been interpreting other events as confirmation of the distressing belief, 'I am stupid'. Collectively, these elements constitute a stable RAF, as indicated by the thick blue line forming an oval around them. The thickness of this line indicates that the RAF has a large impact on Clive's thinking. (**b**) Thera praises Clive's brilliant problem solving ability, which generates a new foodset item, the notion that he is 'brilliant'. Since this is inconsistent with the belief 'I am stupid', it reduces the impact of that RAF, as indicated by the fact that the width of the line forming a blue oval is now thinner. (**c**) Two more foodset items are socially transmitted from Thera to Clive. (**d**) Making use of what Thera modeled for Clive about diffusing negative feelings using humor, he makes a joke. The joke is catalyzed by Thera's prompt to explore alternate explanations for why he received a bad grade. The joke depletes negative feelings associated with the bad grade, such that it is less able to serve as a 'reactant' to support the belief that he is unintelligent, as illustrated by the further dissolution of the oval representing that RAF. His joke constitutes a second RAF. (**e**) Thera's laughter at Clive's joke catalyzes a new belief, 'I am funny', which enhances his self-esteem, and forms a third RAF. These first three RAFs, which are irr-RAFs because they cannot be reduced further, interact with one another, and together form a maxRAF, which encompasses them all. (**f**) A key describing the symbols used in the various panels.

As psychotherapy proceeds, Thera starts to unearth information about Clive's worldview using therapeutic techniques such as reflections, clarifications, and open-ended questions. Clive shares other significant life experiences that affected his self-esteem, such as getting a D- in a high school math class, and not getting a job after working hard to prepare for the interview. By asking questions such as, "Why do you think you received a poor grade?", and "How did that make you feel?", Thera is able to gather information regarding Clive's interpretations of these events. From this, using her preexisting knowledge about social anxiety and mental health, she extrapolates Clive's concerns, and builds a mental model of him in her mind.

While Clive's belief that he is unintelligent is based in real-life adverse experiences—modeled here as 'foodset items'—it involves extrapolation, and possibly distortion. It appears that the triggering incident in which his wife called him a moron served as a 'catalyzing incident' that initiated a tendency toward confirmation bias, such that he reinterprets other past and present events as confirmation of the belief, 'I am stupid'. This, in turn, is damaging his self-esteem. The confirmation bias has thus exacerbated his preexisting concerns; his anxiety in social situations is now more severe due to his belief that he is 'stupid', and he has become isolated and lonely.

We can view what is happening here from the perspective of Clive's worldview as a whole. The self-organizing, self-mending nature of a worldview can create a system of internal feedback combined with external influences that can sustain and amplify an existing (and in this case, negative) narrative [78]. Clive has come to interpret interactions with others as consistent with his negative self-image, and his mistrust and withdrawal have created a positive feedback system that reinforces his belief about social interactions.

To model this using the RAF approach and thereby understand it in more precise terms, the client, Clive, is denoted C, the poor grade is denoted G_C, and not getting the position he applied for is denoted P_C. These memories (and likely others) serve as the raw materials, or reactants, for Clive's confirmation bias. First we note that they are part of his foodset, as follows:

$$G_C, P_C \in F_C. \tag{1}$$

G_C and P_C become reinterpreted as evidence for the belief "I am stupid", denoted $b1_C$. The catalyzing event that initiates this, i.e., the fight with his wife where she called Clive a 'moron', is denoted by m_C. Thus, this process is described as follows:

$$G_C + P_C \xrightarrow{m_C} b1_C \in \neg F_C, \ \neg F_C \mapsto \neg F_C \cup \{b1_C\}. \tag{2}$$

The catalysis of G_C and P_C by m_C is **Step 1a** in Panel (a) of Figure 1. The resulting formation of $b1_C$ is **Step 1b** in Panel (a) of Figure 1. The $\neg$ sign indicates that $b1_C$ is not part of the foodset, i.e., it is a foodset-derived item. The portion of Equation (2) after the comma simply tells us that the set of foodset-derived items has expanded to include $b1_C$.

Note that this little cognitive network now satisfies the conditions for A RAF; (1) all reactions (in this case, there is just $r1$) can proceed, because the needed catalyst is present, and (2) the needed reactants (MRs of events that could be interpreted as confirmation of his lack of intelligence) are also present. The greater the extent to which ambiguous experiences are interpreted as evidence for the foregone conclusion that he is stupid, the greater to which the MR 'I am stupid' constitutes a stable attractor state. (For discussion of attractor states in psychology, see [14,79]). This attractor causes Clive emotional distress, and has an adverse impact on his quality of life.

We now show how Thera extrapolates from what Clive says to build a mental model of him in her mind. Thera interprets Clive's statement as implying something more general about how he views the world and his place in the world: specifically, that his self-concept is increasingly dominated by the belief that he is (as his wife put it), "a moron". We use the subscript T to refer to Thera. Her foodset, denoted F_T, includes knowledge of psychopathology and the treatment options available, as well as her growing knowledge of

Clive, C_T. The formation of this knowledge from Clive's discussion of himself, denoted C_C, is described as follows:

$$F_T \mapsto F_T \cup \{C_T\},\ C_T \in \mathbb{S}_T[C_C]. \tag{3}$$

This equation tells us that her foodset now 'maps to' a foodset that includes knowledge of Clive, and the part after the comma tells us that knowledge of Clive was socially transmitted from Clive himself.

As Thera's understanding of Clive's concerns increases, she feels more emotionally connected to him, which in turn impacts the quality of her responses to him. Clive feels increasingly heard, and begins trusting her. Her reflections, statements, and questions may more readily facilitate adaptive change in Clive that may not have occurred otherwise. Thera's interactions create perturbations in Clive' s cognitive network that disrupt the 'I am stupid' attractor state described above.

Elaborating on the 'Clive and Thera' example, we now show how the RAF model brings to light four distinct ways in which a therapist such as Thera facilitates therapeutic changes in the worldview of a client such as Clive.

5.2. Providing Counter-Evidence to Distressing Belief

During the session, Thera's chair malfunctions. Clive is able to fix the problem by adjusting the various knobs and gears on the chair. Thera says, "You're a brilliant problem solver—you fixed my chair!" This provides counter-evidence to the belief that he is unintelligent. Her observation, which lifts his mood, temporarily decreases his distress, and enhances his self-concept, is socially transmitted to him, resulting in the new MR: 'Someone thinks I'm brilliant'. Transmission of the information that he is brilliant, denoted B_C, is described by Equation (4), as follows:

$$F_C \mapsto F_C \cup \{B_C\},\ \text{where } B_C \in \mathbb{S}_C(B_T) \tag{4}$$

Thera's observation becomes part of Clive's 'mental foodset' because it did not come into existence within Clive's mind; it was 'born' in Thera's mind, and socially transmitted from Thera to Clive. His self-concept now contains both the constellation of experiences and negative beliefs about his intelligence, described as A RAF, consisting of multiple mutually consistent memories that support his belief that he is 'stupid', as well as a new experience that is inconsistent with this RAF. This experience of being described as 'brilliant' by Thera therefore has an *inhibitory* role on that RAF; it weakens the strength of that reaction, thereby diminishing the proclivity to interpret ambiguous events as confirmation of the belief, "I am stupid". This is depicted in Panel (b) of Figure 1, where **Step 2** refers to the social transmission of words of praise from Thera to Clive, and **Step 3** refers to its inhibitory role on the existing RAF.

5.3. Modeling Adaptive Mindset through Self-Disclosure

Thera models how to consciously resist the tendency to interpret ambiguous evidence in a negative manner through the use of self-disclosure regarding how she manages self-critical thoughts in her own life. She also spontaneously models adaptive responses in her interactions with him. She bumps her elbow on the desk, and then laughs at her clumsiness. The laughter enables her to re-frame the thought 'I am clumsy' into something innocuous. Thera's social transmission of this to him is **Step 4** in Panel (c) of Figure 1. Where the socially transmitted laughter is denoted L_C, this process is described as follows:

$$F_C \mapsto F_C \cup \{L_C\},\ \text{where } L_C \in \mathbb{S}_C(L_T) \tag{5}$$

The possibility of responding to one's inadequacies with laughter and/or self-deprecating humor is a new and striking concept for Clive. As such, this experience forms a foodset element in Clive's mind. His decision to use this strategy himself (i.e., to incorporate it as a reactant in subsequent steps) depends on a number of variables, such as the degree to which he trusts and respects the therapist, his motivation to change, and so forth. (Elaboration of these variables is beyond the scope of this discussion).

5.4. Catalyzing Alternative Explanations

Thera not only transmits *existing* knowledge that Clive can import, wholesale, into his worldview, she also prompts the independent formation of *new* information, specifically tailored to his personality, that are conducive to adaptive perspectives and behaviors, and help him resolve or come to terms with the issues he faces. This could take the form of asking Clive questions that prompt him to reconsider existing beliefs, or by challenging Clive's beliefs directly. For example, Thera might as him, "Might there be other reasons that you didn't do so well on that math test?" This question gently challenges Clive's forgone interpretation of the event. It prompts him to explore reasons other than that he is unintelligent. This is depicted in panel (c) of Figure 1, as **Step 5**.

Clive responds, laughing, "Well yeah, I spent a lot of time playing video games". This alternative interpretation does not play into the notion that Clive is unintelligent, and it shows that he has assimilated her proclivity to diffuse a negative conclusion with humor. It generates a *new* interpretation. This is depicted in Panel d of Figure 1, where **Step 6a** shows the catalyzing event (i.e., the search for an alternate explanation), and **Step 6b** shows the product of this 'reaction', Clive's joke.

We describe this in terms of RAFs as follows. Clive's memory of getting a bad grade on the test, denoted G_C, undergoes change, so it serves as a reactant that transforms through RR to the product, the joke about video games, denoted V_C. This new interpretation is provoked, or 'catalyzed' by Thera's question, denoted q_T. We describe this as follows:

$$G_C \xrightarrow{q_T} V_C \in \neg F_C,\ \neg F_C \mapsto \neg F_C \cup \{V_C\}. \tag{6}$$

This conceptual shift transforms one or more element(s) of the foodset F_C into a new foodset-derived MR, V_C, i.e., a member of $\neg F_C$. Clive's self-concept now contains a new MR—the joke he made—represented as a new node in his cognitive network.

5.5. Catalyzing a New Belief That Is Adaptive

When Clive dilutes the potency of this once-distressing memory with humor, Thera laughs. This laughter, denoted as l_T catalyzes a new belief in Clive's mind, 'I am funny', denoted as $b2_C$. This new belief buoys his self-esteem, and reduces his distress. The joke, denoted as J_C serves as a reactant. Thus, the reaction is described as follows:

$$J_C \xrightarrow{l_T} b2_C \in \neg F_C,\ \neg F_C \mapsto \neg F_C \cup \{J_C\}. \tag{7}$$

This is depicted in Panel (e) of Figure 1, where **Step 7a** refers to the catalyzing event, i.e., Thera's laughter, and **Step 7b** refers to the resulting formation of the new belief, 'I am funny'.

Using these four distinct methods, Thera simultaneously reduces the strength of a distressing belief that was damaging his self-concept, and facilitate the creation of a new belief that enhances Clive's self-concept, as illustrated by the relative thickness of the RAFs in Figure 1.

We note there are now hierarchical levels of RAF structure, composed of interacting RAFs that collectively form a maxRAF (panel (e) of Figure 1) (The nodes in the therapist's mind associated with Steps 2, 4, and 5 in Figure 1 can either be included in this maxRAF or not, since from the RAF point of view they are merely copies of the corresponding elements in the client's worldview). We note also that RAF structure extends across the two individuals, providing a means of formally describing the dyadic relationship between therapist and client, and the emergence of a therapeutic alliance between them.

6. Discussion and Conclusions

We presented a RAF model of how a therapist fosters self-esteem and well-being in the client. The model illustrated four distinct ways by which a therapist accomplishes this: (1) providing direct examples/evidence contrary to a client's distressing belief about themselves, (2) challenging the client's existing interpretations of events, (3) using self-disclosure, provide examples of strategies for diffusing the potency of a negative belief, (4) reinforcing the client's attempts to assimilate such strategies in their own ways of thinking.

As discussed in Sections 3 and 4, RAF networks have been used to model the origins of evolutionary processes, biological (the origin of life) as well as cultural (the origin of cumulative innovation). We think this is not coincidental; indeed, elsewhere, we showed that both the evolution of early life and cultural evolution are instantiations of a primitive form of evolution—i.e., cumulative, adaptive, open-ended change—referred to as Self-Other Reorganization (SOR) [34,80,81]. Instead of replication using a self-assembly code, SOR entails internal self-organizing and self-maintaining processes within entities, as well as interaction between entities. The argument for SOR bolsters the argument that they share a deep structure, and thus strengthens the rationale for applying RAFs in both domains. In any case, the RAF approach to modeling therapeutic change is consistent with the theory that humans possess two levels of complex, adaptive, self-organizing structure: an organismal level, and a psychological level [43,45,82]. Psychological research tends to be data rich and theory poor [83], and psychological theorizing remains fragmented [84]. Psychotherapy research relies on momentary snapshots of the perceptions of client and therapist; it is vague about the nature of psychotherapeutic change, i.e., what happens at the level of mental representations and their interrelations and interactions, and how this kind of micro-level change alters the global structure of the client's worldview [5]. We take a first step towards such a global understanding in this paper. We posit that psychotherapeutic processes affect people not just at the individual level but at the society level, by providing a means to the creative transformation and cultural evolution of human worldviews.

Traditional methods for studying psychotherapeutic change have limitations [54] that the complex systems approach is well positioned to overcome [85–88], by enabling psychotherapy to be modeled and understood more precisely, using tools that embed it in a larger framework that includes other systems and disciplines. The above model of the therapeutic process provides a framework for empirical data collection and analysis. A next step is to incorporate into such a model specific factors that affect therapeutic outcomes (such as the degree of trust in the therapist). The impact of the therapeutic alliance between therapist and client on the therapy outcome is well-known [54,89]. It would be interesting to analyze psychotherapy sessions to track cognitive change, and the emergence of a therapeutic alliance, and its impact on this change. Our model accommodates the perspectives of both the therapist and client. While we have chosen to emphasize the client in this interaction, the RAF approach can also model potential changes in the therapist's worldview. The RAF approach could also be used to investigate a number of other issues related to psychopathology and treatment, such as the development of mental illness, the trajectory of various mental health concerns, and whether there are differences in the conceptual frameworks of individuals experiencing depression and those with anxiety. It could also be used to model the impact of different types of psychotherapy on conceptual network structure, and the impact of this structure on mental health and well-being. One promising possibility is to study whether individual differences in reliance on foodset versus foodset-derived information sources (i.e., the propensity to think things through for oneself) culminate in different kinds of conceptual networks, which might differentially affeact therapeutic progress. In addition, using RAF networks to precisely model the psychotherapeutic process could be informative in the design and execution of computerized psychotherapies [90,91], or as an aid to the human psychotherapist for keeping track of, and visually depicting, specific interactions in the psychotherapy process and their outcome. We are a long way from this, but in keeping with the adage "a picture is worth a thousand words", the RAF framework for psychotherapy could form the basis for a software program that enables the therapist to visualize and identify change in RAF structure as it occurs over the course of psychotherapy, and to visualize desired possible future states of their clients' worldviews.

The RAF approach offers an established mathematical framework for integrating research on creative cognition, semantic networks, and the kinds of structures that exhibit cumulative, adaptive, open-ended change, i.e., that evolve, with a similarly dynamic process of psychotherapy. Though still in its infancy, it has the potential to provide a new way of understanding how the therapeutic alliance works, one that embeds psychotherapy research in the formal study of self-organizing structures and their role in evolutionary processes.

Author Contributions: Conceptualization, K.G. and L.G. writing, K.G. and L.G. All authors have read and agreed to the published version of the manuscript.

Funding: L.G. acknowledges funding from Grant 62R06523 from the Natural Sciences and Engineering Research Council of Canada. Funding was also provided by private donors, Susan and Jacques LeBlanc, for which are very grateful.

Institutional Review Board Statement: Not applicable.

Informed Consent Statement: Not applicable.

Conflicts of Interest: The authors declare no conflict of interest. The funders had no role in the writing of the manuscript, or in the decision to publish the results.

Abbreviations

The following abbreviations are used in this manuscript:

CRS	Catalytic reaction system
RAF	Reflexively Autocatalytic foodset-derived network
MR	Mental representation
RR	Representational redescription
SOC	Self-organized criticality

Appendix A. Glossary of Terms

Abstract thought: the processing of internally sourced mental contents.
Autocatalytic: the whole can be reconstituted through interactions amongst its parts.
Catalyst: facilitates a transition that would otherwise be highly unlikely to occur. Here, the role of catalyst is played by a problem, desire, or need, or a realization or external stimuli that trigger a thought that would be highly unlikely to occur otherwise. For example, if a stranger on the street reminds you of a deceased relative, and this triggers a memory of being with that person, the strange (or more precisely, your mental representation of the stranger plays the role of a catalyst.)
Catalytic reaction system: a network of interrelated parts, such as a conceptual network.
Closed RAF: A RAF that is stable unless the foodset changes or the reactions they take part in changes. (Formally, a closed RAF is A RAF that contains every reaction in the network that has each of its reactants and at least one catalyst present either in the foodset or as a product of some reaction in the RAF.) The maxRAF is always closed. The *closure* of any subRAF will contain the original subRAF, and be larger (unless the original subRAF was already closed).
Conceptual network: a web of shared properties, contexts, associations, and relationships of logic, causation, and so forth, that bind them together.
Co-RAF: a nonempty set of reactions that is not A RAF on its own, but that forms A RAF when combined with an existing set of reactions.
Catalytic reaction system: a network of components, such as a network of catalytic molecules, or a conceptual network.

Foodset, F: the elements that are initially present, as opposed to those that are the products of interactions amongst them.
foodset-derived (sometimes called F-generated, or foodset-derived), $\neg F$: an element that can be generated from the foodset F through a series of reactions in $\mathcal{R}'$ itself. That is, an element of the network that is *not* part of the foodset. The term 'foodset-derived' is more often used in the cognitive application of RAFs.
Individual learning: obtaining pre-existing information from the environment by nonsocial means through direct perception.
IrrRAF: A RAF that is irreducible, i.e., cannot be broken down into smaller RAFs.
MaxRAF: the largest RAF in the network. It includes all other RAFs.
Mental representation (MR). Items in declarative or procedural memory composed of one or more concepts or percepts, and which came about through individual learning, social learning, or abstract thought. The set of all mental representations in individual i is denoted X_i. (As mentioned in the text, we emphasize that although we use the terms 'mental representation', we are sympathetic with the view that what we call mental representations do not 'represent', but act as contextually elicited bridges between the mind and the world.)
Reflexive: each part is related to the whole.
Reflexively autocatalytic: each reaction $r \in \mathcal{R}'$ is catalyzed by at least one element type that is either produced by $\mathcal{R}'$ or is present in the foodset F.
Phase transition: rapid transition from one state to another.
Reactant: a mental representation that participates in a given 'reaction', i.e., an event that alters the structure of the conceptual network.
Reaction: a change of state or interaction between existing elements that results in a new element. The set of all possible reactions in individual i is denoted R_i.
Representational redescription (RR): conceptual restructuring that causes a mental representation to change. In the RAF framework this is modeled as a reaction.
Self-organized criticality (SOC): a phenomenon wherein, through simple local interactions, complex systems tend to find a critical state poised at the cusp of a transition between order and chaos, from which a single small perturbation occasionally exerts a disproportionately large effect.
SubRAF: A RAF that is not the maxRAF. It is a component of the maxRAF.
Transient RAF: a subRAF that is not closed. A transient RAF may add additional reactions until it becomes closed.

References

1. Dragioti, E.; Karathanos, V.; Gerdie, B.; Evangelou, E. Does psychotherapy work? An umbrella of meta-analyses of randomized control trials. *Acta Psychiatr. Scand.* **2017**, *136*, 236–246. [CrossRef] [PubMed]
2. Moldovan, R.; Pinetea, S. Mechanisms of change in psychotherapy: Methodological and statistical considerations. *Cogn. Brain Behav.* **2015**, *19*, 299–311.
3. Carey, T.A.; Huddy, V.; Griffiths, R. To mix or not to mix? A meta-method approach to rethinking evaluation practices for improved effectiveness and efficiency of psychological therapies illustrated with the application of perceptual control theory. *Front. Psychol.* **2019**, *10*, 1445. [CrossRef] [PubMed]
4. Kazdin, A.E. Understanding how and why psychotherapy leads to change. *Psychother. Res.* **2009**, *19*, 418–428. [CrossRef] [PubMed]
5. Carey, T.A.; Griffiths, R.; Dixon, J.E.; Hines, S. Identifying functional mechanisms in psychotherapy: A scoping systematic review. *Front. Psychiatry* **2020**, *11*, 291. [CrossRef] [PubMed]
6. Bassett, D.S.; Bullmore, E.D. Small-world brain networks. *Neuroscientist* **2006**, *12*, 512–523. [CrossRef] [PubMed]
7. Betzel, R.F.; Bassett, D.S. Generative models for network neuroscience: Prospects and promise. *J. R. Soc. Interface* **2017**, *14*, 20170623. [CrossRef] [PubMed]
8. Karuza, E.A.; Thompson-Schill, S.L.; Bassett, D.S. Local patterns to global architectures: Influences of network topology on human learning. *Trends Cogn. Sci.* **2016**, *20*, 629–640. [CrossRef]
9. Siew, C.S.; Wulff, D.U.; Beckage, N.M.; Kenett, Y.N. Cognitive network science: A review of research on cognition through the lens of network representations, processes, and dynamics. *Complexity* **2019**, *2019*, 2108423. [CrossRef]
10. Baronchelli, A.; Ferrer-i-Cancho, R.; Pastor-Satorras, R.; Chater, N.; Christiansen, M.H. Networks in cognitive science. *Trends Cogn. Sci.* **2013**, *17*, 348–360. [CrossRef]
11. Borge-Holthoefer, J.; Arenas, A. Semantic networks: Structure and dynamics. *Entropy* **2010**, *12*, 1264–1302. [CrossRef]

12. Kumar, A.; Steyvers, M.; Balota, D.A. A critical review of network-based and distributional approaches to semantic memory structure and processes. *Top. Cogn. Sci.* 2021, *in press*. [CrossRef] [PubMed]
13. Hofmann, S.G.; Curtiss, J.; McNally, R.J. A Complex Network Perspective on Clinical Science. *Perspect. Psychol. Sci.* **2016**, *11*, 597–605. [CrossRef] [PubMed]
14. Orsucci, F. *Mind Force: On Human Attractions*; World Scientific Publishing: Singapore, 2009.
15. Tschacher, W.; Haken, H. *The Process of Psychotherapy*; Springer Nature: Cham, Switzerland, 2019.
16. Gabora, L.; Beckage, N.; Steel, M. Modeling cognitive development with reflexively autocatalytic networks. *Top. Cogn. Sci.* 2022, *in press*.
17. Gabora, L.; Steel, M. From uncertainty to insight: An autocatalytic framework. In *Uncertainty: A Catalyst for Creativity, Learning and Development*; Beghetto, R., Jaeger, G., Eds.; Springer: Berlin/Heidelberg, Germany, 2022.
18. Ganesh, K.; Gabora, L. Modeling Discontinuous Cultural Evolution: The Impact of Cross-domain Transfer. *Front. Psychol.* **2022**, *13*, 786072. [CrossRef]
19. Farmer, J.D.; Kauffman, S.A.; Packard, N.H. Autocatalytic replication of polymers. *Phys. D Nonlinear Phenom.* **1986**, *22*, 50–67. [CrossRef]
20. Kauffman, S.A. *The Origins of Order*; Oxford University Press: Oxford, UK, 1993.
21. Erdös, P.; Rényi, A. On the evolution of random graphs. *Publ. Math. Inst. Hung. Acad. Sci.* **1960**, *5*, 17–61.
22. Gabora, L. Revene of the 'neurds': Characterizing creative thought in terms of the structure and dynamics of human memory. *Creat. Res. J.* **2010**, *22*, 1–13. [CrossRef]
23. Aerts, D.; Aerts, S.; Gabora, L. Experimental evidence for quantum structure in cognition. In Proceedings of the International Symposium on Quantum Interaction, Saarbrücken, Germany, 25–27 March 2009; Bruza, P., Lawless, W., van Rijsbergen, K., Sofge, D., Eds.; Springer: Berlin/Heidelberg, Germany, 2009; pp. 59–70.
24. Aerts, D.; Gabora, L.; Sozzo, S. Concepts and their dynamics: A quantum theoretical model. *Top. Cogn. Sci.* **2013**, *5*, 737–772. [CrossRef]
25. Aerts, D.; Broekaert, J.; Gabora, L.; Sozzo, S. Generalizing prototype theory: A formal quantum framework. *Front. Psychol.* **2016**, *7*, 418. [CrossRef]
26. Buskell, A.; Enquist, M.; Jansson, F. A systems approach to cultural evolution. *Palgrave Commun.* **2019**, *5*, 131. [CrossRef]
27. Gabora, L.; Aerts, D. Contextualizing concepts using a mathematical generalization of the quantum formalism. *J. Exp. Theor. Artif. Intell.* **2002**, *14*, 327–358. [CrossRef]
28. Gabora, L.; Aerts, D. Evolution as context-driven actualisation of potential: Toward an interdisciplinary theory of change of state. *Interdiscip. Sci. Rev.* **2005**, *30*, 69–88. [CrossRef]
29. Wang, Z.; Busemeyer, J.R. Interference effects of categorization on decision making. *Cognition* **2016**, *150*, 133–149. [CrossRef] [PubMed]
30. Steel, M.; Hordijk, W.; Xavier, J.C. Autocatalytic networks in biology: Structural theory and algorithms. *J. R. Soc. Interface* **2019**, *16*, 20180808. [CrossRef] [PubMed]
31. Hordijk, W.; Steel, M. Detecting autocatalytic, self-sustaining sets in chemical reaction systems. *J. Theor. Biol.* **2004**, *227*, 451–461. [CrossRef] [PubMed]
32. Hordijk, W.; Steel, M. Chasing the tail: The emergence of autocatalytic networks. *Biosystems* **2016**, *152*, 1–10. [CrossRef]
33. Mossel, E.; Steel, M. Random biochemical networks and the probability of self-sustaining autocatalysis. *J. Theor. Biol.* **2005**, *233*, 327–336. [CrossRef]
34. Gabora, L.; Steel, M. An evolutionary process without variation and selection. *J. R. Soc. Interface* **2021**, *18*, 20210334. [CrossRef]
35. Hordijk, W.; Steel, M. Autocatalytic sets and boundaries. *J. Syst. Chem.* **2015**, *6*, 1. [CrossRef]
36. Hordijk, W.; Hein, J.; Steel, M. Autocatalytic sets and the origin of life. *Entropy* **2010**, *12*, 1733–1742. [CrossRef]
37. Xavier, J.C.; Hordijk, W.; Kauffman, S.; Steel, M.; Martin, W.F. Autocatalytic chemical networks at the origin of metabolism. *Proc. R. Soc. London Ser. B Biol. Sci.* **2020**, *287*, 20192377. [CrossRef] [PubMed]
38. Gabora, L.; Steel, M. Autocatalytic networks in cognition and the origin of culture. *J. Theor. Biol.* **2017**, *431*, 87–95. [CrossRef] [PubMed]
39. Gabora, L.; Steel, M. Modeling a cognitive transition at the origin of cultural evolution using autocatalytic networks. *Cogn. Sci.* **2020**, *44*, e12878. [CrossRef] [PubMed]
40. Gabora, L.; Steel, M. A model of the transition to behavioral and cognitive modernity using reflexively autocatalytic networks. *Proc. R. Soc. Interface* **2020**, *17*, 20200545. [CrossRef] [PubMed]
41. Steel, M.; Xavier, J.C.; Huson, D.H. Autocatalytic networks in biology: Structural theory and algorithms. *J. R. Soc. Interface* **2020**, *17*, 20200488. [CrossRef] [PubMed]
42. Hordijk, W.; Kauffman, S.A.; Steel, M. Required levels of catalysis for emergence of autocatalytic sets in models of chemical reaction systems. *Int. J. Mol. Sci.* **2011**, *12*, 3085–3101. [CrossRef] [PubMed]
43. Barton, S. Chaos, self-organization, and psychology. *Am. Psychol.* **1994**, *49*, 5–14. [CrossRef] [PubMed]
44. Pribram, K.H. *Origins: Brain and Self-Organization*; Lawrence Erlbaum: Hillsdale, NJ, USA, 1994.
45. Maturana, H.; Varela, F. Autopoiesis and cognition: The realization of the living. In *Boston Studies in the Philosophy of Science*; Cohen, R.S., Wartofsky, M.W., Eds.; Reidel: Dordecht, The Netherlands, 1973; Volume 42.
46. Varela, F.; Thompson, E.; Rosch, E. *The Embodied Mind*; MIT Press: Cambridge, MA, USA, 1991.

47. Mednick, S.A. The associative basis of the creative process. *Psychol. Rev.* **1962**, *69*, 220–232. [CrossRef]
48. Cautin, R.L. A century of psychotherapy, 1860–1960. In *History of Psychotherapy: Continuity and Change*; American Psychological Association: Wahington, DC, USA, 2011; pp. 3–38. [CrossRef]
49. Finsrud, I.; Nissen-Lie, H.A.; Vrabel, K.; Høstmælingen, A.; Wampold, B.E.; Ulvenes, P.G. It's the therapist and the treatment: The structure of common therapeutic relationship factors. *Psychother. Res.* **2022**, *32*, 139–150. [CrossRef] [PubMed]
50. Lambert, M.J. Outcome in psychotherapy: The past and important advances. *Psychotherapy* **2013**, *50*, 42–51. [CrossRef] [PubMed]
51. Caspi, A.; Moffit, T.E. All for one and one for all: Mental disorders in one dimension. *Am. J. Psychiatry* **2018**, *175*, 831–844. [CrossRef] [PubMed]
52. Bernhardt, I.S.; Nissen-Lie, H.A.; Rabu, M. The embodied listener: A dyaduc case study of how therapist and patient reflect on the significance of the therapist's personal experience. *Psychother. Researcg* **2020**, *31*, 682–692. [CrossRef] [PubMed]
53. Carey, T.A.; Tai, S.J.; Mansell, W.; Huddy, V.; Griffiths, R.; Marken, R.S. Improving professional psychological practice through an increased repertoire of research methodologies: Illustrated bu the development of MOL. *Prof. Psychol. Res. Pract.* **2017**, *48*, 175–182. [CrossRef]
54. Hayes, A.M.; Andrews, L.A. A complex systems approach to the study of change in psychotherapy. *BMC Med.* **2020**, *18*, 197. [CrossRef] [PubMed]
55. Tschacher, W.; Meir, D. Physiological synchrony in psychotherapy sessions. *Psychother. Res.* **2019**, *30*, 558–573. [CrossRef] [PubMed]
56. Ramseyer, F.; Tschacher, W. Synchrony in dyadic psychotherapy sessions. In *Simultaneity*; Vrobel, S., Rössler, O., Marks-Tarlow, T., Eds.; World Scientific: Singapore, 2008.
57. Reich, C.M.; Berman, J.S.; Dale, R.; Levitt, H.M. Vocal synchrony in psychotherapy. *J. Soc. Clin. Psychol.* **2015**, *33*, 481–494. [CrossRef]
58. Ramseyer, F.; Tschacher, W. Nonverbal synchrony in psychotherapy. *J. Consult. Clin. Psychol.* **2011**, *79*, 284–295. [CrossRef]
59. Bachelor, A. Clients' and Therapists' Views of the Therapeutic Alliance: Similarities, Differences and Relationship to Therapy Outcome. *Clin. Psychol. Psychother.* **2011**, *20*, 118–135. [CrossRef]
60. Beghetto, R.A. Structured uncertainty: How creativity thrives under constraints and uncertainty. In *Creativity under Duress in Education? Resistive Theories, Practices, and Actions*; Mullen, C.A., Ed.; Springer International Publishing: Cham, Switzerland, 2019; pp. 27–40. [CrossRef]
61. Kauffman, S.A. *Humanity in a Creative Universe*; Oxford University Press: Oxford, UK, 2016.
62. Karmiloff-Smith, A. *Beyond Modularity: A Developmental Perspective on Cognitive Science*; MIT Press: Cambridge, MA, USA, 1992.
63. Bowers, K.S.; Farvolden, P.; Mermigis, L. Intuitive antecedents of insight. In *The Creative Cognition Approach*; Ward, S., Finke, R.A., Eds.; MIT Press: Cambridge, MA, USA, 1995; pp. 27–51.
64. Beghetto, R.; Jaeger, G. *Uncertainty: A Catalyst for Creativity, Learning and Development*; Springer: Berlin/Heidelberg, Germany, 2021.
65. Cabell, K.R.; Valsiner, J. *The Catalyzing Mind: Beyond Models of Causality, Annals of Theoretical Psychology*; Springer: Berlin/Heidelberg, 2016; Volume 11.
66. Barsalou, L.W. Context-independent and context-dependent information in concepts. *Mem. Cogn.* **1982**, *10*, 82–93. [CrossRef]
67. Hampton, J.A. Disjunction of natural concepts. *Mem. Cogn.* **1988**, *16*, 579–591. [CrossRef] [PubMed]
68. Osherson, D.N.; Smith, E.E. On the adequacy of prototype theory as a theory of concepts. *Cognition* **1981**, *9*, 35–58. [CrossRef]
69. Veloz, T.; Tëmkin, I.; Gabora, L. A conceptual network-based approach to inferring the cultural evolutionary history of the Baltic psaltery. In Proceedings of the 34th Annual Meeting of the Cognitive Science Society, Sapporo, Japan, 1–4 August 2012; Miyake, N., Peebles, D., Cooper, R.P., Eds.; Cognitive Science Society: Austin, TX, USA, 2012; pp. 2487–2492.
70. Gabora, L.; Leijnen, S.; Veloz, T.; Lipo, C. A non-phylogenetic conceptual network architecture for organizing classes of material artifacts into cultural lineages. In Proceedings of the 33rd annual meeting of the Cognitive Science Society, Boston, MA, USA, 20–23 July 2011; Carlson, L., Holscher, C., Shipley, T.F., Eds.; Cognitive Science Society: Austin, TX, USA, 2011; pp. 2923–2928.
71. Gabora, L.; Tseng, S. The social benefits of balancing creativity and imitation: Evidence from an agent-based model. *Psychol. Aesthet. Creat. Arts* **2017**, *11*, 457–473. [CrossRef]
72. Gabora, L.; Smith, C. Two cognitive transitions underlying the capacity for cultural evolution. *J. Anthropol. Sci.* **2018**, *96*, 27–52. [PubMed]
73. Sousa, F.; Hordijk, W.; Steel, M.; Martin, W. Autocatalytic sets in *E. coli* metabolism. *J. Syst. Chem.* **2015**, *6*, 4. [CrossRef]
74. Steel, M. The emergence of a self-catalyzing structure in abstract origin-of-life models. *Appl. Math. Lett.* **2000**, *13*, 91–95. [CrossRef]
75. Vasas, V.; Fernando, C.; Santos, M.; Kauffman, S.; Szathmáry, E. Evolution before genes. *Biol. Direct* **2012**, *7*, 1. [CrossRef] [PubMed]
76. Cazzolla Gatti, R.; Fath, B.; Hordijk, W.; Kauffman, S.; Ulanowicz, R. Niche emergence as an autocatalytic process in the evolution of ecosystems. *J. Theor. Biol.* **2018**, *454*, 110–117. [CrossRef]
77. Rogers, C. The necessary and sufficient conditions of therapeutic personality change. *J. Couns. Psychol.* **1956**, *21*, 95–103. [CrossRef]
78. O'Connor, B.; Gabora, L. Applying complexity theory to a dynamical process model of the development of pathological belief systems. *Chaos Complex. Lett.* **2009**, *4*, 75–96.

79. Hiver, P. Attractor States. In *Motivational Dynamics in Language Learning*; Dornyei, Z., MacIntyre, P., Henry, A., Eds.; Multilingual Matters: Bristol, UK, 2015; pp. 20–28.
80. Gabora, L. From deep learning to deep reflection: Toward an appreciation of the integrated nature of cognition and a viable theoretical framework for cultural evolution. In Proceedings of the 2019 Annual Meeting of the Cognitive Science Society, Montreal, QC, Canada, 24–27 July 2019; Nadel, L., Stein, D., Eds.; Cognitive Science Society: Austin, TX, USA, 2019; pp. 1801–1807.
81. Gabora, L. Creativity: Linchpin in the quest for a viable theory of cultural evolution. *Curr. Opin. Behav. Sci.* **2019**, *27*, 77–83. [CrossRef]
82. Gabora, L. Ideas are not replicators but minds are. *Biol. Philos.* **2004**, *19*, 127–143. [CrossRef]
83. Fried, E.I. Lack of theory building and testing impedes progress in the factor and network literature. *Psychol. Inq.* **2020**, *31*, 271–288. [CrossRef]
84. Teo, T. *Outline of Theoretical Psychology: Critical Investigations*; Springer: Berlin/Heidelberg, Germany, 2018.
85. Orsucci, F. Introduction. In *Human Dynamics: A Complexity Science Open Handbook*; Orsucci, F., Ed.; Nova Science Publishers: New York, NY, USA, 2016; pp. 119–133.
86. de Felice, G.; Giuliani, A.; Halfon, S.; Andreassi, S.; Paoloni, G.; Orsucci, F. The misleading Dodo Bird verdict. How much of the outcome variance is explained by common and specific factors? *New Ideas Psychol.* **2019**, *54*, 50–55. [CrossRef]
87. de Felice, G.; Giuliani, A.; Gelo, O.C.G.; Mergenthaler, E.; De Smet, M.M.; Meganck, R.; Paoloni, G.; Andreassi, S.; Schiepek, G.K.; Scozzari, A.; et al. What Differentiates Poor- and Good-Outcome Psychotherapy? A Statistical-Mechanics-Inspired Approach to Psychotherapy Research, Part Two: Network Analyses. *Front. Psychol.* **2019**, *11*, 788. [CrossRef] [PubMed]
88. de Felice, G.; Orsucci, F.; Scozzari, A.; Gelo, O.C.G.; Serafini, G.; Andreassi, S.; Vegni, N.; Paoloni, G.; Lagetto, G.; Mergenthaler, E.; et al. What Differentiates Poor and Good OutcomePsychotherapy? A Statistical-Mechanics-InspiredApproach to Psychotherapy Research. *Systems* **2019**, *7*, 22. [CrossRef]
89. Fluckiger, C.; Del, A.C.; Wampold, B.E.; Horvath, A.O. The alliance in adult psychotherapy: A meta-analytic synthesis. *Psychotherapy* **2018**, *55*, 316–340. [CrossRef]
90. Liu, Z.; Qiao, D.; Xu, Y.; Zhao, W.; Yang, Y.; Wen, D.; Li, X.; Nie, X.; Dong, Y.; Tang, S.; et al. The efficacy of computerized cognitive behavioral therapy for depressive and anxiety symptoms in patients With COVID-19: Randomized controlled trial. *J. Med. Internet Res.* **2021**, *23*, e26883. [CrossRef]
91. Foroushani, P.; Schneider, J.; Assareh, N. Meta-review of the effectiveness of computerised CBT in treating depression. *BMC Psychiatry* **2011**, *11*, 131. [CrossRef]

MDPI

Article

Studying Physiological Synchrony in Couple Therapy through Partial Directed Coherence: Associations with the Therapeutic Alliance and Meaning Construction

Evrinomy Avdi [1,*], Evangelos Paraskevopoulos [2], Christina Lagogianni [1], Panagiotis Kartsidis [3] and Fotis Plaskasovitis [1]

[1] Department of Psychology, Aristotle University of Thessaloniki, 541 24 Thessaloniki, Greece; clagogia@psy.auth.gr (C.L.); fplaskas@psy.auth.gr (F.P.)
[2] Department of Psychology, University of Cyprus, Nicosia 1678, Cyprus; paraskevopoulos.evangelso@ucy.ac.cy
[3] School of Medicine, Aristotle University of Thessaloniki, 541 24 Thessaloniki, Greece; panos.kartsidis@auth.gr
* Correspondence: avdie@psy.auth.gr; Tel.: +30-231-099-7363

Abstract: In line with the growing recognition of the role of embodiment, affect and implicit processes in psychotherapy, several recent studies examine the role of physiological synchrony in the process and outcome of psychotherapy. This study aims to introduce Partial Directed Coherence (PDC) as a novel approach to calculating psychophysiological synchrony and examine its potential to contribute to our understanding of the therapy process. The study adopts a single-case, mixed-method design and examines physiological synchrony in one-couple therapy in relation to the therapeutic alliance and a narrative analysis of meaning construction in the sessions. Interpersonal Physiological Synchrony (IPS) was calculated, via a windowed approach, through PDC of a Heart Rate Variability-derived physiological index, which was measured in the third and penultimate sessions. Our mixed-method analysis shows that PDC quantified significant moments of IPS within and across the sessions, modeling the characteristics of interpersonal interaction as well as the effects of therapy on the interactional dynamics. The findings of this study point to the complex interplay between explicit and implicit levels of interaction and the potential contribution of including physiological synchrony in the study of interactional processes in psychotherapy.

Keywords: physiological synchrony; heart rate; therapeutic alliance; psychotherapy process; couple therapy

Citation: Avdi, E.; Paraskevopoulos, E.; Lagogianni, C.; Kartsidis, P.; Plaskasovitis, F. Studying Physiological Synchrony in Couple Therapy through Partial Directed Coherence: Associations with the Therapeutic Alliance and Meaning Construction. *Entropy* **2022**, *24*, 517. https://doi.org/10.3390/e24040517

Academic Editors: Franco Orsucci and Wolfgang Tschacher

Received: 31 January 2022
Accepted: 2 April 2022
Published: 6 April 2022

1. Introduction

This study rests on the assumption that psychotherapy relies on both implicit and explicit processes and that both need to be taken into account when studying clinical process [1,2]. It focuses on one aspect of implicit interaction, interpersonal physiological synchrony (IPS), and introduces the use of Partial Directed Coherence as a metric for operationalizing IPS in psychotherapy sessions. Using a single-case, mixed-method design on one couple therapy, physiological synchrony is examined in relation to the therapeutic alliance and a qualitative analysis that draws upon narrative principles of meaning reconstruction in the sessions.

Synchrony is observed in many complex biological systems and is assumed to occur through nonlinear dynamic processes rather than simple causal links. In social interaction, synchrony concerns the temporal covariation of behavior or internal states in interacting partners and can be broadly defined as 'the social coupling of two (or more) individuals in the here-and-now of a communication context that emerges alongside, and in addition to, their verbal exchanges' [3] (p. 558).

A key concept in the literature on interactional synchrony is interpersonal coordination, which refers to the degree to which the behaviors of interacting partners are nonrandom, patterned or synchronized in timing and form [4]. There is ample evidence that behavioral matching and interactional synchrony are ubiquitous features of human interaction, on both verbal (e.g., vocal tone, word choice, laughter, speech accent, syntax, intonation) and nonverbal (e.g., posture, gesture, facial expression, orientation, etc.) levels. Interpersonal coordination emerges early in life and is an automatic, non-conscious process that is associated with liking, affiliation, rapport, cooperation, self–other merging, perspective taking, empathy, smoothness of interaction, prosocial behaviors, compassion and increased performance in tasks that rely on joint actions [5–7].

In the context of psychotherapy, 'being in sync' has been examined primarily in relation to nonverbal behaviors and has been shown to be associated with important psychotherapy processes, such as rapport [8], therapist empathy, the therapeutic alliance [6,7], session quality and therapy outcome [9–11], as well as mental state in relation to attachment [12–14]. Drawing upon developmental research, several authors have proposed that synchronous behaviors between therapist and client are crucial for the formation of the therapeutic alliance, which in turn promotes affect regulation in the client and fosters therapeutic change [7]. Similarly, research on infant development suggests that repeated experiences of biobehavioral synchrony between infants and their parents are central to the development of affect regulation capacities in the infant and security of attachment [15–18]. There is some evidence that synchrony is associated with affect regulation in adulthood as well, as interacting partners in close relationships coregulate their arousal around a homeostatic optimal level [19,20].

1.1. Interpersonal Physiological Synchrony

In addition to studying synchrony in observable behavior, in recent years, there has been a growing interest in the role of synchrony in physiological arousal in psychotherapy. This is in line with the recognition that psychological and social processes cannot be isolated from embodiment and affect [21,22]. The inclusion of affective and embodied aspects of interaction is arguably particularly relevant to psychotherapy, given that affect is intimately linked with meaning construction and forms an integral part of the work of therapy [23]. The Autonomic Nervous System (ANS) plays a key role in cognition, emotion and behavior [24], and although ANS activation is not specific to affect, most emotions are associated with increased physiological arousal [25,26]. As such, several recent studies include psychophysiological measures in psychotherapy process research and treat physiological activation, and particularly its arousal component, as an index of affect [27,28]. In this literature, it is assumed that measures of psychophysiology enable the study of aspects of the therapy process that may not be accessible through self-report or observation, and can therefore add another layer of information on clinical process [29]. In other words, psychophysiological measures may reflect non-conscious, implicit affective processes and can then be used as correlates of implicit intra- and interpersonal processes in therapy [23,30]. In addition to these theoretical developments, technological advances make the continuous recording of physiological states in therapy relatively easy and unobtrusive.

Research on interpersonal physiology concerns the temporal coregulation of physiological activation in interacting partners, using continuous measures of physiological activity. The indices of ANS arousal most commonly used include electrodermal activity (EDA), considered to reflect sympathetic arousal, and variables associated with heart rate (e.g., heart rate variability), which are associated with both sympathetic and parasympathetic activity. Due to the sufficient time resolution of these variables [31], their outcome may be used to estimate the influence that one person's physiological indices exert over another's, through a model of physiological interactions or coupling [32,33]. In this context, interpersonal physiological synchrony (IPS) is defined as 'any interdependent or associated activity identified in the physiological processes of two or more individuals' [34] (p. 2).

Recent reviews of studies of IPS in different interactional contexts suggest that physiological synchrony is a robust phenomenon identifiable through different methods [11,34].

In the context of psychotherapy, physiological synchrony between therapists and clients was first examined in a series of studies in the 1950s in relation to rapport and empathy [35]. More recently, the role of IPS in the psychotherapy process has been examined in several studies of psychotherapy sessions [3,23,30,36–40], as well as simulated sessions [13,38,41]. In a recent review of this literature, Kleinbub [14] concluded that physiological synchrony in psychotherapy is an established fact, although its clinical meaning is far from known.

Physiological synchrony in psychotherapy has primarily been associated with empathy [11,16,42,43]. However, research in interactional contexts other than psychotherapy suggest that physiological synchrony is not uniquely associated with empathy and is not necessarily positive for interactions. For example, research on infant development [17,18] shows that attachment security is associated with medium-range synchrony in parent–infant interaction and that 'too much' synchrony is predictive of attachment insecurity. Similarly, findings regarding the role of physiological linkage in the quality of adult romantic relationships are mixed, with several studies showing that increased physiological linkage in couples tends to be associated with poorer relationship satisfaction and the escalation of negative affect [44]. The evidence to date suggests that, in the context of negative interactions, IPS is associated with relationship dissatisfaction and conflict, whereas in positive interactions, it is primarily associated with empathy and rapport [34]. In addition to the *affective valence* of interactions, the *degree of emotional arousal* may also moderate physiological synchrony; for example, in studies of mother–infant interactions, higher maternal heart rate, thought to reflect increased affective arousal, has been associated with lower physiological synchrony with her infant [45]. Drawing upon these findings, it seems important for future research to take into account the characteristics of the relational context when studying the role of IPS in psychotherapy.

A related issue concerns the way IPS is conceptualized, operationally defined and calculated. The majority of studies to date of IPS in psychotherapy examine only positive correlations, i.e., in-phase synchrony, where the therapist's and client's arousal covary in the same direction, and assume that negative correlations, or anti-phase synchrony, reflect lack of synchrony. Other studies, however, suggest that anti-phase synchrony, where one partner's physiological arousal decreases as the other partner's increases, reflects processes of coregulation or complementarity [46,47]. For example, in one study implicating a storytelling task, it was found that the narrator's autonomic arousal decreased when the listener's increased and he or she displayed affiliation; this was interpreted as reflecting a process of 'sharing the emotional load', whereby the listener's engagement regulated the teller's physiological arousal [48]. Similarly, in a study of ANS activation in psychoanalytic therapy, the therapists' empathic displays were associated with increased arousal in the therapist and decreased arousal in the client, whereas sequences of the therapists' challenges were associated with increases in both participants' arousal [49]. In line with these findings, Butler & Randell [19] suggest that asynchrony may be associated with stress buffering, whereby one individual moderates the stress level of another. Based on the above, including both in-phase and anti-phase synchrony in studies of IPS in psychotherapy is likely to provide a more nuanced approach to understanding this multifaceted interactional phenomenon.

The metric employed to estimate interaction is also of importance. Most studies investigating IPS use correlation-derived estimates, which are sensitive to spurious correlations and do not address causality or directionality in the interaction [14]. In order to overcome this issue, approaches that employ specific causality tests, such as Granger causality, adjusted for estimating the information flow between multivariate time series can be used in the frequency domain [50]. Combined with surrogate testing of the parameters used to estimate interactions [11], such approaches may be combined with a windowed analysis to reach a stable and fine-grained temporal resolution that can also provide directionality.

Another important issue when examining physiological synchrony in psychotherapy relates to the timescales employed in the analysis. Most studies calculate IPS over whole sessions, despite the fact that IPS is likely to be a transient phenomenon that fluctuates through sessions [1,34]. Similarly, recent studies approach the therapeutic alliance as a dynamic phenomenon and show that therapy sessions contain several periods characterized by ruptures in the alliance, often followed by interactive repair [51]. Indeed, several authors suggest that it is precisely such repairs that are important for optimal development and therapeutic change [51–53]. Therefore, examining synchrony on a more micro-level of interaction can shed light on processes that may not be apparent at the session level.

In sum, research on physiological synchrony in psychotherapy suggests that it can add important information regarding the psychotherapy process; given that IPS may reflect different interactional processes—including empathy, affect coregulation and conflict—caution is needed when interpreting findings. Moreover, the field is fragmented on both conceptual and methodological levels, as reflected in the prevalent lack of agreement on terminology, data collection methods, research designs and statistical analyses [11,34,54]. Recent reviews suggest that, given how little we know about the context-specific factors that affect IPS, it may be preferable to use idiographic designs and theoretically informed analyses of the therapy process. Since the publication of these reviews, a few such studies have been published that shed light on the different functions of physiological synchrony in psychotherapy [23,30,37,42,55–57].

Before turning to the current study, we briefly discuss the concept of the therapeutic alliance, with a focus on couple therapy, given that it is a key clinical concept that has been associated with physiological synchrony.

1.2. The therapeutic Alliance in Couple Therapy

Several contemporary approaches to psychotherapy adopt a discursive and narrative perspective and conceptualize the process of change in psychotherapy in terms of meaning reconstruction [58]. In this framework, psychotherapy is described as a semantic process that relies on the creation of a dialogical space, which facilitates the reconstruction of clients' life narratives so that they become more complex, polyphonic, emotionally salient, inclusive and flexible [59]. The therapist's receptive and relationally responsive attitude towards the clients' storytelling and expression of affect are considered crucial elements in this process [60]. There is ample evidence that different therapist actions associated with responsiveness play an important role for the process and outcome of psychotherapy [61], with the therapeutic alliance being a key relational aspect in this process.

The therapeutic alliance is a pan-theoretical concept that is associated with the collaborative aspects of the therapeutic relationship and has been extensively studied as an important process variable in psychotherapy. It is usually conceptualized as comprising three interlinked aspects: a strong emotional bond between clients and therapists, and agreement and collaboration on the goals and the tasks of therapy [62]. The quality of the therapeutic alliance has consistently been shown to be a predictor of outcome in individual psychotherapy across different modalities [63], as well as couple and family therapy (CFT) [64–69]

In conjoint treatments, such as couple therapy, the therapeutic alliance consists of a web of interlinked relationships between participants and the various subsystems thus formed [63,66]. Several factors—such as power dynamics, conflict, trust, loyalties and secrets in the couple or family—affect the formation of the alliance in CFT [69–71]. A strong overall alliance in couple therapy requires a balanced alliance between the therapist and each partner, as well as agreement in the couple on the problems, goals and values of therapy; as such, the therapist is encouraged to foster an alliance with each partner, avoiding 'split alliances', and to promote within-couple alliance [66].

The current study is a mixed-method, single-case study aiming to illustrate the potential of the PDC metric as a useful way of examining IPS in relation to the therapy process; it assumes a theoretically driven idiographic design and examines whether the therapeutic alliance maps onto IPS findings.

2. Materials and Methods

The research material in this study is drawn from one-couple therapy, conducted in an outpatient Family Therapy Department in Greece, in the context of a wider naturalistic, multisite research study [30,39,72]. The treatment in this service follows systemic principles and includes the use of reflective conversations with a co-therapist. In usual clinical practice, sessions are provided monthly; a second therapist watches the session between the primary therapist and the couple behind a one-way mirror and joins them for a reflective conversation towards the end of each session [73]. Participating couples were informed about the study by a graduate researcher at the end of their first session. Participation in the project was voluntary, and ethical approval was granted by the Family Therapy Department's Scientific Board. Both clients and therapists gave permission for the data to be used for research purposes.

2.1. The Case

This therapy consisted of 15 sessions spanning 14 months. The couple, Costas and Demetra, is a white heterosexual couple in their mid-thirties. Demetra is a law graduate with a successful professional career. Costas has no university education; he worked as a technician in the past and is currently unemployed. The couple had been in a long-term relationship of over 10 years when they came to therapy. They sought therapy because of increasing tension in their relationship following the birth of their baby 10 months earlier. Two experienced female clinical psychologists and systemic family therapists in their fifties participated in this therapy. The therapy centered on Demetra's distress in her role as a mother, the expression of anger and conflict between the spouses, and Costas' low self-esteem associated with periods of unemployment. At the end of treatment, the couple reported an improvement in their personal lives and their relationship.

2.2. Procedure

All sessions were video-recorded in split-screen mode with four web-cameras. In addition, in two sessions (sessions 3 and 14), physiological measures of the participants' heart rates were recorded for the duration of the session. Within 24 h of the measurement sessions, a graduate researcher conducted separate Stimulated Recall interviews [74,75] with each client and therapist, each lasting approximately 30 min.

2.3. Measures

2.3.1. Autonomic Nervous System Responses

The participants' autonomic nervous system (ANS) responses were recorded via Firstbeat Bodyguard (Firstbeat Technologies, Jyväskylä, Finland) [76] mobile heart rate (HR) monitors. Ag/AgCl electrodes, connected to the Firstbeat Bodyguard, were attached on two sites on the skin of the chest before the start of each measurement session and were removed the next day, with the guidance of a graduate researcher. HR was continuously recorded during this period.

2.3.2. Clinical Outcomes in Routine Evaluation–Outcome Measure (CORE-OM)

The outcome of therapy was examined using the CORE-OM, administered at the start and end of therapy. The CORE-OM is a widely used, 34-item self-report measure that examines psychological distress in four domains: wellbeing, problems, functioning and risk [77].

2.3.3. Session Rating Scale (SRS)

The SRS is a four-item, ultra-brief visual analogue instrument to assess the global strength of the alliance, designed to be used in routine outcome monitoring [78]. The four items measure the therapist–client emotional bond, agreement on goals, agreement on tasks and overall rating of the alliance. It is scored by summing the marks measured to the nearest centimeter on each of the four lines. Based on a total possible score of 40, any score lower than 36 overall, or 9 on any scale, could be a source of concern.

2.3.4. System for Observing Family Therapy Alliances (SOFTA-o)

The SOFTA-o is an observer-based measure developed to study the therapeutic alliance in couple and family therapy [79]. It examines the contribution of each participant to the alliance by coding specific behaviors in four dimensions: Emotional connection, Engagement in the therapeutic process, Safety within the therapeutic system and Shared sense of purpose. The first three dimensions concern the therapist(s)–clients relationship, whereas the fourth concerns the couple sub-system. Following the coding of specific items, global ratings are provided for each dimension on a 7-point ordinal scale, ranging from -3 (extremely problematic) to +3 (extremely strong), with 0 denoting an unremarkable or neutral alliance. These dimensions are conceptually interdependent and moderately correlated and can be combined in a composite score [66].

2.4. Data Analysis

Interpersonal Physiological Synchrony

Data from the ANS were analyzed using Firstbeat PRO Wellness Analysis Software® version 1.4.1. This software uses neural network modeling to calculate Heart Rate Variability (HRV) indices second-by-second. This is achieved using a short-time Fourier Transform method (STFT) combining data from HR- and HRV-derived variables that describe respiration rate and oxygen consumption (VO2). In addition, the absolute stress vector (ASV) is calculated from the HR, high-frequency power (HFP), low-frequency power (LFP) and HRV-derived respiratory variables, as an index of the activity of the sympathetic nervous system. The ASV grounds on detecting sympathetic reactivity that exceeds the momentary metabolic requirements of the ANS. Hence, the ASV is high when the heart rate is elevated, HRV is low and respiration rate is low relative to HR and HRV [80]. The ASV is calculated at a 1 Hz rate.

2.5. Partial Directed Coherence within Sessions

Within-session, directed, interpersonal physiological synchrony based on ASV was estimated using Partial Directed Coherence (PDC) [50]. PDC analysis transforms the ASV time series into the frequency domain and provides time-lagged associations between two participants' multivariate signals, assessing their statistical independence or predictability [50]. Specifically, grounded on instantaneous Granger causality, it implies that, knowing the previous states of the first signal (the leading signal), one may achieve a better prediction of the second signal (the pacing signal), than just knowing the previous states of the second signal. Hence, it describes the direction of information flow between isolated pairs of time series, in a frequency-domain representation of the notion of Granger causality. This approach has recently been proposed as a method of choice for estimating IPS in psychotherapy by Kleinbub [54], due to its ability to establish direction, and thus causality, in interactions. Due to the time-varying conditional variance of HRV signals [81], PDC as a frequency-domain method for identifying causal interactions between the signals was preferred over the classical Granger causality, which estimates interactions in the time domain. In addition, PDC has previously been used to successfully estimate the frequency-domain causality in cardiovascular time series with Instantaneous Interactions [82].

The second-by-second ASV data of the measurement sessions were imported into Matlab (MathWorks Inc., Natick, MA, USA) as time series. The ASV time series were segregated into time-windows of 50 s, and the PDC for each window was estimated independently for each pair of participants in each session via an in-house script based on the work of Baccalá and Sameshima [50]. The length of the time-window was empirically determined on the basis of a series of tests comparing the number of significant PDC time-windows within independent sets of surrogate data generated via Matlab, aiming to achieve the best possible balance between the resolution of the analysis (i.e., smallest time-window) and the absence of false-positive significant PDC time-windows. Hence, for each 50 s time-window of the session, we retrieved two PDC values for each pair of participants (one for each direction, i.e., one in which participant 1 leads and participant 2 paces, and one in which participant 2 leads and participant 1 paces). Additionally, a statistical test based on Monte Carlo iterations of the corresponding data was performed for each pair, in order to identify time-windows with a significant PDC. The threshold of significance was defined as $p = 0.05/3$, accounting for the total number of comparisons in which the same set of data participated, thereby effectively controlling for multiple comparisons. Only significant PDC values were taken into account.

2.6. *Partial Directed Coherence between Sessions*

The number of significant PDC time-windows for each pair of participants was compared between sessions 3 and 14 as an index of the overall effect of therapy on interpersonal physiological synchronization. The aim was to identify differences in the global characteristics of IPS between sessions at the start and end of therapy.

2.7. *Qualitative Analysis of the Therapy Process*

2.7.1. Topical Episodes

The measurement sessions were segmented into topical episodes, i.e., periods of time during which a specific topic was discussed [83]. This coding was initially carried out by two graduate researchers and was checked by third researcher, and any discrepancies were resolved through discussion. This initial thematic coding provides a description of the main themes discussed in a session. Session 3 was segmented into 14 topical episodes, ranging from 2 to 15 min' duration, and session 14 was segmented into 12 topical episodes, ranging from approximately 1 to 9 min' duration.

2.7.2. Therapeutic Alliance

Two graduate psychologists, trained in using the SOFTA-o, coded each session. The raters coded the sessions independently and then discussed any discrepancies until consensus was reached. Next, in order to gain a more fine-grained coding of the development of the alliance through the session, the strength of the alliance was coded for each topical episode.

3. Findings and Discussion

With regards to the outcome of therapy, the clients' CORE-OM scores decreased significantly over the course of therapy, suggesting a clinically significant reduction in psychological distress (Table 1). At the onset of therapy, both partners reported a medium level of distress, and, importantly, Costas scored on items concerning the risk of harming himself. At the end of therapy, Demetra's CORE-OM score decreased to the cut-off point for clinical distress (<10), and Costas' showed clinically significant change (>5 clinical score points) [77]. In terms of the therapeutic alliance, Costas' scores indicated a positive alliance in session 3, which further increased in the penultimate session, whereas Demetra's scores indicated a problematic alliance in session 3, which improved in the penultimate session.

Table 1. Clients' CORE-OM and SRS scores.

	CORE-OM		**CORE-OM RISK**		**SRS**	
	Session 1	**Session 15**	**Session 1**	**Session 15**	**Session 3**	**Session 14**
Demetra	12	10	0	1.6	5.6	8.0
Costas	19	11	5	0	8.9	9.8

Next, we present the key quantitative findings regarding interpersonal physiological synchrony (IPS) within and across the two measurement sessions. Then, the potential of PDC analysis as a useful way of examining the process of therapy is explored through a mixed-method analysis of session 3.

The physiological activity of the couple, as reflected in their ASV, in the two sessions is presented in Figure 1. In both sessions, Demetra's autonomic arousal decreased as the session progressed, whereas Costas' remained relatively constant through. It is worth noting that Demetra's mean ASV score in the penultimate session was significantly higher than in the third session, and her arousal shows higher variance. The clinical relevance of this observation would require further investigation and lies beyond the scope of this study.

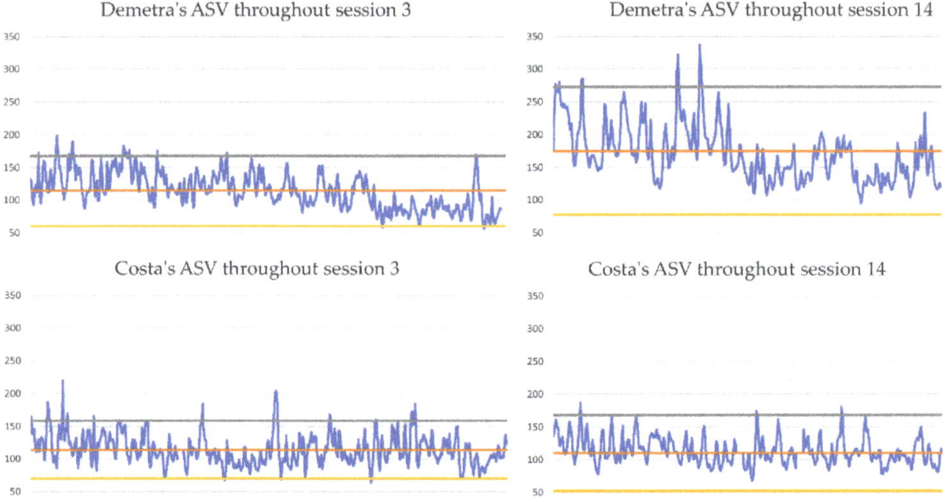

Figure 1. Demetra and Costa's ASV values in the sessions, plotted against the mean ASV value for the session (red line), 2 Standard Deviations below (yellow line) and 2 SD above (grey line) the mean.

3.1. Interpersonal Physiological Synchrony

3.1.1. IPS in Session 3

The PDC analysis identified 29 time-windows in which the participants' ASV were synchronized in session 3, out of a total of 93 time-windows (Table 2 and Figure 2). This corresponds to at least two participants' physiological arousal being synchronized in 31,2% of the total session time. More specifically, Demetra's ASV values led Costa's ASV in one time-window, and the therapist's ASV in four. In contrast, Costa's ASV led Demetra's ASV in eight time-windows, and the therapist's ASV in nine. Lastly, the therapist's ASV led Demetra's ASV in six time-windows, and Costas' in eight. Overall, in session 3, Costas' autonomic arousal was found to lead IPS to a greater degree than Demetra's; moreover, the therapist had a leading role in several parts of the session, while Demetra primarily had a pacing role.

Table 2. Number of time-windows showing significant PDC synchronization between clients and therapist in session 3.

		Leading Role		
		Demetra	**Costas**	**Therapist**
Pacing role	Demetra		8	6
	Costas	1		8
	Therapist	4	9	

Note: Number of time-windows in session = 93. Time-windows in which at least two participants show significant PDC = 29.

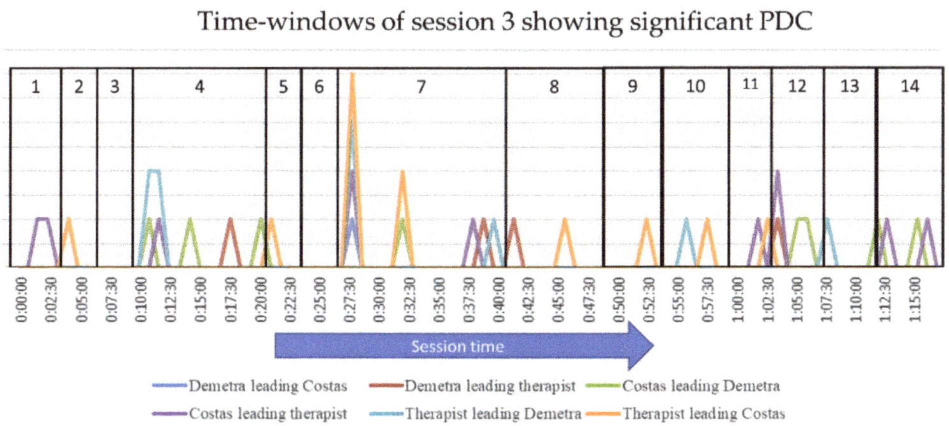

Time-windows of session 3 showing significant PDC

Session time

Demetra leading Costas — Demetra leading therapist — Costas leading Demetra
Costas leading therapist — Therapist leading Demetra — Therapist leading Costas

Figure 2. Time-windows showing significant PDC in session 3. The height of each line indicates the number of pairs that show significant PDC within each time-window. The vertical lines denote the boundaries of the Topical Episodes.

In addition, in session 3, several time-windows showed increased IPS; we use this term to describe time-windows in which more than one of the six possible directed synchronizations were observed. We consider these time-windows as particularly significant. Specifically, four time-windows showed increased physiological synchrony. Notably, in one time-window, all three participants were physiologically synchronized, with a mutual IPS between the clients and both clients' arousal also leading the therapist's ASV. Moreover, as can be seen in Figure 2, the time-windows with IPS tended to cluster around specific points in the session. We consider this clustering of IPS as reflecting time periods in the session that are significant for the process of therapy.

3.1.2. IPS in Session 14

The PDC analysis of the penultimate session identified 10, out of a total of 58, time-windows in which the participants' ANS arousal was synchronized (Table 3 and Figure 3). This corresponds to at least two participants' physiological arousal being synchronized in 17.2% of the total session time. More specifically, Demetra's arousal led Costas' ASV in four time-windows, and the therapist's ASV in one. Costas' arousal led Demetra's ASV in one time-window, and the therapist's in two. Lastly, the therapist's ASV led Demetra's ASV in four time-windows, and Costas' in one. Overall, IPS in the penultimate session was equally led by Demetra and the therapist, and both clients had similar pacing roles, with Demetra pacing the therapist's ASV and Costas pacing Demetra's. Three time-windows showed increased physiological synchrony in this session, and again, time-windows with PDC tended to cluster together.

Table 3. Number of time-windows showing significant PDC synchronization between clients and therapist in the penultimate session.

		Leading Role		
		Demetra	**Costas**	**Therapist**
Pacing role	Demetra		1	4
	Costas	4		1
	Therapist	1	2	

Note: Number of time-windows in session = 58. Time-windows in which at least two participants show significant PDC = 10.

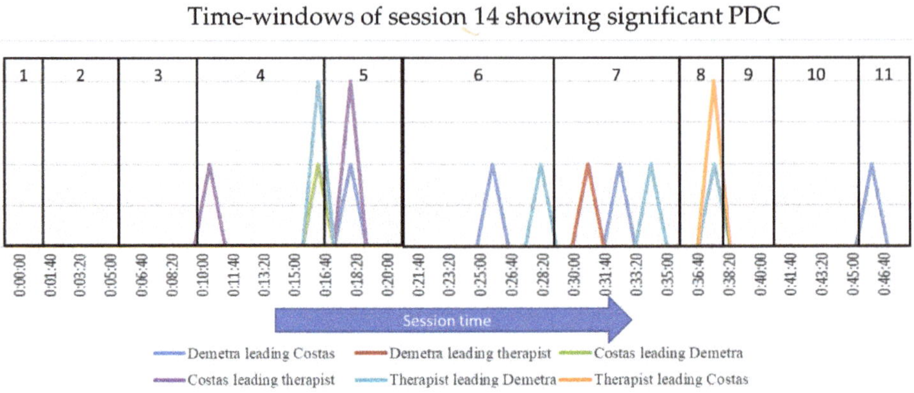

Time-windows of session 14 showing significant PDC

Demetra leading Costas — Demetra leading therapist — Costas leading Demetra — Costas leading therapist — Therapist leading Demetra — Therapist leading Costas

Figure 3. Time-windows showing significant PDC in the penultimate session (session 14). The height of each line indicates the number of pairs with significant PDC within each time-window. The vertical lines denote the boundaries of the Topical Episode.

3.1.3. IPS between Sessions

A comparison of the PDC values between the two measurement sessions shows that (i) the total time spent in interpersonal physiological synchrony was significantly lower in session 14 as compared to session 3, and (ii) the global architecture of the interpersonal physiological synchrony network was reorganized to become more balanced as therapy progressed (Figure 4). As mentioned above, the percentage of the total session time with IPS decreased from 31.2% in session 3 to 17.2% in session 14. The IPS between the therapist and each of the clients showed the most marked decrease, from twenty-seven time-windows (28.1% of the session time) in session 3 to eight time-windows (13.7% of total session time) in session 14. This reduction in IPS over the course of therapy can be seen to reflect the clients' reduced affective arousal and their gradual disengagement from the process. As therapy progressed, the clients' difficult feelings and conflicts were expressed, elaborated upon and gradually reconstructed, and the physiological synchrony between the clients and the therapist decreased. This is in line with the characteristics of the closing stages of therapy, which entail less affectively charged and more reflective conversations, as well as a process of gradual disengagement from the therapeutic relationship and the work of therapy.

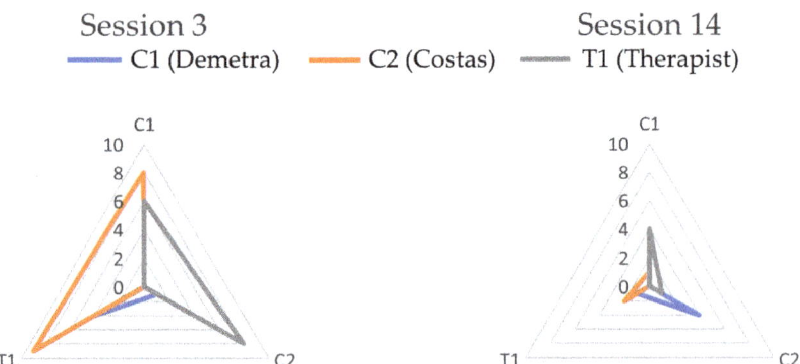

Figure 4. Time-windows showing significant PDC per leading participant.

Furthermore, IPS in a multi-actor setting such as couple therapy is more complex, as there are six possible pairs of participants. A shift was observed in how the IPS was distributed between participants; in session 3, the IPS was mainly driven by Costas, who led Demetra's and the therapist's ASV in seventeen time-windows and paced the therapist in eight. In contrast, in session 14, the overall synchrony was more equally driven by all participants, producing a more balanced or 'democratic' network structure (Figure 4). This finding points to the co-creation of a more equally distributed and balanced embodied relatedness between participants as therapy was reaching termination.

3.2. Physiological Synchrony and Clinical Process

In order to deepen our understanding of the relational meaning of IPS as it fluctuated through a session, the clinical process in session 3 was qualitatively analyzed drawing upon narrative principles, and the findings were subsequently examined in relation to the PDC analysis. In brief, the session was segmented into topical episodes [83]; this thematic coding allows researchers to identify key themes in a session and track the process of meaning co-construction, thus obtaining a relatively fine-grained description of meaning making through the session. Next, a qualitative analysis was performed to identify the significant moments in the session, which were defined as those topical episodes where: (a) important issues in the couple's life were introduced and narratively elaborated; (b) associated emotions were recognized, explored and expressed; and (c) the meaning of these key issues began to be reconstructed. The central theme in this session concerned Demetra's low mood and her strong ambivalent feelings regarding her role as a mother. Two topical episodes were identified through the qualitative analysis as entailing the elaboration of this theme, accompanied by intense emotional expression; these are briefly described below.

The theme of Demetra's conflicts in her role as a mother was first introduced in TE4, approximately ten min into the session (duration 10′40″). This episode started with the therapist asking how the couple would choose to spend their time together if they had the finances and caretaking support. Costas made several suggestions that Demetra firmly rejected as she felt 'bored' with everything. Through the therapist's gentle curiosity and empathic questioning, Demetra's boredom was gradually reconstructed as entailing intense sadness; Demetra cried as she described her low mood, exhaustion, and sense of suffocation in her role as a mother. Towards the end of the episode, Costas gently talked about Demetra's lack of interest in sex, and this led to the expression of more sadness by Demetra. This episode contained the elaboration of the key theme of the session along with nonverbal displays of affect, as well as several markers of a moderately strong therapeutic alliance; this was particularly evident in the relationship between Demetra and the therapist, as well as within the couple (Table 4). The PDC analysis for this episode shows a cluster of five time-windows with IPS, accounting for 39% of the episode time; two of these time-windows show increased IPS, whereby more than two participants are in-sync (Figure 3). In other words, the findings from the PDC analysis concur with those of the narrative analysis and with the coding of the therapeutic alliance, suggesting that this topical episode was important for the therapy process on both semantic and embodied levels.

The same theme was further elaborated with increased emotional expression in TE7. This long episode (duration 14′40″) took place in the middle of the session. It started with Demetra crying as she described feeling trapped and suffocating in her relationship with their baby; she vividly described her frustration and rage towards their baby, the wish to hurt him that she sometimes experienced, her angry outbursts towards him, and the intense guilt that she felt after such outbursts. As she listened to Demetra's emotional narrative, the therapist displayed many nonverbal signs of affiliation and empathy. She also introduced the hypothesis that Demetra's difficulties and sense of failure result from comparing herself to an idealized version of motherhood; this was followed by a productive conversation that challenged Demetra's idealization of her own mother, as well as the dominant discourse of ideal motherhood [60]. In terms of the alliance, this episode contained markers of a moderately strong alliance between Demetra and the therapist, whereas Costas displayed some markers of difficulty in the alliance. Based on the PDC analysis, there were five time-windows with IPS in this episode (including two windows with increased IPS), and these account for 28% of the episode time. Notably, this topical episode contained one time-window during which all three participants were physiologically synchronized. This corresponds to the point in the session where Demetra cried as she talked about the rage and guilt she felt towards their baby. A cluster of time-windows with IPS can be seen at the end of the episode, as Demetra's sadness and sense of suffocation in her maternal role were expressed. Once again, in this topical episode, the PDC analysis identified a period in the session that entailed intense affective expression by Demetra, empathic responsiveness by the therapist and a strong therapeutic alliance.

Table 4. Comparison of SOFTA scores and number of PDC identified significant time-windows for each Topical Episode.

TE	SOFTA Scores					Time-Windows with PDC	% Episode Time in PDC
	Costas	Demetra	Therapist	SSP	Composite Score		
1	2	1	0	0	3	2	47.6
2	1	1	2	0	4	1	26.3
3	−1	0	1	2	3-1	0	0
4	1	2	1	2	6	5	39
5	0	0	0	1	1	1	15.4
6	1	0	0	0	1	0	0
7	−1	2	2	0	4-1	5	28
8	0	1	1	2	4	2	19.2
9	1	0	1	2	4	1	17.9
10	0	1	1	1	3	2	28.6
11	0	0	0	1	1	1	31.2
12	0	1	1	0	2	5	66
13	1	2	1	2	6	1	34.5
14	1	0	1	1	3	0	0

Next, we examined a topical episode identified as significant through the PDC, but not through the qualitative analysis: TE12 contained a cluster of time-windows with IPS that account for 66% of the episode time. The episode took place towards the end of the session (duration 5′20″) and focused on Demetra's lack of sexual desire, a delicate and affectively charged issue in the couple's relationship. In this instance, the PDC analysis identified a part of the session that was not identified as important from a qualitative perspective, but which proved to be significant later in the treatment.

In sum, the two topical episodes that were identified as important for the process of therapy through the qualitative analysis also entailed increased IPS and increased ratings of the therapeutic alliance, as compared to the rest of the session. These findings are in line with the literature that points to the role of IPS in empathy, affiliation, rapport and the therapeutic alliance [7,11,43]. At the same time, PDC analysis proved useful in identifying significant moments in the session.

In order to explore the relational significance of IPS, we examine the findings from the PDC analysis in relation to the interactions between participants in session 3. With regards to the therapist–client(s) interaction, periods of physiological synchrony between the therapist and Demetra account for 10.1% of the session time (corresponding to 27.8% of the total time with IPS); IPS between the therapist and Costas is significantly higher, accounting for 18.3% of the total session time (corresponding to 47% of the total time with IPS). This points to the presence of a more intense affective connection between Costas and the therapist, and this is in line with the clients' respective SRS scores (Table 1). This finding reflects the complex interplay between explicit and implicit aspects of interaction in psychotherapy. More specifically, the therapist was very responsive to Demetra's distress on an explicit level, as she openly expressed empathy and affiliation with Demetra's painful conflicts regarding motherhood. At the same time, she was significantly more in-sync with Costas on an embodied level. This is in line with findings that behavioral and physiological synchrony seem to be independent processes that do not always co-occur [84]. The therapist was able to maintain a balanced therapeutic alliance with both members of the couple, and this was achieved through different modalities. In other words, implicit and explicit modalities of interaction were used to manage different interactional goals [30]. This is important, as split alliances, i.e., situations where the therapist takes sides by colluding with one partner, have a negative impact on the outcome of therapy [79]. In addition, there is some evidence that the therapeutic alliance with the male partner is critical for therapy outcome in heterosexual couples [70].

With regards to the couple's relationship, Costas' physiological arousal led Demetra's ASV significantly more than the opposite. Thus, although Demetra's difficulties were central on the level of talk, on an implicit embodied level, Costas had a more powerful influence on the interaction. Again, this is an observation that illustrates the complex interplay between the verbal and embodied aspects of psychotherapeutic interaction. A possible interpretation of this observation is that of affect co-regulation, where Costas' presence could be seen to regulate Demetra's affective arousal, as often happens in complementary, i.e., homeostatic, couple relationships [36]. A closer examination of the level of both partners' arousal, the valence of their affective displays, and an examination of the talk in these episodes would help contextualize and interpret these observations more fully.

4. Conclusions

The findings of this study point to the complex interplay between explicit and implicit levels of interaction and the potential added value of including physiological synchrony in the study of interactional processes in couple therapy [36,55,72,85,86]. In line with contemporary theories of therapeutic change, a key assumption of this work is that psychotherapy entails processes of intersubjective meaning making that take place through different modalities and, presumably, with different degrees of conscious awareness [23]. From this perspective, including measures of physiological activation in the study of psychotherapy sessions can help examine implicit, embodied interactional processes that contribute significantly to the formation of the therapeutic alliance, the co-creation of new meanings and, ultimately, therapeutic change. Although several research methods have been developed to study the talk in interaction [48,58], these methods generally fail to grasp the implicit, procedural level of interaction. Our attempt to include measures of autonomic arousal in studies of the therapy process and to operationalize implicit interactional processes of embodied responsiveness are in the spirit of exploring ways to include the implicit realm when studying the psychotherapy process [23].

Research to date suggests that IPS reflects different interactional processes, and these need to be disentangled for the field to progress [11,34,54]. In our study, we propose the use of a windowed Partial Directed Coherence-based approach as a metric to calculate physiological synchrony, as this allows a more nuanced examination of the dynamic nature of IPS in psychotherapy sessions. PDC analysis allows us to examine the therapy process in specific interactional events in the session, and this *micro-focus* provides a more fine-grained description of interactional dynamics as they develop, thus allowing a more nuanced interpretation of the role of IPS in the therapy process. Importantly, PDC analysis allows us to examine the *directionality* of synchronous interactions, which again adds another layer of complexity to our understanding of the role of physiological synchrony in the therapy process. Therefore, the proposed approach models the couple's interactions within the setting of a therapy as a self-organizing system, a system that is both open and complex, exchanging energy and information between its component parts and with its surroundings [87]. This exchange may be synchronic and diachronic, in spatial distribution and time transitions, therefore demanding multidimensional theoretical models to represent its hybrid nature [88].

A key aim of this study was to explore the links between a quantitative approach to the study of IPS and the characteristics of interactional dynamics and the clinical process, and this mixed-method analysis produced promising initial findings. More specifically, it examined shifts in IPS between the start and end of therapy in a successful couple therapy and identified a reduction in IPS as therapy progressed. This decrease primarily concerned the therapist–client(s) interaction and was interpreted as a reflection of progress, in the sense of a decrease in the intensity of negative affects expressed by the clients and the need for therapist empathy, as well as the couple's gradual disengagement from the process of therapy in line with the termination phase. In addition, the network of IPS between the three participants became more balanced. Both these findings are in line with a good therapy outcome, and as such, they provide support for the clinical validity of PDC analysis.

The main limitation, inherent in this approach, is that only one couple is included; hence, the descriptive outcome of the study cannot be generalized. Nonetheless, we propose that this detailed analysis provides a necessary step for evaluating the usefulness of employing PDC analysis to examine IPS in therapy sessions, which can now be further elaborated. Studying IPS via a windowed PDC approach may lead to an even more detailed identification of the underlying processes if the characteristics of the ANS signals during significant time-windows are further investigated. In addition, calculating positive vs. negative correlations of ANS activity or specific patterns of ANS reactivity within the significant time-windows may be used in future studies to examine their associations with different intersubjective processes, such as empathy, alliance or affect contagion. We hope that future work in this field will exploit the strengths of the PDC analysis and further our understanding of the embodied, relational aspects of the therapy process.

Author Contributions: Conceptualization, E.A. and E.P.; literature review: E.A., E.P., C.L. and F.P.; methodology, E.A., E.P. and P.K.; formal analysis, E.A., E.P., P.K. and C.L.; original draft preparation, E.A.; writing—review and editing, E.A., E.P., C.L. and F.P. All authors have read and agreed to the published version of the manuscript.

Funding: This research received no external funding.

Institutional Review Board Statement: The study was conducted in accordance with the Declaration of Helsinki, and approved by the Ethics Committee of the Scientific Board of the Psychiatric Hospital of Thessaloniki in June 2013.

Informed Consent Statement: Informed consent was obtained from all subjects involved in the study.

Data Availability Statement: The ANS data presented in this study are available on request from the corresponding author, and the data from the video material are not publicly available due to ethical restrictions.

Acknowledgments: We would like to thank Vasileia Lerou, Katerina Lazaridou and Kalliopi Papadopoulou for participating in the coding of the sessions with the SOFTA and the segmentation into Topical Episodes. We would also like to thank V. Escudero for useful comments on the use of the SOFTA analysis of the topical episodes.

Conflicts of Interest: The authors declare no conflict of interest.

References

1. Stern, D.N.; Sander, L.W.; Nahum, J.P.; Harrison, A.M.; Lyons-Ruth, K.; Morgan, A.C.; Bruschweiler-Stern, N.; Tronick, E.Z. Non-interpretive mechanisms in psychoanalytic therapy. The 'something more' than interpretation. The Process of Change Study Group. *Int. J. Psychoanal.* **1998**, *79*, 903–921. [PubMed]
2. Boston Change Process Study Group. Explicating the implicit: The local level and the microprocesses of change in the analytic situation. *Int. J. Psychoanal.* **2002**, *83*, 1051–1062. [CrossRef]
3. Tschacher, W.; Meier, D. Physiological synchrony in psychotherapy sessions. *Psychother. Res.* **2020**, *30*, 558–573. [CrossRef] [PubMed]
4. Bernieri, F.J.; Rosenthal, R. Interpersonal coordination: Behavior matching and interactional synchrony. In *Fundamentals of Nonverbal Behavior*; Rime, B., Feldman, R.S., Eds.; Cambridge University Press: Cambridge, UK, 1991; pp. 401–432.
5. Chartrand, T.L.; van Baaren, R. Human mimicry. *Adv. Exp. Soc. Psychol.* **2009**, *41*, 219–274.

6. Lakin, J.L.; Chartrand, T.L. Using nonconscious behavioral mimicry to create affiliation and rapport. *Psychol. Sci.* **2003**, *14*, 334–339. [CrossRef]
7. Koole, S.L.; Tschacher, W. Synchrony in psychotherapy: A review and an integrative framework for the therapeutic alliance. *Front. Psychol.* **2016**, *7*, 862. [CrossRef]
8. Raingruber, B.J. Three perspectives regarding what works and does not work in therapy: A comparison of judgments of clients, nurse-therapists, and uninvolved evaluators. *J. Am. Psychiatr. Nurses Assoc.* **2001**, *7*, 13–21. [CrossRef]
9. Ramseyer, F.; Tschacher, W. Synchrony: A core concept for a constructivist approach to psychotherapy. *Constr. Hum. Sci.* **2006**, *11*, 150–171.
10. Ramseyer, F.; Tschacher, W. Nonverbal synchrony in psychotherapy: Coordinated body movement reflects relationship quality and outcome. *J. Consult. Clin. Psychol.* **2011**, *79*, 284–295. [CrossRef]
11. Wiltshire, T.J.; Philipsen, J.S.; Trasmundi, S.B.; Jensen, T.W.; Steffensen, S.V. Interpersonal coordination dynamics in psychotherapy: A systematic review. *Cogn. Ther. Res.* **2020**, *44*, 752–773. [CrossRef]
12. Jaffe, J.; Beebe, B.; Feldstein, S.; Crown, C.L.; Jasnow, M.D.; Rochat, P.; Stern, D. Rhythms of dialogue in infancy: Coordinated timing in development. *Monogr. Soc. Res. Child Dev.* **2001**, *66*, i–viii, 1–132. [PubMed]
13. Palmieri, A.; Kleinbub, J.R.; Calvo, V.; Benelli, E.; Messina, I.; Sambin, M.; Voci, A. Attachment-security prime effect on skin-conductance synchronization in psychotherapists: An empirical study. *J. Couns. Psychol.* **2018**, *65*, 490–499. [CrossRef] [PubMed]
14. Kleinbub, J.R.; Talia, A.; Palmieri, A. Physiological synchronization in the clinical process: A research primer. *J. Couns. Psychol.* **2020**, *67*, 420–437. [CrossRef] [PubMed]
15. Feldman, R. Child development parent—Infant synchrony: A biobehavioral model of mutual influences in the formation of affiliative bonds. *Monogr. Soc. Res. Child Dev.* **2005**, *77*, 42–51. [CrossRef]
16. Ham, J.; Tronick, E. Relational psychophysiology: Lessons from mother-infant physiology research on dyadically expanded states of consciousness. *Psychother. Res.* **2009**, *19*, 619–632. [CrossRef]
17. Beebe, B.; Lachmann, F.M. *Infant Research and Adult Treatment*; The Analytic Press: London, UK, 2002.
18. Beebe, B.; Lachmann, F.M. *The Origins of Attachment: Infant Research and Adult Treatment*; Routledge: New York, NY, USA, 2014.
19. Butler, E.A.; Randall, A.K. Emotional coregulation in close relationships. *Emot. Rev.* **2013**, *5*, 202–210. [CrossRef]
20. Timmons, A.C.; Margolin, G.; Saxbe, D.E. Physiological linkage in couples and its implications for individual and interpersonal functioning: A literature review. *J. Fam. Psychol.* **2015**, *29*, 720–731. [CrossRef]
21. Cromby, J. Feeling the way: Qualitative clinical research and the affective turn. *Qual. Res. Psychol. Spec. Issue Qual. Clin. Res.* **2012**, *9*, 88–98. [CrossRef]
22. Fuchs, T.; Koch, S.C. Embodied affectivity: On moving and being moved. *Front. Psychol.* **2014**, *5*, 508. [CrossRef]
23. Avdi, E.; Evans, C. Exploring conversational and physiological aspects of psychotherapy talk. *Front. Psychol.* **2020**, *11*, 3001. [CrossRef]
24. Cacioppo, J.T.; Berntson, G.G.; Andersen, B.L. Psychophysiological approaches to the evaluation of psychotherapeutic process and outcome, 1991: Contributions from social psychophysiology. *Psychol. Assess.* **1991**, *3*, 321–336. [CrossRef]
25. Kreibig, S.D. Autonomic nervous system activity in emotion: A review. *Biol. Psychol.* **2010**, *84*, 394–421. [CrossRef] [PubMed]
26. Soma, C.S.; Baucom, B.R.W.; Xiao, B.; Butner, J.E.; Hilpert, P.; Narayanan, S.; Atkins, D.C.; Imel, Z.E. Coregulation of therapist and client emotion during psychotherapy. *Psychother. Res.* **2020**, *30*, 591–603. [CrossRef] [PubMed]
27. Russell, J.A. A circumplex model of affect. *J. Personal. Soc. Psychol.* **1980**, *39*, 1161–1178. [CrossRef]
28. Gennaro, A.; Carola, V.; Ottaviani, C.; Pesca, C.; Palmieri, A.; Salvatore, S. Affective saturation index: A lexical measure of affect. *Entropy* **2021**, *23*, 1421. [CrossRef]
29. Deits-Lebehn, C.; Baucom, K.J.W.; Crenshaw, A.O.; Smith, T.W.; Baucom, B.R.W. Incorporating physiology into the study of psychotherapy process. *J. Couns. Psychol.* **2020**, *67*, 488–499. [CrossRef]
30. Avdi, E.; Seikkula, J. Studying the process of psychoanalytic psychotherapy: Discursive and embodied aspects. *Br. J. Psychother.* **2019**, *35*, 217–232. [CrossRef]
31. Shaffer, F.; Ginsberg, J.P. An overview of heart rate variability metrics and norms. *Front. Public Health* **2017**, *5*, 258. [CrossRef]
32. Friston, K.J. Functional and effective connectivity: A review. *Brain Connect.* **2011**, *1*, 13–36. [CrossRef]
33. Mc Craty, R. New frontiers in heart rate variability and social coherence research: Techniques, technologies, and implications for improving group dynamics and outcomes. *Front. Public Health* **2017**, *5*, 267. [CrossRef]
34. Palumbo, R.V.; Marraccini, M.E.; Weyandt, L.L.; Wilder-Smith, O.; McGee, H.A.; Liu, S.; Goodwin, M.S. Interpersonal autonomic physiology: A systematic review of the literature. *Personal. Soc. Psychol. Rev.* **2017**, *21*, 99–141. [CrossRef] [PubMed]
35. Dimascio, A.; Boyd, R.W.; Greenblatt, M. Physiological correlates of tension and antagonism during psychotherapy; a study of interpersonal physiology. *Psychosom. Med.* **1957**, *19*, 99–104. [CrossRef] [PubMed]
36. Kykyri, V.-L.; Karvonen, A.; Wahlström, J.; Kaartinen, J.; Penttonen, M.; Seikkula, J. Soft prosody and embodied attunement in therapeutic interaction: A multimethod case study of a moment of change. *J. Constr. Psychol.* **2017**, *30*, 211–234. [CrossRef]
37. Kodama, K.; Tanaka, S.; Shimizu, D.; Hori, K.; Matsui, H. Heart Rate Synchrony in Psychological Counseling: A Case Study. *Psychology* **2018**, *9*, 1858–1874. [CrossRef]
38. Marci, C.D.; Orr, S.P. The effect of emotional distance on psychophysiologic concordance and perceived empathy between patient and interviewer. *Appl. Psychophysiol. Biofeedback* **2006**, *31*, 115–128. [CrossRef]
39. Seikkula, J.; Karvonen, A.; Kykyri, V.L.; Penttonen, M.; Nyman-Salonen, P. The relational mind in couple therapy: A bateson-inspired view of human life as an embodied stream. *Fam. Process* **2018**, *57*, 855–866. [CrossRef] [PubMed]

40. Villmann, T.; Liebers, C.; Bergmann, B.; Gumz, A.; Geyer, M. Investigation of psycho-physiological interactions between patient and therapist during a psychodynamic therapy and their relation to speech using in terms of entropy analysis using a neural network approach. *New Ideas Psychol.* **2008**, *26*, 309–325. [CrossRef]
41. Messina, I.; Palmieri, A.; Sambin, M.; Kleinbub, J.R.; Voci, A.; Calvo, V. Somatic underpinnings of perceived empathy: The importance of psychotherapy training. *Psychother. Res.* **2013**, *23*, 169–177. [CrossRef]
42. Gennaro, A.; Kleinbub, J.R.; Mannarini, S.; Salvatore, S.; Palmieri, A. Training in psychotherapy: A call for embodied and psychophysiological approaches. *Res. Psychother. Psychopathol. Process Outcome* **2019**, *22*, 395. [CrossRef]
43. Palmieri, A.; Pick, E.; Grossman-Giron, A.; Tzur Bitan, D. Oxytocin as the neurobiological basis of synchronization: A research proposal in psychotherapy settings. *Front. Psychol.* **2021**, *12*, 628011. [CrossRef]
44. Levenson, R.W.; Ruef, A.M. Empathy: A physiological substrate. *J. Personal. Soc. Psychol.* **1992**, *63*, 234–246. [CrossRef]
45. Creaven, A.M.; Skowron, E.A.; Hughes, B.M.; Howard, S.; Loken, E. Dyadic concordance in mother and preschooler resting cardiovascular function varies by risk status. *Dev. Psychobiol.* **2014**, *56*, 142–152. [CrossRef] [PubMed]
46. Dale, R.; Fusaroli, R.; Duran, N.D.; Richardson, D.C. The Self-Organization of human interaction. *Psychol. Learn. Motiv.* **2013**, *59*, 43–95. [CrossRef]
47. Reed, R.G.; Barnard, K.; Butler, E.A. Distinguishing emotional coregulation from codysregulation: An investigation of emotional dynamics and body weight in romantic couples. *Emotion* **2015**, *15*, 45–60. [CrossRef]
48. Peräkylä, A.; Henttonen, P.; Voutilainen, L.; Kahri, M.; Stevanovic, M.; Sams, M.; Ravaja, N. Sharing the emotional load: Recipient affiliation calms down the storyteller. *Soc. Psychol. Q.* **2015**, *78*, 301–323. [CrossRef]
49. Voutilainen, L.; Henttonen, P.; Kahri, M.; Ravaja, N.; Sams, M.; Peräkylä, A. Empathy, challenge, and psychophysiological activation in therapist–client interaction. *Front. Psychol.* **2018**, *9*, 530. [CrossRef]
50. Baccalá, L.A.; Sameshima, K. Partial directed coherence: A new concept in neural structure determination. *Biol. Cybern.* **2001**, *84*, 463–474. [CrossRef]
51. Safran, J.D.; Muran, J.C. *Negotiating the Therapeutic Alliance: A Relational Treatment Guide*; Guilford Press: New York, NY, USA, 2000.
52. Tronick, E.Z. Emotions and emotional communication in infants. *Am. Psychol.* **1989**, *44*, 112–119. [CrossRef]
53. Tronick, E.Z.; Bruschweiler-Stern, N.; Harrison, A.M.; Lyons-Ruth, K.; Morgan, A.C.; Nahum, J.P.; Sander, L.; Stern, D.N. Dyadically expanded states of consciousness and the process of therapeutic change. *Infant Ment. Health J.* **1998**, *19*, 290–299. [CrossRef]
54. Kleinbub, J.R. State of the art of interpersonal physiology in psychotherapy: A systematic review. *Front. Psychol.* **2017**, *8*, 2053. [CrossRef]
55. Laitila, A.; Vall, B.; Penttonen, M.; Karvonen, A.; Kykyri, V.L.; Tsatsishvili, V.; Kaartinen, J.; Seikkula, J. The added value of studying embodied responses in couple therapy research: A case study. *Fam. Process* **2019**, *58*, 685–697. [CrossRef] [PubMed]
56. Mylona, A.; Avdi, E. Alliance ruptures and embodied arousal in psychodynamic psychotherapy: An exploratory study. *Hell. J. Psychol.* **2021**, *18*, 226–248. [CrossRef]
57. Tschacher, W.; Haken, H. Causation and chance: Detection of deterministic and stochastic ingredients in psychotherapy processes. *Psychother. Res.* **2020**, *30*, 1075–1087. [CrossRef] [PubMed]
58. Smoliak, O.; Strong, T. *Therapy as Discourse: Practice and Research*; Springer International Publishing: London, UK, 2018; ISBN 978-3-319-93066-4.
59. Avdi, E.; Georgaca, E. Discourse analysis and psychotherapy: A critical review. *Eur. J. Psychother. Couns.* **2007**, *9*, 157–176. [CrossRef]
60. Avdi, E.; Georgaca, E. Researching the discursive construction of subjectivity. In *Therapy as Discourse: Practice and Research*; Smoliak, O., Strong, T., Eds.; Palgrave: London, UK, 2018; pp. 45–70.
61. Norcross, J.C.; Lambert, M.J. *Psychotherapy Relationships That Work: Evidence-Based Therapist Contributions*; Oxford University Press: Oxford, UK, 2019.
62. Bordin, E.S. The generalizability of the psychoanalytic concept of the working alliance. *Psychotherapy* **1979**, *16*, 252–260. [CrossRef]
63. Horvath, A.O.; Bedi, R.P. The alliance. In *Psychotherapy Relationships that Work: Therapist Contributions and Responsiveness to Patients*; Norcross, J.C., Ed.; Oxford University Press: Oxford, UK, 2002; pp. 37–69.
64. Larner, G. Towards a common ground in psychoanalysis and family therapy: On knowing not to know. *J. Fam. Ther.* **2000**, *22*, 61–82. [CrossRef]
65. Anderson, H. Postmodern collaborative and person-centred therapies: What would Carl Rogers say? *J. Fam. Ther.* **2001**, *23*, 339–360. [CrossRef]
66. Friedlander, M.L.; Escudero, V.; Heatherington, L.; Diamond, G.M. Alliance in couple and family therapy. *Psychotherapy* **2011**, *48*, 25–33. [CrossRef]
67. Kuhlman, I.; Tolvanen, A.; Seikkula, J. The therapeutic alliance in couple therapy for depression: Predicting therapy progress and outcome from assessments of the alliance by the patient, the spouse, and the therapists. *Contemp. Fam. Ther.* **2013**, *35*, 1–13. [CrossRef]
68. Speed, B. All aboard in the NHS: Collaborating with colleagues who use different approaches. *J. Fam. Ther.* **2004**, *26*, 260–279. [CrossRef]
69. Tilden, T.; Johnson, S.U.; Hoffart, A.; Zahl-Olsen, R.; Wampold, B.E.; Ulvenes, P.; Håland, Å.T. Alliance predicting progress in couple therapy. *Psychotherapy* **2021**, *58*, 391–400. [CrossRef] [PubMed]
70. Glebova, T.; Bartle-Haring, S.; Gangamma, R.; Knerr, M.; Delaney, R.O.; Meyer, K.; McDowell, T.; Adkins, K.; Grafsky, E. Therapeutic alliance and progress in couple therapy: Multiple perspectives. *J. Fam. Ther.* **2011**, *33*, 42–65. [CrossRef]
71. Halford, T.C.; Owen, J.; Duncan, B.L.; Anker, M.G.; Sparks, J.A. Pre-therapy relationship adjustment, gender and the alliance in couple therapy. *J. Fam. Ther.* **2016**, *38*, 18–35. [CrossRef]

72. Seikkula, J.; Karvonen, A.; Kykyri, V.-L.; Kaartinen, J.; Penttonen, M. The embodied attunement of therapists and a couple within dialogical psychotherapy: An introduction to the relational mind research project. *Fam. Process* **2015**, *54*, 703–715. [CrossRef] [PubMed]
73. Andersen, T. *The Reflecting Team: Dialogues and Dialogues about the Dialogues*; Norton: New York, NY, USA, 1991.
74. Kagan, N.; Krathwohl, D.R.; Miller, R. Stimulated recall in therapy using video tape: A case study. *J. Couns. Psychol.* **1963**, *10*, 237–243. [CrossRef]
75. Larsen, D.; Flesaker, K.; Stege, R. Qualitative interviewing using interpersonal process recall: Investigating internal experiences during professional-client conversations. *Int. J. Qual. Methods* **2008**, *7*, 18–37. [CrossRef]
76. Firstbeat Technologies, Oy. Available online: https://www.firstbeat.com/en/ (accessed on 28 March 2022).
77. Evans, C.; Connell, J.; Barkham, M.; Margison, F.; McGrath, G.; Mellor-Clark, J.; Audin, K. Towards a standardised brief outcome measure: Psychometric properties and utility of the CORE-OM. *Br. J. Psychiatry* **2002**, *180*, 51–60. [CrossRef]
78. Miller, S.D.; Duncan, B.L.; Brown, J.; Sparks, J.; Claud, D. The outcome rating scale: A preliminary study of the reliability, validity, and feasibility of a brief visual analog measure. *J. Br. Ther.* **2003**, *2*, 91–100.
79. Friedlander, M.L.; Escudero, V.; Horvath, A.O.; Heatherington, L.; Cabero, A.; Martens, M.P. System for observing family therapy alliances: A tool for research and practice. *J. Couns. Psychol.* **2006**, *53*, 214–225. [CrossRef]
80. Kettunen, J.; Saalasti, S. Procedure for Detection of Stress by Segmentation and Analyzing A Heart Beat Signal. US Patent 7,330,752, 12 February 2008.
81. Leite, A.; Silva, M.E.; Rocha, A.P. Modeling volatility in heat rate variability. In Proceedings of the 2016 38th Annual International Conference of the IEEE Engineering in Medicine and Biology Society (EMBC), Orlando, FL, USA, 16–20 August 2016; pp. 3582–3585. [CrossRef]
82. Faes, L.; Nollo, G. Assessing frequency domain causality in cardiovascular time series with instantaneous interactions. *Methods Inf. Med.* **2010**, *49*, 453–457. [CrossRef]
83. Seikkula, J.; Laitila, A.; Rober, P. Making sense of multi—Actor dialogues in family therapy and network meetings. *J. Marital Fam. Ther.* **2012**, *38*, 667–687. [CrossRef] [PubMed]
84. Codrons, E.; Bernardi, N.F.; Vandoni, M.; Bernardi, L. Spontaneous group synchronization of movements and respiratory rhythms. *PLoS ONE* **2014**, *9*, e107538. [CrossRef] [PubMed]
85. Päivinen, H.; Holma, J.; Karvonen, A.; Kykyri, V.L.; Tsatsishvili, V.; Kaartinen, J.; Penttonen, M.; Seikkula, J. Affective arousal during blaming in couple therapy: Combining analyses of verbal discourse and physiological responses in two case studies. *Contemp. Fam. Ther.* **2016**, *38*, 373–384. [CrossRef]
86. Tourunen, A.; Kykyri, V.-L.; Seikkula, J.; Kaartinen, J.; Tolvanen, A.; Penttonen, M. Sympathetic nervous system synchrony: An exploratory study of its relationship with the therapeutic alliance and outcome in couple therapy. *Psychotherapy* **2020**, *57*, 160–173. [CrossRef]
87. Haken, H.; Portugali, J. Information and selforganization: A unifying approach and applications. *Entropy* **2016**, *18*, 197. [CrossRef]
88. Orsucci, F. Human synchronization maps—The hybrid consciousness of the embodied mind. *Entropy* **2021**, *23*, 1569. [CrossRef]

MDPI

Article

Recurrence-Based Synchronization Analysis of Weakly Coupled Bursting Neurons under External ELF Fields

Aissatou Mboussi Nkomidio [1,†], Eulalie Ketchamen Ngamga [2], Blaise Romeo Nana Nbendjo [1], Jürgen Kurths [2,3] and Norbert Marwan [2,*]

1 Laboratory of Modelling and Simulation in Engineering, Biomimetics and Prototypes, Faculty of Sciences, University of Yaoundé I, Yaoundé 812, Cameroon; nananbendjo@yahoo.com

2 Potsdam Institute for Climate Impact Research (PIK), Member of the Leibniz Association, Telegraphenberg, 14473 Potsdam, Germany; eulaliejoelle@yahoo.com (E.K.N.); juergen.kurths@pik-potsdam.de (J.K.)

3 Department of Physics, Humboldt University of Berlin, 12489 Berlin, Germany

* Correspondence: marwan@pik-potsdam.de

† This paper is dedicated to the memory of A. Mboussi Nkomidio (1980–2016) who passed away during the work at this paper.

Abstract: We investigate the response characteristics of a two-dimensional neuron model exposed to an externally applied extremely low frequency (ELF) sinusoidal electric field and the synchronization of neurons weakly coupled with gap junction. We find, by numerical simulations, that neurons can exhibit different spiking patterns, which are well observed in the structure of the recurrence plot (RP). We further study the synchronization between weakly coupled neurons in chaotic regimes under the influence of a weak ELF electric field. In general, detecting the phases of chaotic spiky signals is not easy by using standard methods. Recurrence analysis provides a reliable tool for defining phases even for noncoherent regimes or spiky signals. Recurrence-based synchronization analysis reveals that, even in the range of weak coupling, phase synchronization of the coupled neurons occurs and, by adding an ELF electric field, this synchronization increases depending on the amplitude of the externally applied ELF electric field. We further suggest a novel measure for RP-based phase synchronization analysis, which better takes into account the probabilities of recurrences.

Keywords: neuron; electric field; weak coupling; gap junction; synchronization; recurrence plot

Citation: Nkomidio, A.M.; Ngamga, E.K.; Nbendjo, B.R.N.; Kurths, J.; Marwan, N. Recurrence-Based Synchronization Analysis of Weakly Coupled Bursting Neurons under External ELF Fields. *Entropy* **2022**, *24*, 235. https://doi.org/10.3390/e24020235

Academic Editors: Franco Orsucci and Wolfgang Tschacher

Received: 30 December 2021
Accepted: 1 February 2022
Published: 3 February 2022

1. Introduction

Action potentials, or spikes, are responsible for the transmission of information through the nervous system [1]. A neuron can generate various temporal patterns of spike signals when it is driven by stimuli or noise from both internal or external environments. Therefore, analyzing spiking patterns of neurons under different stimulations plays an important role in the exploration of the encoding and decoding mechanism of information for neurons. External environmental stimuli in the brain can be of various origins, such as a wide utilization of power lines or electrical equipment. Electromagnetic exposure in the environment today is nearly one hundred times stronger than in previous centuries and many neuronal diseases are probably caused by electromagnetic exposure, as reported by Huang et al. [2]. Experiments with transcranial electrical stimulation have shown that electric field magnitudes in the cortex can be as high as 0.4 mV/mm for a 1 mA stimulation current. For typical electrode configurations used in clinical trials, maximal field intensities of up to 0.8 mV/mm were found when applying 2 mA. More extended areas can reach values of 0.28 mV/mm (95th percentile) under 2 mA stimulation [3–5].

An electromagnetic field can affect the neuron sensibility [6–9]. It also exhibits the excitability of many nerve cells, such as hippocampal cells, or cortical neurons [10,11]. Neurons exposed to an electromagnetic field can change the normal firing properties, which may lead to many neural diseases such as amyotrophic lateral sclerosis, senile dementia, Parkinson's disease, and Alzheimer's disease [7,12–14].

On the other hand, neurons are strongly coupled in the brain, and they need to synchronize information to encode and decode. Synchronization is a universal concept of nonlinear dynamics [15]. In the brain system, synchronization is a typical form of group motion rhythm, which means the neurons discharge at the same time or their discharge rhythms have at least some kind of relationship [16,17]. Neuronal synchrony activities can be found not only among coupled neuron groups in the same brain region but also among uncoupled neuron groups in the same brain region or among different cortical areas; moreover, synchronization can cross over two hemispheres of the brain [18]. Synchronization processes are crucially important for the neuronal system, and well-coordinated synchrony within and between neuronal populations appears to play an important role in neuronal signaling and information processing.

To study synchronization between neurons, different models of neuron dynamics have been developed, such as the Hodgkin–Huxley (HH) model and all the models derived from it. One of the derived models is the Morris–Lecar (ML) model [19,20]. It has the advantage of exhibiting class I and II neurons. Most studies on neuron synchronization use the Morris–Lecar model under an external electric field. For example, Kitajima and Kurths [21] investigated forced synchronization of electrically coupled class I and class II neurons under different coupling strengths. It was found that class II neurons have a wide parameter region of forced synchronization. However, in general, such studies did not consider the effect of small variations of the coupling strength between neurons.

The assumption of weak neuronal connection is based on the observation that the typical size of a postsynaptic potential is less than 1 mV, which is small in comparison to the mean potential necessary to discharge a cell or the average value of the action potential [22]. In a study of synaptic organization and dynamical properties of weakly connected neuronal oscillators, Hoppensteadt and Izhikevich [23] showed some phase synchronization between neurons in this range of coupling. Moreover, Izhikevich [24] studied the synchronization of elliptic bursters in a range of weak connectivity and found that such weakly connected bursters need few bursts to synchronize and synchronization is possible for bursters having quite different quantitative features. These phenomena were found in different neuron models, such as the FitzHugh–Rinzel, ML, and HH models.

The important question is if, even in the range of small coupling strength, a pair of neurons weakly coupled with gap junction are able to synchronize under the effect of an electric field (EF). Because electromagnetic stimulation can cause many disorders in the neural system, the theoretical investigation of the impact of an external EF on the synchronization of weakly coupled neurons is an important step to understand what happens in the brain during this exposure. Thus, in this work, we study the synchronization of a pair of ML neurons weakly connected with gap junction under an externally applied extremely low frequency (ELF) EF. Here, extremely low frequency means a frequency range between 0 and 10 Hz. Mammalian neurons show intrinsic resonance with frequency selectivity for inputs within the range from 4 to 10 Hz [25–30]. Gap junctions (channels that physically connect adjacent cells) provide an efficient and extremely fast way to propagate those signals between neurons [31,32]. In contrast, signal transmissions via chemical synapses have a significant delay (in the order of milliseconds) [33] and are not fast enough to respond to the EF. Therefore, we consider here coupling via gap junction, allowing direct response to the ELF EF. Using recurrence plot-based time series analysis, we investigate how the applied EF affects the condition of synchronization of the coupled neurons. This specific method has the advantage of being able to compare the phases of chaotic weakly coupled systems, even within noncoherent regimes or for spiky signals [34].

Recurrence plots (RPs) represent manifold recurrence features of a dynamical system in phase space [35] and are widely applied in the field of neuroscience [36–42]. For example, RPs can differentiate the stochastic and deterministic dynamics of irregularly firing cortical neurons [43] and show the average dynamics within a network of synchronized neurons [44] or spontaneous activity in neuronal in vitro cultures [45]. They are also powerful tools to study inter-relationships, coupling directions, phase synchronization, and generalized synchronization [34,46–48] and have been applied in different fields, such as chemistry, engineering, physiology, financial markets, and climatology [41,49–53]. Based on EEG measurements, the joint recurrence and the correlation of probability of recurrence were used to reconstruct brain networks [54,55]. A similar approach was used to study the synchronization between neurons based on the Hindmarsh–Rose model [39].

The correlation of probability of recurrence is a commonly used measure for recurrence-based phase synchronization analysis [34,50]. However, this measure is based on Pearson correlation and, thus, has a methodological concern because of the spiky nature of the probability of recurrence.

In this study, we first formulate the mathematical modeling of a single ML neuron and present typical neuron bursting patterns under varying ELF EFs and their corresponding recurrence features as obtained by RPs. We then study the synchronization of two weakly coupled chaotic bursting neurons with and without the influence of an ELF EF. We present the effect of EFs on the mismatch of the mean frequencies of both neurons, even when they are weakly connected. For this purpose, we suggest a slight modification of the recurrence-based phase synchronization measure.

2. Model

2.1. Morris–Lecar Neuron Model under an Extremely Low Frequency Electric Field

The ML neuron model is a model for electrical activity in the barnacle muscle fiber [19]. It is a simplified version of the HH neuron model for describing the discharge and the refractory properties of real neurons. It can explain the dynamical and biophysical mechanisms of the action potential initiation. This model is chosen as a compromise between a realistic representation of neuronal dynamics and an analytically tractable system. Furthermore, it has an advantage in that the excitability of types I and II can be obtained with a single parameter change. It can also exhibit a variety of bursting types involving regular bursting or irregular bursting and complex bifurcation structures [20,56–58].

The ML model has a fast activation variable v (membrane voltage) and a slower recovery variable w. v represents voltage (expressed in mV) and controls the instantaneous activation of fast currents (i_{fast}); w is a function of v and controls the activation of slower currents (i_{slow}). $c\frac{dv}{dt}$ is the current flowing through the capacitor related to the variation of ionic density between external and internal faces of the membrane. i_{fast}, i_{slow}, and i_{leak} are ionic currents characterizing the movement of charged particles through the ion channels. This movement of charged particles is due to the opening and closing of each ion channel. i_{stim} and c are the external input current and the membrane capacity, respectively. Finally, this model is given by the following equations:

$$c\frac{dv}{dt} = i_{\text{stim}} - i_{\text{fast}} - i_{\text{leak}} - i_{\text{slow}} \quad (1)$$

$$\frac{dw}{dt} = \varphi\frac{m_2(v) - w}{b(v)} \quad (2)$$

with the currents

$$\begin{aligned} i_{\text{fast}} &= g_{\text{fast}} m_1(v)(v - e_{\text{Na}}) \\ i_{\text{slow}} &= g_{\text{slow}} w(v - e_{\text{K}}) \\ i_{\text{leak}} &= g_{\text{leak}}(v - e_{\text{leak}}). \end{aligned}$$

The parameters e_{Na}, e_{K}, and e_{leak} represent the equilibrium potentials of Na^+, K^+, and leak ions, respectively, and g_{fast}, g_{slow}, and g_{leak} are the maximal conductances of the corresponding ion currents. They reflect the ion channels' densities distributed over the membranes. Control parameter φ is used to control the rate of change of w. The steady states m_1 and m_2 are nonlinear functions of v, given by

$$m_1(v) = 0.5\left(1 + \tanh\frac{v - u_1}{u_2}\right) \tag{3}$$

$$m_2(v) = 0.5\left(1 + \tanh\frac{v - u_3}{u_4}\right) \tag{4}$$

$$b(v) = \frac{1}{\cosh\frac{v-u_3}{2u_4}}. \tag{5}$$

u_1 and u_3 are the activation midpoint potentials at which the corresponding currents are half activated. u_2 and u_4 denote the slope factors of the activation. The time constant of the potassium activation is b. When a time-varying ELF EF is applied to the brain, it can induce a charge movement in the brain tissue; in which case, the current flow occurs mostly in the extracellular medium [59]. Therefore, an external EF will induce a membrane depolarization Δv which will modulate neuronal bursting behavior. For the sake of simplicity, we consider a steady external sinusoidal electrical field

$$v_e = \frac{A}{\omega}\sin\omega t + V_E \tag{6}$$

where V_E is the direct voltage, A the amplitude, and ω the frequency of the ELF EF. The field-induced membrane depolarization Δv can be expressed by [60]

$$\Delta v = \frac{A}{\omega}\frac{\sin\omega t - \cos\omega t}{1 + (\omega t_1)^2} + V_E \tag{7}$$

with t_1 significantly small and the frequency in the extremely low frequency area $\omega t_1 \ll 1$. Thereby, Equation (7) can be simplified to

$$\Delta v = \frac{A}{\omega}\sin\omega t + V_E. \tag{8}$$

According to Equation (8), the sinusoidal EF v_e equals its field-induced membrane depolarization Δv. Considering that Δv acts as an additive perturbation to the membrane potential, the dynamics of a neuron during exposure can be described by [61]

$$c\frac{dv}{dt} = i_{\mathrm{stim}} - \frac{d\Delta v}{dt} - i_{\mathrm{fast}} - i_{\mathrm{leak}} - i_{\mathrm{slow}} \tag{9}$$

$$\frac{dw}{dt} = \varphi\frac{m_2(v) - w}{b(v)} \tag{10}$$

with

$$\begin{aligned} i_{\mathrm{fast}} &= g_{\mathrm{fast}} m_1(v)(v + \Delta v - e_{\mathrm{Na}}) \\ i_{\mathrm{slow}} &= g_{\mathrm{slow}} w(v + \Delta v - e_{\mathrm{K}}) \\ i_{\mathrm{leak}} &= g_{\mathrm{leak}}(v + \Delta v - e_{\mathrm{leak}}). \end{aligned}$$

We assume that the synaptic input current $i_{\mathrm{stim}} = 0$ in order to study the response of a cortical neuron model exposed to an external sinusoidal field. Throughout this paper, we use the same parameter values for the ML model as explained in Table 1 [62].

Table 1. Parameters used for the ML model.

u_1	−1.2 mV	g_{fast}	20 mS/cm^2	e_{Na}	50 mV	φ	0.15
u_2	18 mV	g_{slow}	20 mS/cm^2	e_K	−100 mV	c	2 μ
u_3	−13 mV	g_{leak}	2 mS/cm^2	e_{leak}	−70 mV	V_E	−17.63 mV
u_4	10 mV						

2.2. Bursting Patterns of a Neuron

To explore how the neuron model responds to the externally applied ELF EF, we study the dynamics described by Equations (9) and (10) under the sinusoidal stimulus v_e, Equation (6). The simulations are implemented using the 4th-order Runge–Kutta method with a time step of 0.01 ms. Initial conditions are chosen as the resting values of membrane voltage in the absence of stimuli, that is, $v(0) = -65$ mV and $w(0) = 0$. The length of the time series is 2000 ms. The response of a neuron induced by an EF depends on the EF's frequency ω (Figure 1 for ω in the range $0 \leq \omega \leq 0.5$ rad/ms). The amplitude of the external EF is set very small, $A = 0.1$.

The firing pattern of a neuron stimulated by an external EF varies when changing the frequency ω (Figures 1 and 2). For an ELF EF with very low frequency, the neuron fires periodically. We find n spike bursting states, and the number n can be large. n spike bursting means that we have n action potentials in every stimulus period (Figures 1A–D and 2) for $\omega = [0.001, 0.120]$ rad/ms. After this range ω of n-periodic bursting, the neuron bursts synchronously to the stimulus $\omega = [0.121, 0.280]$ rad/ms. After this range of ω, the neuron exhibits a chaotic response (Figure 1E) with $\omega = [0.281, 0.320]$ rad/ms where the membrane potential responses are aperiodic and irregular. As ω is further increased, a mode locking pattern of bursting appears (Figure 1F), finally followed by synchronized firing with only one action potential in every stimulus, which can maintain this state for a long-term frequency band (Figure 1G). Neuron dynamics are obviously very sensitive to the frequency of the stimulus by the ELF EF.

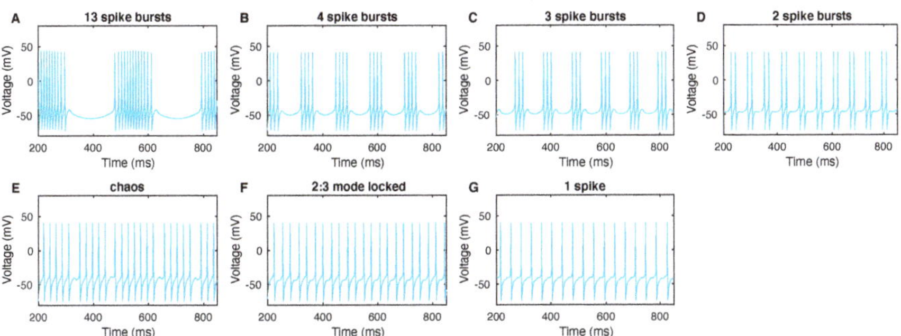

Figure 1. Spiking patterns of ML neuron membrane voltage under an external EF for different frequencies: (**A**) $\omega = 0.02$ rad/ms, (**B**) $\omega = 0.05$ rad/ms, (**C**) $\omega = 0.06$ rad/ms, (**D**) $\omega = 0.10$ rad/ms, (**E**) $\omega = 0.286$ rad/ms, (**F**) $\omega = 0.32$ rad/ms, and (**G**) $\omega = 0.5$ rad/ms.

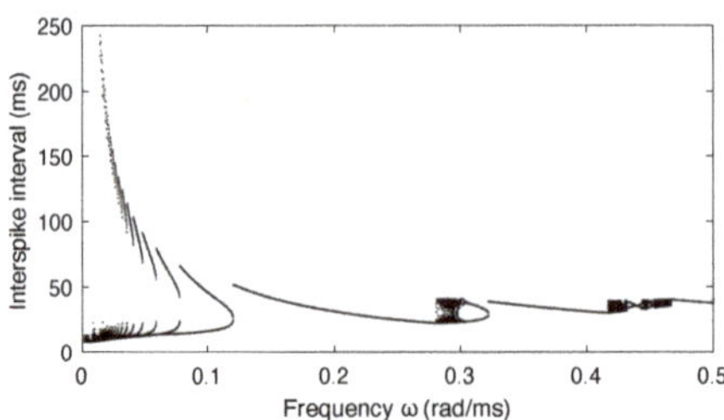

Figure 2. Bifurcation diagram of ML neuron dynamics under an external EF for varying frequencies ω (based on interspike intervals of the membrane voltage). For better visibility of the dynamics for larger ω, the y-axis was bounded to 250 ms.

3. Recurrence Quantification Analysis (RQA)

In the following, neuron dynamics will be studied using recurrence quantification analysis (RQA). This method quantifies certain recurrence features of the dynamical system in its corresponding phase space [35,63]. We define a recurrence of a trajectory $\vec{x}(t) \in \mathbb{R}^m$ (with m the dimension of the system) of a dynamical system by saying that the trajectory has returned at time $t = j$ to the former point in phase space visited at $t = i$ (with $i \in [1, N]$ and N the length of time series) if

$$R_{i,j} = \Theta\Big(\varepsilon - \big\|\vec{x}(i) - \vec{x}(j)\big\|\Big) \tag{11}$$

where ε is a pre-defined threshold and $\Theta(\cdot)$ is the Heaviside function. We have a matrix of (0, 1), where 1 at (i, j) means that $\vec{x}(i)$ and $\vec{x}(j)$ are neighbors and 0 means that they are not. The resulting black and white representation of this binary matrix is called a recurrence plot (RP). For the selection of the recurrence threshold ε, different strategies are available, depending on the research question [64–69]. Here, we use an approach to select ε in a way that ensures a certain recurrence point density. This allows a better comparability between RPs of different systems [68].

The RP method has been intensively studied and applied in the last years. Different measures of complexity have been proposed that can classify different dynamics, identify dynamical transitions, or detect couplings, causality, or synchronization [35].

If not all state variables of the state vector $\vec{x}$ are available, a phase space reconstruction has to be applied. Here, we use the recently proposed PECUZAL method to reconstruct the phase space trajectories [70]. This method allows us to use multiple embedding delays τ. The embedding parameters are listed in Table 2.

Table 2. Embedding parameters indicated by the PECUZAL algorithm.

Time Series	Dimension	Delay
4 spike burst ($\omega = 0.05$ rad/ms)	3	17, 22
2 spike burst ($\omega = 0.10$ rad/ms)	2	16, 20
chaos ($\omega = 0.286$ rad/ms)	2	20
1 spike ($\omega = 0.5$ rad/ms	2	19

RPs of the different bursting neurons represent a typical pattern (Figure 3, using an ε that ensures a recurrence point density of 0.15). Each "dashed-dotted" diagonal line in the RPs corresponds to a spike. For the alternating spiking behavior, we have a set of dashed lines followed by an extended black region (Figure 3A,B). The set of n spikes is well distinguished by the number of dashed lines (see some orange boxes marked in the figure). The block-like black region represents the silent state between each stimulus, which is a period for which the neuron cannot respond to a stimulus. On the small scale, the diagonal lines show some additional patterns, i.e., small structures sitting perpendicularly at these lines or thickenings, similar to bumps or knobs. This is a typical feature of slow–fast systems [71].

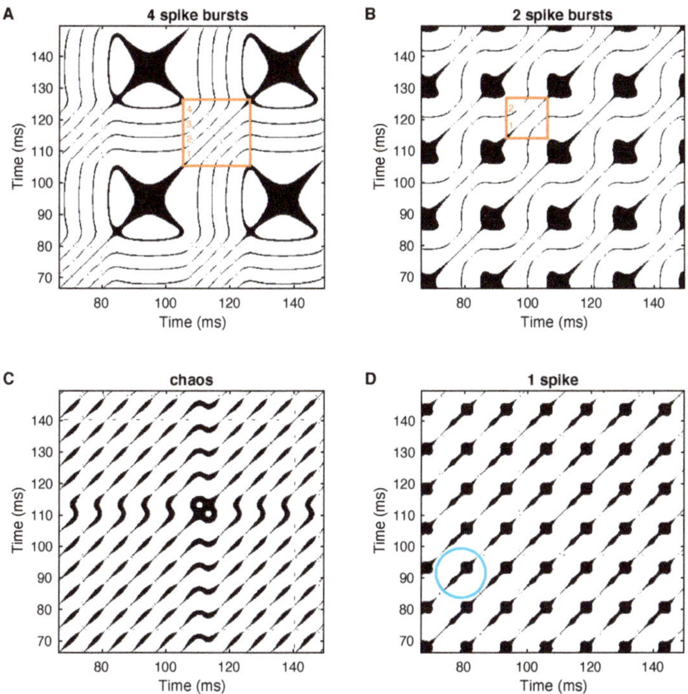

Figure 3. RPs of the membrane voltage v of selected bursting neurons: (**A**) 4 spike burst (ω = 0.05 rad/ms), (**B**) 2 spike burst (ω = 0.10 rad/ms), (**C**) chaos (ω = 0.286 rad/ms), and (**D**) 1 spike (ω = 0.5 rad/ms). Diagonal lines represent the spikes, the larger extended structures represent the "silent" epochs, and the structures perpendicular to the diagonal lines and small thickenings represent the slow–fast dynamics (blue circle in (**D**)). The orange boxes in (**A**,**B**) mark a sequence of diagonal lines. The number of diagonal lines counted from the main diagonal of such a box towards the corner of this box represents the number of spikes within this period. Recurrence threshold ε is selected to ensure a recurrence point density of 0.15.

In order to go beyond the visual impression of the RP, we use recurrence quantification analysis (RQA) [35,72]. The RQA measures are based on the recurrence point density and the diagonal and vertical line structures of the RP. For example, the recurrence point density $\frac{1}{N^2}\sum R_{i,j}$ corresponds to the probability that a state will recur. The calculation of this measure can also be restricted to a diagonal-wise calculation, i.e., the recurrence point density along a diagonal with distance τ from the main diagonal $R_{i,i} = 1$ [35]. This gives us an estimator of the probability that the system returns to a previous state after time τ and is called the τ-recurrence rate,

$$RR_\tau = \frac{1}{N-\tau}\sum_{i=0}^{N-\tau} R_{i,i+\tau} \tag{12}$$

where τ is the set time and N the total number of points in the phase space. The distance between the peaks in an RR_τ plot corresponds to the period length of oscillations or the interspike intervals of spike trains similar to the neuron's spiking/bursting patterns.

The spike trains of 4 spikes, 2 spikes, chaos, and 1 spike have their specific probability distributions for recurrence after lag τ (Figure 4). Where the 1-periodic spike occurrence is clearly visible for 1 spike (Figure 4D), the 2 and 4 spikes produce more subtle probability distributions, revealing different periodicities and large blocks between the bursting periods (Figure 4A,B). The RR_τ of the chaotic bursting exhibits a more complicated distribution of peaks corresponding to the unpredictable occurrence of spikes (Figure 4C).

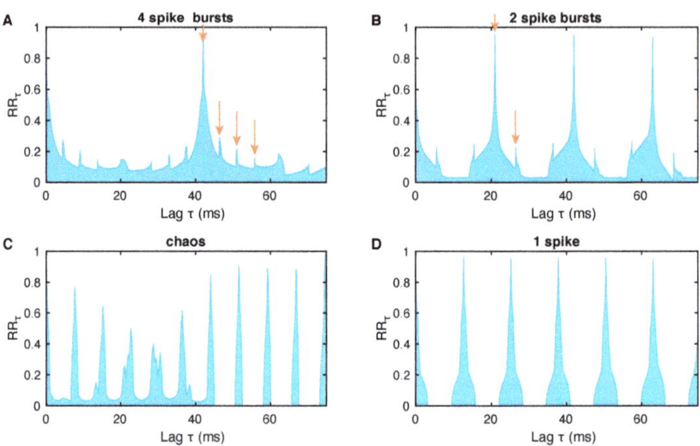

Figure 4. Probability of recurrence after time τ (τ-recurrence rate) for the bursting neurons as shown in Figure 3: (**A**) 4 spike burst, (**B**) 2 spike burst, (**C**) chaos, and (**D**) 1 spike. The n bursts are visible as the rather thin side peaks of the main peaks (in addition to the main peak).

4. Coupling of Two Bursting Neurons

4.1. Model and Numerical Simulation

A coupling between ML neurons is realized by a gap junction. We suppose that the two neurons are slightly different by considering different values of u_2 and u_3 in Equations (4) and (5), i.e., $u_{2,1} = -18.0$ mV and $u_{2,2} = -18.1$ mV, and $u_{3,1} = -12.8$ mV and $u_{3,2} = -10$ mV for neurons 1 and 2, respectively. Moreover, both neurons start using different initial conditions $v_1(0) = -65.6$ mV and $v_2(0) = -60$ mV. We integrate the model for 50,000 time steps (with $dt = 0.05$) and remove the first 10,000 points as transients. Using the ELF EF frequency that leads to chaotic bursting $\omega = 0.286$ rad/ms, the coupled chaotic bursting ML neurons under ELF EF exposure can be expressed as

$$c\frac{dv_1}{dt} = i_{\text{stim}} - \frac{d\Delta v_1}{dt} - i_{1,\text{fast}} - i_{1,\text{slow}} - i_{1,\text{leak}} - g(v_1 - v_2) \tag{13}$$

$$\frac{dw_1}{dt} = \varphi\frac{m_2(v_1) - w_1}{b(v_1)} \tag{14}$$

$$c\frac{dv_2}{dt} = i_{\text{stim}} - \frac{d\Delta v_2}{dt} - i_{2,\text{fast}} - i_{2,\text{slow}} - i_{2,\text{leak}} - g(v_2 - v_1) \tag{15}$$

$$\frac{dw_2}{dt} = \varphi\frac{m_2(v_2) - w_2}{b(v_2)}, \tag{16}$$

where g is the gap which represents the electrical junction between the neurons. With these two different chaotic neurons, we will now study the phase synchronization between them and focus on the range of weak coupling, i.e., with 500 values of g within the interval $g = [0, 0.15]$ (Figure 5).

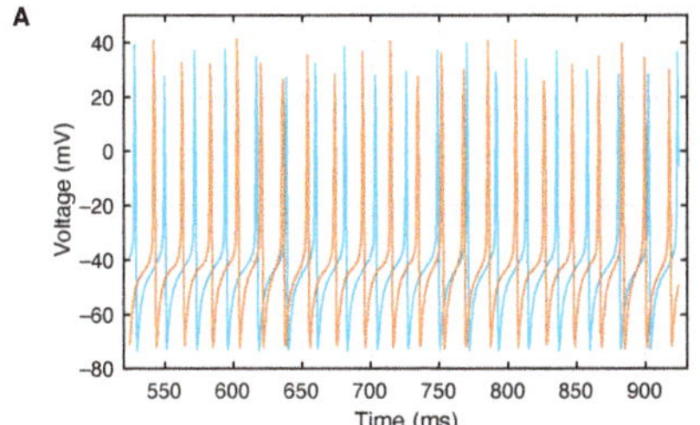

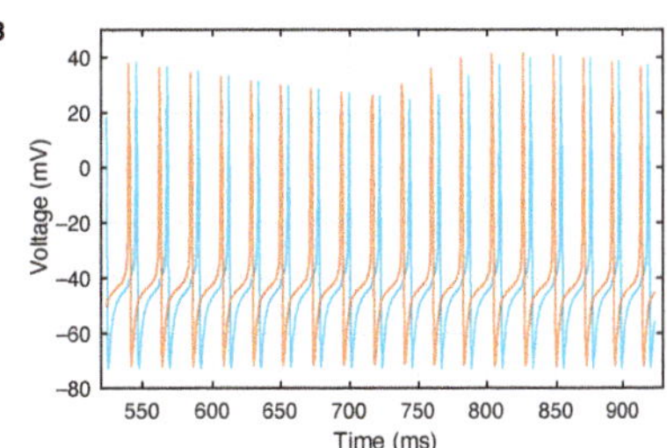

Figure 5. Spiking pattern in the membrane voltage of two weakly coupled neurons in an ELF EF in chaotic regime ($\omega = 0.286$) with (**A**) no synchronization with $g = 0.01$ and (**B**) phase synchronization with $g = 0.04$.

When bursting begins at the same time in the coupled neurons, we have bursting synchronization irrespective of the neurons' spiking behaviors within a given burst event. From a dynamical point of view, since we assign a phase that increases by 2π at each burst event, we regard bursting synchronization as a kind of phase synchronization [73]. Thus, first we will determine the phase of each chaotic bursting neuron. A frequently used approach to calculate the phase of a signal is using the Hilbert transformation [15]

$$\phi(t) = \arctan 2(v_H(t), v(t)) \tag{17}$$

where v_H is the complex part of the Hilbert transform of the membrane voltage $v(t)$ and $\phi(t)$ increases continually with time. Since chaotic neurons have chaotic spikes, the phase of chaotic neurons changes also chaotically. Unfortunately, this approach does not work well for spiky signals and can cause slipping of the instantaneous phases. Nevertheless, for long-term averages, it provides useful results.

To detect phase synchronization of chaotic coupled neurons and to evaluate the effect of an ELF EF on this synchronization, we first consider the absolute phase difference between the membrane voltage of both neurons without and with applied EF. Phase synchronization occurs if the difference $\phi_1(t) - \phi_2(t)$ between the phases of the two neurons does not grow with time [74]. This means that the two neurons, on average, generate spikes almost simultaneously. With the knowledge of the phase $\phi(t)$, the frequency $\bar{\omega}(t) = \frac{d\phi(t)}{dt}$ and the mean frequency $\Omega = \left\langle \frac{d\phi(t)}{dt} \right\rangle$ can be defined. A weaker form of synchronization is frequency locking. Frequency locking between coupled systems can be measured by the mismatch between the average frequencies $\Delta\Omega = \Omega_1 - \Omega_2$, with $\Delta\Omega \to 0$ for phase locking.

The weakly coupled neurons show frequency locking without ELF EF when the coupling exceeds a critical value (Figure 6). The frequency mismatch $\Delta\Omega$ between both neurons is constant between $g = 0$ (no coupling) and $g = 0.025$. After this value, $\Delta\Omega$ is decreasing and vanishes around $g = 0.066$, indicating the onset of synchronization frequency locking between the neurons.

With ELF EF applied, the frequency difference is smaller, even for $g = 0$, and decreases much faster than without EF; the neurons become frequency-locked for $g = 0.037$ (Figure 6). Thus, in a range of weakly connected neurons, applying an external ELF EF on the chaotic coupled ML neurons enhances frequency-locked synchronization. This confirms earlier findings of synchronized neurons using a different model of weakly connected bursters [24].

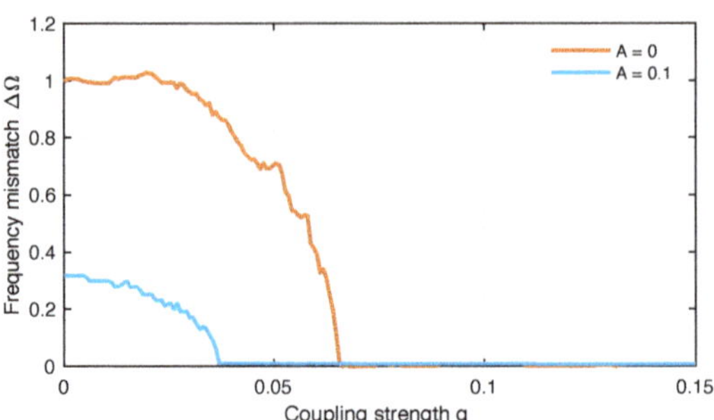

Figure 6. Frequency mismatch between chaotic coupled neurons for increasing coupling g without external EF (red) and with external EF where $A = 0.1$ and $\omega = 0.286$ (blue).

Since the firing pattern strongly depends on the amplitude of the ELF EF, we expect that the occurrence of frequency locking also depends on this external stimulus amplitude. In fact, we find that an amplitude value of $A = 0.15$ is strong enough to cause a complete synchronization of two neurons even without coupling (Figure 7). Therefore, we select a lower amplitude value of $A = 0.1$, where we still have a significant frequency mismatch between the uncoupled neurons. A weak coupling between the neurons leads, finally, to frequency-locked synchronization for lower ELF EF amplitudes.

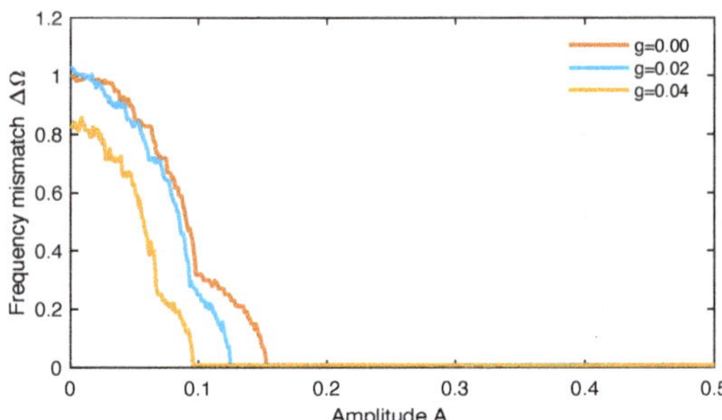

Figure 7. Frequency mismatch between chaotic coupled neurons for increasing amplitude A and for different coupling strengths g.

To test for phase synchronization, i.e., whether the difference $\phi_1(t) - \phi_2(t)$ remains constant, we will use an alternative method which can derive the phases of spiky signals in a more reliable way.

4.2. Phase Synchronization Analysis Using Recurrence Features

Phase synchronization is related to recurrence of states. Therefore, RPs are a natural tool to study phase synchronization [35]. The spiking pattern causes regular and almost periodic occurrence of diagonal line structures in the RPs (Figure 8). Here, we use a recurrence threshold ε to ensure a recurrence point density of 0.1. Although we notice a certain amount of similarity between the RPs of neuron 1 and neuron 2 in the nonsynchronized regime, we still see deviations in the line patterns of the RP of neuron 2 (Figure 8A,B). In contrast, the RPs of neuron 1 and neuron 2 for the in-phase synchronized regime show a striking similarity (Figure 8C,D).

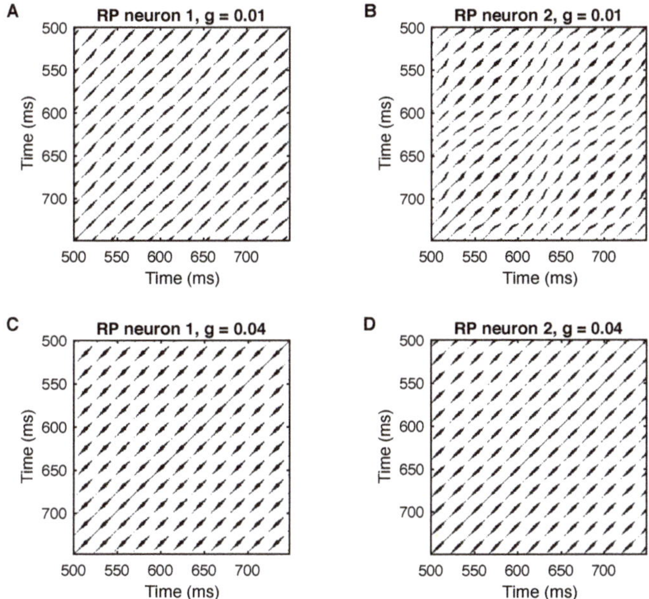

Figure 8. Recurrence plots of the membrane voltage for weakly coupled neurons as shown in Figure 5 for (**A**,**B**) no synchronization, $g = 0.01$, and (**C**,**D**) phase synchronization, $g = 0.04$. Embedding parameters were estimated using the PECUZAL method [70]; the recurrence threshold is selected to ensure a recurrence rate of $RR = 0.1$.

The vertical distance between these diagonal line structures is related to the phase. Therefore, we can use the density of recurrence points along diagonals parallel to the main diagonal, the τ-recurrence rate RR_τ, Equation (12), as an estimator of the phase distribution and compare it between different systems. For two nonsynchronized systems, the recurrence probabilities should differ significantly (Figure 9A). During phase synchronization, RR_τ should have high probabilities at the same τ values; thus, the shape of RR_τ should be very similar (Figure 9B). Therefore, RR_τ has been used to construct a measure for phase synchronization between two signals x_1 and x_2 by calculating the Pearson correlation of probability of recurrence (CPR) between $RR_\tau^{x_1}$ and $RR_\tau^{x_2}$ [34]

$$CPR^{\mathrm{P}} = \frac{\mathrm{cov}(RR_\tau^{x_1}, RR_\tau^{x_2})}{\sigma_{RR_\tau^{x_1}} \sigma_{RR_\tau^{x_2}}}, \tag{18}$$

with σ_{RR_τ} the standard deviation of the corresponding RR_τ series. CPR values of 1 would then correspond to phase synchronization and 0 to no synchronization. Here, it is important to remove the first peak in RR_τ close to $\tau = 0$ because these values correspond to the main diagonal in the RP present in all systems [50]. Therefore, this first peak would indicate some kind of similarity between $RR_\tau(x_1)$ and $RR_\tau(x_2)$ even for completely desynchronized systems. Such exclusion of the first part of the RR_τ series corresponds to applying a Theiler window [75]. Here we used a Theiler window of 25 mS.

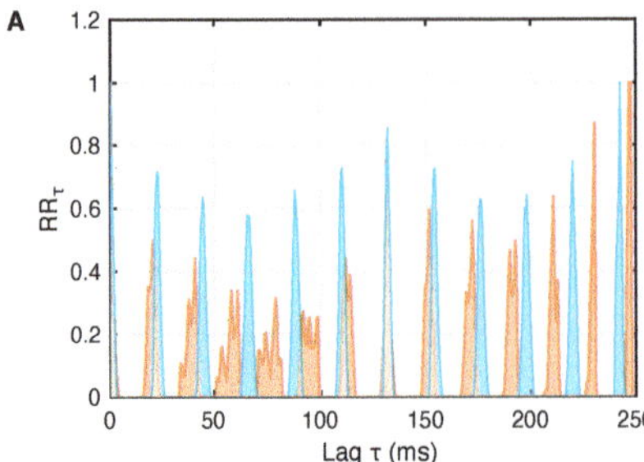

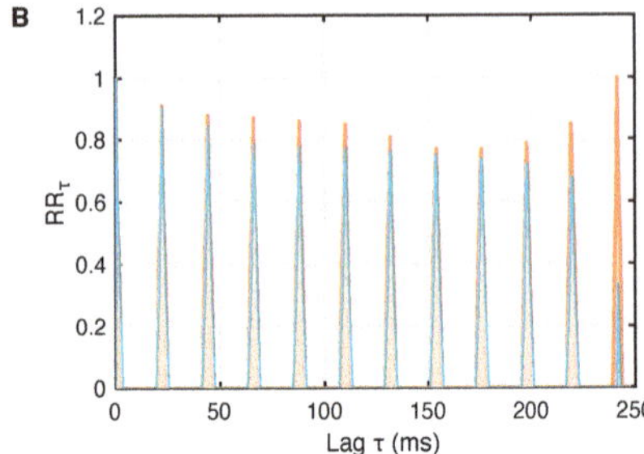

Figure 9. τ-recurrence rate for weakly coupled neurons as shown in Figures 5 and 8 for (**A**) no synchronization, $g = 0.01$, and (**B**) phase synchronization, $g = 0.04$. For phase synchronization, the τ-recurrence rate series for both neurons are almost completely overlapping.

Another concern on calculating the CPR measure using Equation (18) is the spiky shape of the RR_τ series, biasing the Pearson correlation estimation. As an alternative, we could use the Spearman rank correlation instead of the Pearson correlation,

$$CPR^{S} = \frac{\mathrm{cov}\left(\mathrm{R}(RR_\tau^{x_1}), \mathrm{R}(RR_\tau^{x_2})\right)}{\sigma_{\mathrm{R}(RR_\tau^{x_1})}\sigma_{\mathrm{R}(RR_\tau^{x_2})}}, \tag{19}$$

with the RR_τ series converted to the ranks $\mathrm{R}(RR_\tau)$. This correlation measure is expected to work better for non-normal distributed data, as the RR_τ series would be.

Both CPR measures clearly show the onset of phase synchronization at $g = 0.066$ for neurons without ELF EF and at $g = 0.037$ for neurons with ELF EF (Figure 10). There are some differences between CPR^{P} and CPR^{S}. During phase synchronization, CPR^{P} is almost 1, but CPR^{S} is slightly below 1, even more obvious for the coupled neurons without ELF EF. However, for phase synchronization, we would expect to have a CPR value of 1. Moreover, the transition from a nonsynchronized regime to a synchronized regime is not as abrupt as indicated by $\Delta\Omega$ (Figure 6), but CPR^{S} changes almost abruptly from very low values to very large values, whereby CPR^{P} shows a more gradual increase (and even step-wise increase for $A = 0$). This finding indicates that the Spearman-based CPR is obviously not a better choice than the Pearson-based CPR measure.

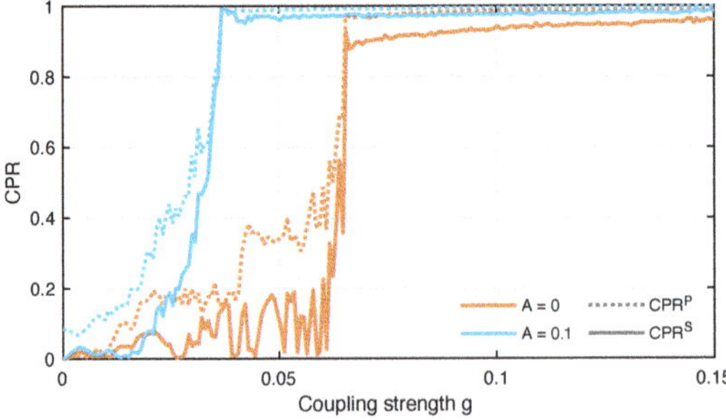

Figure 10. Correlation of probability of recurrence CPR based on Pearson (dotted) and Spearman (line) correlations indicating the onset of phase synchronization between chaotic coupled neurons without external EF (red) and with external EF where $A = 0.1$ and $\omega = 0.286$ (blue).

The RR_τ series represents probabilities of recurrence. Therefore, it seems more natural to use a measure that can directly quantify the difference between probability populations, such as Kullback–Leibler distance [76] or Hellinger distance [77]. Here we test the use of the Hellinger distance

$$H(RR_\tau^{x_1}, RR_\tau^{x_2}) = \frac{1}{\sqrt{2}} \left\| \sqrt{RR_\tau^{x_1}} - \sqrt{RR_\tau^{x_2}} \right\|, \tag{20}$$

which corresponds to the Euclidean norm of the square root distances between the RR_τ series of the two signals. Values of H close to 0 indicate phase transition, whereas values close to 1 indicate nonsynchronized regimes.

To assess whether the variation of H indeed reveals phase synchronization, we use a simple block shuffling approach to test the null hypothesis that the signals are not synchronized. Block shuffling splits a time series into a number of blocks (here, we used five blocks) of equal width at random indices and randomly concatenates these blocks to create a new surrogate time series. Such surrogates preserve short-term temporal properties but destroy long-term dynamical information and, thus, correlations when compared with another signal. The distribution $p(H)$ derived from the ensemble of surrogates is then used to define the confidence limit of 95% (simply by using the 95% quantile of this test distribution $p(H)$). Considering $A = 0$, we find the confidence limit by $H_{0.95}$ as 0.17. Values of H below this value can be considered to represent phase synchronization.

The measure H indicates the transition from the nonsynchronous to the phase synchronization regime of the two weakly coupled neurons (Figure 11). The change in H is significant. Moreover, the variation in H over increasing coupling strength g reveals the more gradual change to the phase synchronization as well as the step-like transition to phase synchronization for the situation without ELF EF caused by phase jumps.

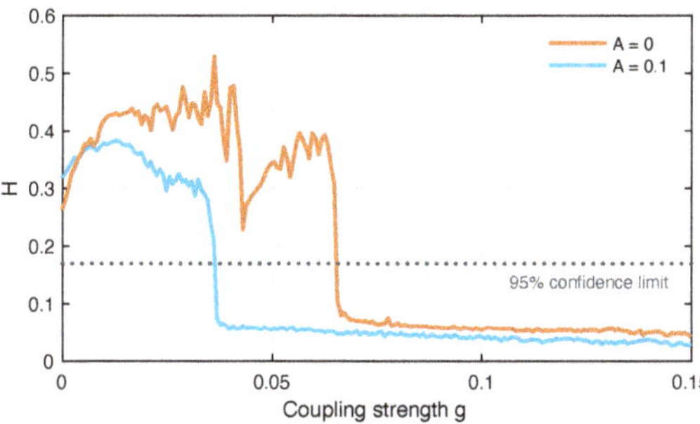

Figure 11. Hellinger distance of the τ-recurrence rate indicating the onset of phase synchronization between chaotic coupled neurons without external EF (red) and with external EF where $A = 0.1$ and $\omega = 0.286$ (blue). A drop of H below the confidence limit of 95% (dotted line) represents the significance of this finding.

5. Conclusions

The synchronization of weakly coupled Morris–Lecar neurons under common external forcing has been studied previously. For example, Kitajima and Kurths [21] used interspike intervals (to study frequency locking) and Yi et al. [62] considered the average firing rate. In general, the numerical calculation of the phases of spiky signals using the Hilbert transform is problematic. An alternative way to identify the phases in dynamical systems is to use recurrence plots [34]. This method can find phases for noncoherent and spiky signals. We, therefore, used a recurrence-based approach, which decodes the phase in terms of specific recurrence patterns in the recurrence plot, and demonstrated its potential for the study of spiking patterns of neurons.

In this work, the spiking patterns of Morris–Lecar neurons under ELF sinusoidal EF and the synchronization of two neurons weakly coupled with gap junction under ELF EF were investigated using this recurrence-based approach. The representation of the dynamics of the neurons' membrane voltages by recurrence plots provided a convenient approach to compare the recurrence features of their spiking patterns. Various spiking patterns, such as periodic and chaotic bursting and periodic spikes, were observed. The spiking patterns were found to be very sensitive to changes of the stimulus frequency.

Moreover, the recurrence approach allows us to consider phase differences between the spiking patterns in a more robust way than the frequently used Hilbert transform. We have introduced an alternative measure for testing phase synchronization using recurrences. Instead of comparing the probabilities of recurrences (as represented by the τ-recurrence rate) by correlation coefficients, we suggest to use the Hellinger distance as a more natural measure because it quantifies the differences between probabilities. The typically used Pearson correlation is biased because the τ-recurrence rate does not follow a normal distribution. The Spearman rank correlation could be an alternative, but we found additional bias due the large number of zeros in the τ-recurrence rate series.

By using recurrence-based synchronization measures, we found that even without external EF, phase synchronization of two ML neurons can occur for a range of values of coupling strength. Moreover, phase synchronization can be enhanced by an additional external EF. This physiological behavior might be of importance for the functioning of the brain when exposed to electromagnetic fields, such as by power lines, electrical equipment, or cellular radio towers.

Author Contributions: Conceptualization, A.M.N., J.K. and N.M.; software, A.M.N., E.K.N. and N.M.; methodology, investigation, and writing—original draft preparation, A.M.N. and N.M.; evaluation, writing—review and editing, J.K., B.R.N.N. and E.K.N.; visualization, N.M.; supervision, N.M. and J.K. All authors have read and agreed to the published version of the manuscript.

Funding: This work was made possible by the financial support from the partnership between Deutsche Forschungsgemeinschaft (DFG) and the Academy of Science for the Developing World (TWAS) under the grant number 31405812. E.J.N. and J.K. acknowledge the Volkswagen Foundation (Grant No. 85391).

Data Availability Statement: Code used to perform the analysis of this study is available via Zenodo, doi:10.5281/zenodo.5910812 (accessed on 29 December 2022).

Acknowledgments: We thank Paul Woafo for support and supervision. We further acknowledge the help of K. Hauke Kraemer and Frank Hollmann for support in the implementation of the model in Julia.

Conflicts of Interest: The authors declare no conflict of interest.

Abbreviations

The following abbreviations are used in this manuscript:

CPR	correlation of probability of recurrence
EF	electric field
ELF	extremely low frequency
HH	Hodgkin–Huxley model
ML	Morris–Lecar model
RP	recurrence plot
RQA	recurrence quantification analysis
RR	recurrence rate

References

1. Adrian, E.D. *The Mechanism of Nervous Action, Electrical Studies of the Neurone*; University of Pennsylvania Press: Philadelphia, PA, USA, 1932.
2. Huang, K.; Li, Y.; Yang, C.; Gu, M. The dynamic principle of interaction between weak electromagnetic fields and living system–Interference of electromagnetic waves in dynamic metabolism. *Chin. J. Med. Phys.* **1996**, *14*, 205–207.
3. Reato, D.; Rahman, A.; Bikson, M.; Parra, L.C. Low-Intensity Electrical Stimulation Affects Network Dynamics by Modulating Population Rate and Spike Timing. *J. Neurosci.* **2010**, *30*, 15067–15079. [CrossRef]

4. Opitz, A.; Falchier, A.; Yan, C.G.; Yeagle, E.M.; Linn, G.S.; Megevand, P.; Thielscher, A.; Deborah A.R.; Milham, M.P.; Mehta, A.D.; et al. Spatiotemporal structure of intracranial electric fields induced by transcranial electric stimulation in humans and nonhuman primates. *Sci. Rep.* **2016**, *6*, 31236. [CrossRef]
5. Huang, Y.; Liu, A.A.; Lafon, B.; Friedman, D.; Dayan, M.; Wang, X.; Bikson, M.; Doyle, W.K.; Devinsky, O.; Parra, L.C. Measurements and models of electric fields in the in vivo human brain during transcranial electric stimulation. *eLife* **2017**, *6*, e18834. [CrossRef]
6. Savitz, D.A.; Loomis, D.P.; Tse, C.K.J. Electrical occupations and neurodegenerative disease: Analysis of US mortality data. *Arch. Environ. Health Int. J.* **1998**, *53*, 71–74. [CrossRef]
7. Johansen, C. Exposure to electromagnetic fields and risk of central nervous system disease in utility workers. *Epidemiology* **2000**, *11*, 539–543. [CrossRef]
8. Radman, T.; Su, Y.; An, J.H.; Parra, L.C.; Bikson, M. Spike timing amplifies the effect of electric fields on neurons: Implications for endogenous field effects. *J. Neurosci.* **2007**, *27*, 3030–3036. [CrossRef]
9. Nkomidio, A.M.; Woafo, P. Effects of imperfection of ionic channels and exposure to electromagnetic fields on the generation and propagation of front waves in nervous fibre. *Commun. Nonlinear Sci. Numer. Simul.* **2010**, *15*, 2350–2360. [CrossRef]
10. Eichwald, C.; Kaiser, F. Model for external influences on cellular signal transduction pathways including cytosolic calcium oscillations. *Bioelectromagnetics* **1995**, *16*, 75–85. [CrossRef]
11. Huang, C.; Xu, B.; Lin, J. Effects of extremely low frequency magnetic fields on hormone-induced cytosolic calcium oscillations. *Shengwu Wuli Xuebao* **1998**, *15*, 543–546.
12. Wertheimer, N.; Leeper, E. Electrical wiring configurations and childhood cancer. *Am. J. Epidemiol.* **1979**, *109*, 273–284. [CrossRef] [PubMed]
13. Moulder, J.E. Power-frequency fields and cancer. *Crit. Rev. Biomed. Eng.* **1998**, *26, 1–116.* [CrossRef] [PubMed]
14. Stuchly, M.A.; Dawson, T.W. Interaction of low-frequency electric and magnetic fields with the human body. *Proc. IEEE* **2000**, *88*, 643–664. [CrossRef]
15. Pikovsky, A.; Rosenblum, M.; Kurths, J. *Synchronization—A Universal Concept in Nonlinear Sciences*; Cambridge University Press: Cambridge, UK, 2001.
16. Golomb, D.; Rinzel, J. Clustering in globally coupled inhibitory neurons. *Phys. D Nonlinear Phenom.* **1994**, *72*, 259–282. [CrossRef]
17. Dayan, P.; Abbott, L. *Theoretical Neuroscience: Computational and Mathematical Modeling of Neural Systems*; MIT Press: Cambridge, MA, USA, 2001.
18. Gray, C.M.; König, P.; Engel, A.K.; Singer, W. Oscillatory responses in cat visual cortex exhibit inter-columnar synchronization which reflects global stimulus properties. *Nature* **1989**, *338*, 334–337. [CrossRef]
19. Morris, C.; Lecar, H. Voltage oscillations in the barnacle giant muscle fiber. *Biophys. J.* **1981**, *35*, 193. [CrossRef]
20. Rinzel, J.; Ermentrout, G.B. Analysis of Neural Excitability and Oscillations. In *Methods in Neuronal Modeling*; Koch, C., Segev, I., Eds.; MIT Press: Cambridge, MA, USA, 1989; pp. 251–291.
21. Kitajima, H.; Kurths, J. Forced synchronization in Morris–Lecar neurons. *Int. J. Bifurc. Chaos* **2007**, *17*, 3523–3528. [CrossRef]
22. Hoppensteadt, F.C.; Izhikevich, E.M. *Weakly Connected Neural Networks*; Springer: New York, NY, USA, 1997.
23. Hoppensteadt, F.C.; Izhikevich, E.M. Synaptic organizations and dynamical properties of weakly connected neural oscillators II. Learning phase information. *Biol. Cybern.* **1996**, *75*, 129–135. [CrossRef]
24. Izhikevich, E.M. Synchronization of elliptic bursters. *Siam Rev.* **2001**, *43*, 315–344. [CrossRef]
25. Gutfreund, Y.; Yarom, Y.; Segev, I. Subthreshold oscillations and resonant frequency in guinea-pig cortical neurons: Physiology and modelling. *J. Physiol.* **1995**, *483*, 621–640. [CrossRef]
26. Hutcheon, B.; Miura, R.M.; Puil, E. Subthreshold membrane resonance in neocortical neurons. *J. Neurophysiol.* **1996**, *76*, 683–697. [CrossRef] [PubMed]
27. Hu, H.; Vervaeke, K.; Storm, J.F. Two forms of electrical resonance at theta frequencies, generated by M-current, h-current and persistent Na + current in rat hippocampal pyramidal cells. *J. Physiol.* **2002**, *545*, 783–805. [CrossRef] [PubMed]
28. Giocomo, L.M.; Zilli, E.A.; Fransén, E.; Hasselmo, M.E. Temporal Frequency of Subthreshold Oscillations Scales with Entorhinal Grid Cell Field Spacing. *Science* **2007**, *315*, 1719–1722. [CrossRef] [PubMed]
29. Vera, J.; Pezzoli, M.; Pereira, U.; Bacigalupo, J.; Sanhueza, M. Electrical Resonance in the θ Frequency Range in Olfactory Amygdala Neurons. *PLoS ONE* **2014**, *9*, e85826. [CrossRef]
30. Fischer, L.; Leibold, C.; Felmy, F. Resonance Properties in Auditory Brainstem Neurons. *Front. Cell. Neurosci.* **2018**, *12*, 8. [CrossRef]
31. Bennett, M.V.; Zukin, R. Electrical Coupling and Neuronal Synchronization in the Mammalian Brain. *Neuron* **2004**, *41*, 495–511. [CrossRef]
32. Dong, A.; Liu, S.; Li, Y. Gap Junctions in the Nervous System: Probing Functional Connections Using New Imaging Approaches. *Front. Cell. Neurosci.* **2018**, *12*, 320. [CrossRef]
33. Sabatini, B.L.; Regehr, W.G. Timing of neurotransmission at fast synapses in the mammalian brain. *Nature* **1996**, *384*, 170–172. [CrossRef]
34. Romano, M.C.; Thiel, M.; Kurths, J.; Kiss, I.Z.; Hudson, J.L. Detection of synchronization for non-phase-coherent and non-stationary data. *Europhys. Lett.* **2005**, *71*, 466–472. [CrossRef]

35. Marwan, N.; Romano, M.C.; Thiel, M.; Kurths, J. Recurrence Plots for the Analysis of Complex Systems. *Phys. Rep.* **2007**, *438*, 237–329. [CrossRef]
36. Marwan, N. A Historical Review of Recurrence Plots. *Eur. Phys. J. Spec. Top.* **2008**, *164*, 3–12. [CrossRef]
37. Babloyantz, A. Evidence for slow brain waves: A dynamical approach. *Electroencephalogr. Clin. Neurophysiol.* **1991**, *78*, 402–405. [CrossRef]
38. Song, I.H.; Lee, D.S.; Kim, S.I. Recurrence quantification analysis of sleep electoencephalogram in sleep apnea syndrome in humans. *Neurosci. Lett.* **2004**, *366*, 148–153. [CrossRef]
39. Bergner, A.; Romano, M.C.; Kurths, J.; Thiel, M. Synchronization analysis of neuronal networks by means of recurrence plots. In *Lectures in Supercomputational Neurosciences*; Beim Graben, P., Zhou, C., Thiel, M., Kurths, J., Eds.; Understanding Complex Systems; Springer: Berlin/ Heidelberg, Germany, 2008; pp. 177–191. [CrossRef]
40. Ouyang, G.; Li, X.; Dang, C.; Richards, D.A. Using recurrence plot for determinism analysis of EEG recordings in genetic absence epilepsy rats. *Clin. Neurophysiol.* **2008**, *119*, 1747–1755. [CrossRef] [PubMed]
41. Budzinski, R.C.; Boaretto, B.R.R.; Prado, T.L.; Lopes, S.R. Phase synchronization and intermittent behavior in healthy and Alzheimer-affected human-brain-based neural network. *Phys. Rev.* **2019**, *99*, 022402. [CrossRef] [PubMed]
42. Rodriguez-Sabate, C.; Rodriguez, M.; Morales, I. Studying the functional connectivity of the primary motor cortex with the binarized cross recurrence plot: The influence of Parkinson's disease. *PLoS ONE* **2021**, *16*, e0252565. [CrossRef] [PubMed]
43. Mendonça, P.R.F.; Vargas-Caballero, M.; Erdélyi, F.; Szabó, G.; Paulsen, O.; Robinson, H.P.C. Stochastic and deterministic dynamics of intrinsically irregular firing in cortical inhibitory interneurons. *eLife* **2016**, *5*, e16475. [CrossRef] [PubMed]
44. Boaretto, B.R.R.; Budzinski, R.C.; Prado, T.L.; Kurths, J.; Lopes, S.R. Neuron dynamics variability and anomalous phase synchronization of neural networks. *Chaos* **2018**, *28*, 106304. [CrossRef]
45. Tibau, E.; Soriano, J. Analysis of spontaneous activity in neuronal cultures through recurrence plots: Impact of varying connectivity. *Eur. Phys. J. Spec. Top.* **2018**, *227*, 999–1014. [CrossRef]
46. Marwan, N.; Thiel, M.; Nowaczyk, N.R. Cross Recurrence Plot Based Synchronization of Time Series. *Nonlinear Process. Geophys.* **2002**, *9*, 325–331. [CrossRef]
47. Hirata, Y.; Aihara, K. Identifying hidden common causes from bivariate time series: A method using recurrence plots. *Phys. Rev. E* **2010**, *81*, 016203. [CrossRef] [PubMed]
48. Astakhov, S.V.; Dvorak, A.; Anishchenko, V.S. Influence of chaotic synchronization on mixing in the phase space of interacting systems. *Chaos* **2013**, *23*, 013103. [CrossRef] [PubMed]
49. Konvalinka, I.; Xygalatas, D.; Bulbulia, J.; Schjodt, U.; Jegindø, E.M.; Wallot, S.; Van Orden, G.C.; Roepstorff, A. Synchronized arousal between performers and related spectators in a fire-walking ritual. *Proc. Natl. Acad. Sci. USA* **2011**, *108*, 8514–8519. [CrossRef] [PubMed]
50. Goswami, B.; Ambika, G.; Marwan, N.; Kurths, J. On interrelations of recurrences and connectivity trends between stock indices. *Physica A* **2012**, *391*, 4364–4376. [CrossRef]
51. Ramos, A.M.T.; Builes-Jaramillo, A.; Poveda, G.; Goswami, B.; Macau, E.E.N.; Kurths, J.; Marwan, N. Recurrence measure of conditional dependence and applications. *Phys. Rev. E* **2017**, *95*, 052206. [CrossRef]
52. Hobbs, B.; Ord, A. Nonlinear dynamical analysis of GNSS data: Quantification, precursors and synchronisation. *Prog. Earth Planet. Sci.* **2018**, *5*, 36. [CrossRef]
53. Godavarthi, V.; Pawar, S.A.; Unni, V.R.; Sujith, R.I.; Marwan, N.; Kurths, J. Coupled interaction between unsteady flame dynamics and acoustic field in a turbulent combustor. *Chaos* **2018**, *28*, 113111. [CrossRef]
54. Schinkel, S.; Zamora-López, G.; Dimigen, O.; Sommer, W.; Kurths, J. Functional network analysis reveals differences in the semantic priming task. *J. Neurosci. Methods* **2011**, *197*, 333–339. [CrossRef]
55. Rangaprakash, D. Connectivity analysis of multichannel EEG signals using recurrence based phase synchronization technique. *Comput. Biol. Med.* **2014**, *46*, 11–21. [CrossRef]
56. Izhikevich, E.M. Neural exciability, spiking and bursting. *Int. J. Bifurc. Chaos* **2000**, *10*, 1171–1266. [CrossRef]
57. Prescott, S.A.; De Koninck, Y.; Sejnowski, T.J. Biophysical basis for three distinct dynamical mechanisms of action potential initiation. *PLoS Comput. Biol.* **2008**, *4*, e1000198. [CrossRef] [PubMed]
58. Orr, D.; Ermentrout, B. Synchronization of oscillators via active media. *Phys. Rev. E* **2019**, *99*, 052218. [CrossRef] [PubMed]
59. Attwell, D. Interaction of low frequency electric fields with the nervous system: The retina as a model system. *Radiat. Prot. Dosim.* **2003**, *106*, 341–348. [CrossRef] [PubMed]
60. Bédard, C.; Kröger, H.; Destexhe, A. Model of low-pass filtering of local field potentials in brain tissue. *Phys. Rev. E* **2006**, *73*, 051911. [CrossRef] [PubMed]
61. Modolo, J.; Thomas, A.; Stodilka, R.; Prato, F.; Legros, A. Modulation of neuronal activity with extremely low-frequency magnetic fields: Insights from biophysical modeling. In Proceedings of the IEEE Fifth International Conference on Bio-Inspired Computing: Theories and Applications (BIC-TA 2010), Changsha, China, 23–26 September 2010.
62. Yi, G.S.; Wang, J.; Han, C.X.; Deng, B.; Wei, X.L. Spiking patterns of a minimal neuron to ELF sinusoidal electric field. *Appl. Math. Model.* **2012**, *36*, 3673–3684. [CrossRef]
63. Eckmann, J.P.; Oliffson Kamphorst, S.; Ruelle, D. Recurrence Plots of Dynamical Systems. *Europhys. Lett.* **1987**, *4*, 973–977. [CrossRef]

64. Matassini, L.; Kantz, H.; Hołyst, J.A.; Hegger, R. Optimizing of recurrence plots for noise reduction. *Phys. Rev. E* **2002**, *65*, 021102. [CrossRef]
65. Marwan, N. How to avoid potential pitfalls in recurrence plot based data analysis. *Int. J. Bifurc. Chaos* **2011**, *21*, 1003–1017. [CrossRef]
66. beim Graben, P.; Hutt, A. Detecting Recurrence Domains of Dynamical Systems by Symbolic Dynamics. *Phys. Rev. Lett.* **2013**, *110*, 154101. [CrossRef]
67. Vega, I.; Schütte, C.; Conrad, T.O.F. Finding metastable states in real-world time series with recurrence networks. *Phys. A* **2016**, *445*, 1–17. [CrossRef]
68. Kraemer, K.H.; Donner, R.V.; Heitzig, J.; Marwan, N. Recurrence threshold selection for obtaining robust recurrence characteristics in different embedding dimensions. *Chaos* **2018**, *28*, 085720. [CrossRef] [PubMed]
69. Andreadis, I.; Fragkou, A.; Karakasidis, T. On a topological criterion to select a recurrence threshold. *Chaos* **2020**, *30*, 013124. [CrossRef] [PubMed]
70. Kraemer, K.H.; Datseris, G.; Kurths, J.; Kiss, I.Z.; Ocampo-Espindola, J.L.; Marwan, N. A unified and automated approach to attractor reconstruction. *New J. Phys.* **2021**, *23*, 033017. [CrossRef]
71. Kasthuri, P.; Pavithran, I.; Krishnan, A.; Pawar, S.A.; Sujith, R.I.; Gejji, R.; Anderson, W.; Marwan, N.; Kurths, J. Recurrence analysis of slow–fast systems. *Chaos* **2020**, *30*, 063152. [CrossRef]
72. Zbilut, J.P.; Webber, C.L., Jr. Embeddings and delays as derived from quantification of recurrence plots. *Phys. Lett. A* **1992**, *171*, 199–203. [CrossRef]
73. Ivanchenko, M.V.; Osipov, G.V.; Shalfeev, V.D.; Kurths, J. Phase synchronization in ensembles of bursting oscillators. *Phys. Rev. Lett.* **2004**, *93*, 134101. [CrossRef]
74. Rosenblum, M.G.; Pikovsky, A.S.; Kurths, J. Phase synchronization of chaotic oscillators. *Phys. Rev. Lett.* **1996**, *76*, 1804. [CrossRef]
75. Theiler, J. Spurious dimension from correlation algorithms applied to limited time-series data. *Phys. Rev. A* **1986**, *34*, 2427–2432. [CrossRef]
76. Amari, S. *Information Geometry and Its Applications*; Springer: Tokyo, Japan, 2016; Volume 194. [CrossRef]
77. Pollard, D. Densities and derivatives. In *A User's Guide to Measure Theoretic Probability*; Cambridge University Press: Cambridge, UK, 2001; pp. 53–76. [CrossRef]

MDPI

Article

Alternate Entropy Computations by Applying Recurrence Matrix Masking

Charles L. Webber, Jr.

Department of Cell and Molecular Physiology, Health Sciences Campus, Loyola University Chicago, 2160 South First Avenue, Maywood, IL 60153, USA; cwebber@luc.edu; Tel.: +1-708-638-7497

Abstract: In practicality, recurrence analyses of dynamical systems can only process short sections of signals that may be infinitely long. By necessity, the recurrence plot and its quantifications are constrained within a truncated triangle that clips the signals at its borders. Recurrence variables defined within these confining borders can be influenced more or less by truncation effects depending upon the system under evaluation. In this study, the question being asked is what if the boundary borders were tilted, what would be the effect on all recurrence variables? This question was prompted by the observation that line entropy values are maximized for highly periodic systems in which the infinitely long line elements are truncated to different unique lengths. However, by redefining the recurrence plot area to a 45-degree tilted box within the triangular area, the diagonal lines would consequently be truncated to identical lengths. Such masking would minimize the line entropy to 0.000 bits/bin. However, what new truncation influences would be imposed on the other recurrence variables? This question is examined by comparing recurrence variables computed with the triangular recurrence area versus boxed recurrence area. Examples include the logistic equation (mathematical series), the Dow Jones Industrial Average over a decade (real-word data), and a square wave pulse (toy series). Good agreement among the variables in terms of timing and amplitude was found for most, but not all variables. These important results are discussed.

Keywords: nonlinear dynamics; recurrence quantifications; line entropy; recurrence matrix masking

Citation: Webber, C.L., Jr. Alternate Entropy Computations by Applying Recurrence Matrix Masking. *Entropy* **2022**, *24*, 16. https://doi.org/10.3390/e24010016

Academic Editors: Franco Orsucci and Wolfgang Tschacher

Received: 31 August 2021
Accepted: 20 December 2021
Published: 23 December 2021

1. Introduction

As an extension of recurrence plots [1], recurrence quantification analysis (RQA) was introduced by Zbilut and Webber [2,3] almost three decades ago and Marwan et al. [4] two decades ago. These quantifications include eight recurrence variables extracted from recurrence plots [5] which have proven to have utility in general-purpose data analyses for linear and nonlinear systems alike [6]. The most challenging aspect of recurrence analyses is the setting of the multiple recurrence parameters [7]. What may be less appreciated is the effect border truncations have on all recurrence variables.

The fundamental feature of recurrence plots, which distinguishes deterministic from stochastic systems, is the presence of diagonal line structures. Since the traditional recurrence plot is symmetrical on either side of the line of identity (LOI), only the upper-triangular half of the plot is utilized. The ubiquitous LOI is excluded. Three recurrence quantification variables are directly derived from these diagonals. First, percent determinism (DETERM) is defined as the ratio of points forming diagonal line structures to the total number of recurrent points. Second, the diagonal maximum (DMAX) is defined as the integer number of points constructing the longest diagonal line. Third, information or line entropy (ENT) is defined as the Shannon entropy [8] of the histogram distribution of all diagonal line lengths within the triangular recurrence window. Because long diagonal lines are necessarily truncated at the borders of the recurrence plot, the question arises how much the truncation effect influences the accuracy of DETERM, DMAX and ENT computations, not to mention the remaining recurrence variables. This is the same question formerly asked by Kraemer and Marwan [9].

This paper will focus on the effect recurrence borders have on the computation of line structures and other features of the recurrence plot. Two types of borders are examined: traditional triangular recurrence borders and novel tilted box recurrence borders. The latter represents a smaller area masked off within the larger triangular area. System studied include a mathematical system, a real-world financial system, and a contrived toy system to clearly illustrate time and amplitude shifts in all compared variables.

2. Disparity between Recurrence Line Entropy Values and Lyapunov Exponents

When this author [10] was studying the logistic equation by recurrence analysis, it was noted that when the equation was in period 1, period 2, period 4, etc., periodic modes, the line entropy values were maximized when the Lyapunov exponents were low. This counterintuitive relationship is easily explained by the differing lengths of diagonal lines which were truncated by the triangular borders into unique lengths. Censi et al. [11] also realized this effect and introduced a correction by assigning all diagonals to the same length as the central LOI. This method effectively minimized the line entropy values but carried with it the assumption that all lines are of equal length. Eroglu et al. [12] took a different approach using weighted recurrence plots. Entropic measures were redefined according to the distribution of return times, not line lengths. Indeed, Shannon entropy values became positively correlated with the largest Lyapunov exponent. More sophisticated approaches for quantifying recurrence entropy are based on matrix microstates [13] and categorical time series [14]. Finally, Kraemer and Marwan [9] described another method similar to the one introduced in this paper. That is they recomputed entropy by masking the recurrence plot such that the square recurrence matrix was windowed within a diamond pattern overlaying the traditional plot. Thus, for fully periodic systems, all diagonal lines were truncated to identical lengths, minimizing line entropy values to 0.000 bits/bin.

Because the auto-recurrence plot is symmetrical, in this paper, only the upper triangle was masked with the best fitting square (not half-diamond or rectangle as in the Kraemer-Marwan case). The box-masking will be fully described. The masking gives low entropy values for periodic systems and high entropy values for chaotic systems. It was necessary to compare these boxed entropy values against the traditional triangular entropy values for several different cases and situations. Indeed, performance testing was accomplished for all eight recurrence variables. The necessary comparisons include amplitude and timing characteristics of the recurrence variables for both triangular and boxed recurrences.

3. Redefinition of Recurrence Plot Boundary Conditions

The boxed recurrence masking can be best be illustrated graphically. Figure 1 presents the 400-point recurrence plot of a simple repeating sinewave consisting of 16 cycles or waves as computed by traditional recurrence programs RQD and RQC, and new boxed programs RQDB and RQDC (see Appendix A). Because all 16 sinewaves are identical across the time series (green), 15 border-to-border lines excluding the LOI (black) are inscribed within the triangular recurrence plot (blue and red). Each one of these deterministic lines is truncated to a unique (different) length spanning from the west border to the north border. That is, the further the line is removed from the LOI, the shorter that line is. This has ramifications for line entropy calculations as described below.

Another approach to this sinewave truncation issue is to fashion a tilted box within the recurrence triangle parallel to the LOI (45-degree, clockwise tilt) as is illustrated by the black square (Figure 1). All recurrent points within this smaller boxed area (red) are simply shorter parts of the longer lines within the larger triangular area (blue). The advantage of this system is that the 10 diagonal line structures within the box are each truncated to identical lengths.

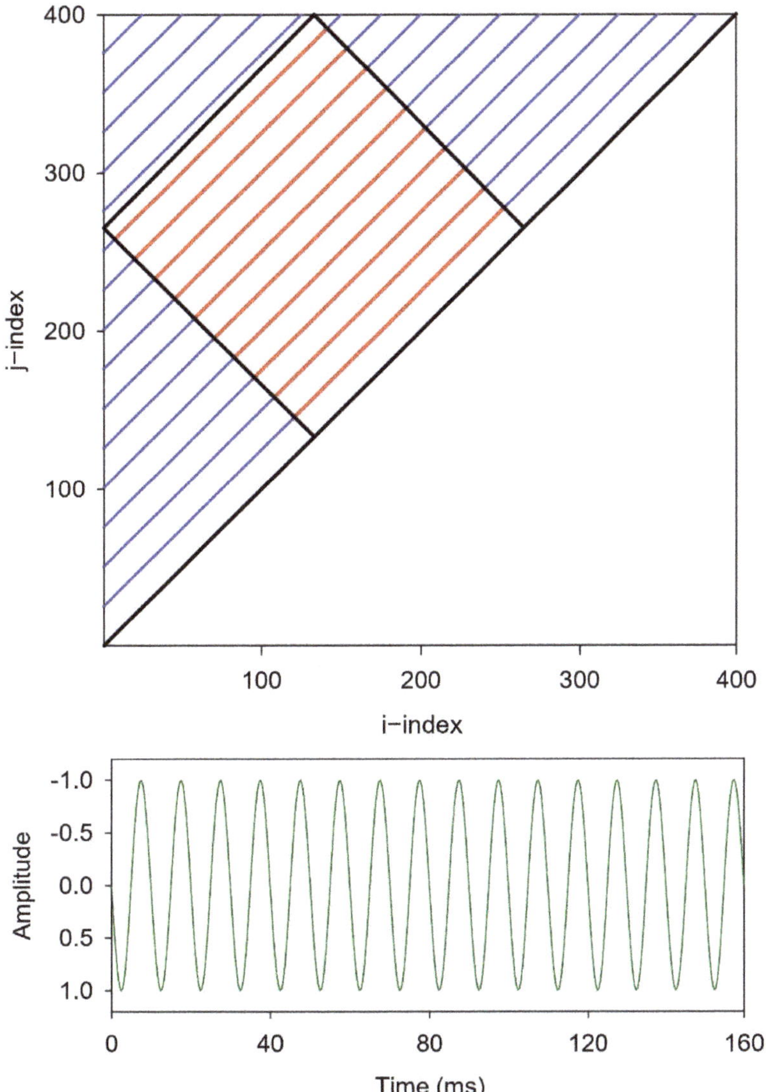

Figure 1. Recurrence plot of a sinewave with no added noise. The 16 noise-free sinewave cycles represent a 100 Hz sinewave digitized at 2500 Hz (25 points/cycle). The boxed recurrences (red lines) are constrained within the tilted box. The triangular recurrences (blue lines + red lines) are constrained with the upper half of the recurrence plot (above the LOI). Parameter settings: DELAY = 1; EMBED = 2; NORM = Euclid; WINDOW = 400; RESCALE = max distance; RADIUS = 1%; LINE = 2.

Then, when 5% random noise is added to the pure sinewave, the long diagonal line structures are chopped up into shorter segments as shown in Figure 2. In this case, the noise mitigates the influence of the truncation effects of the triangular borders insofar that the line structures within the tilted box and triangle appear very similar. Are they in reality? By the way, the embedding dimension was set to 2 for both the pure sinewave with a radius of 1% of the maximum distance. Using these parameter settings there was no false thickening of the diagonal lines as discussed by Thiel et al. [15]. That is, the selected radius was high enough to include the noisy points, but not so high as to thicken the lines.

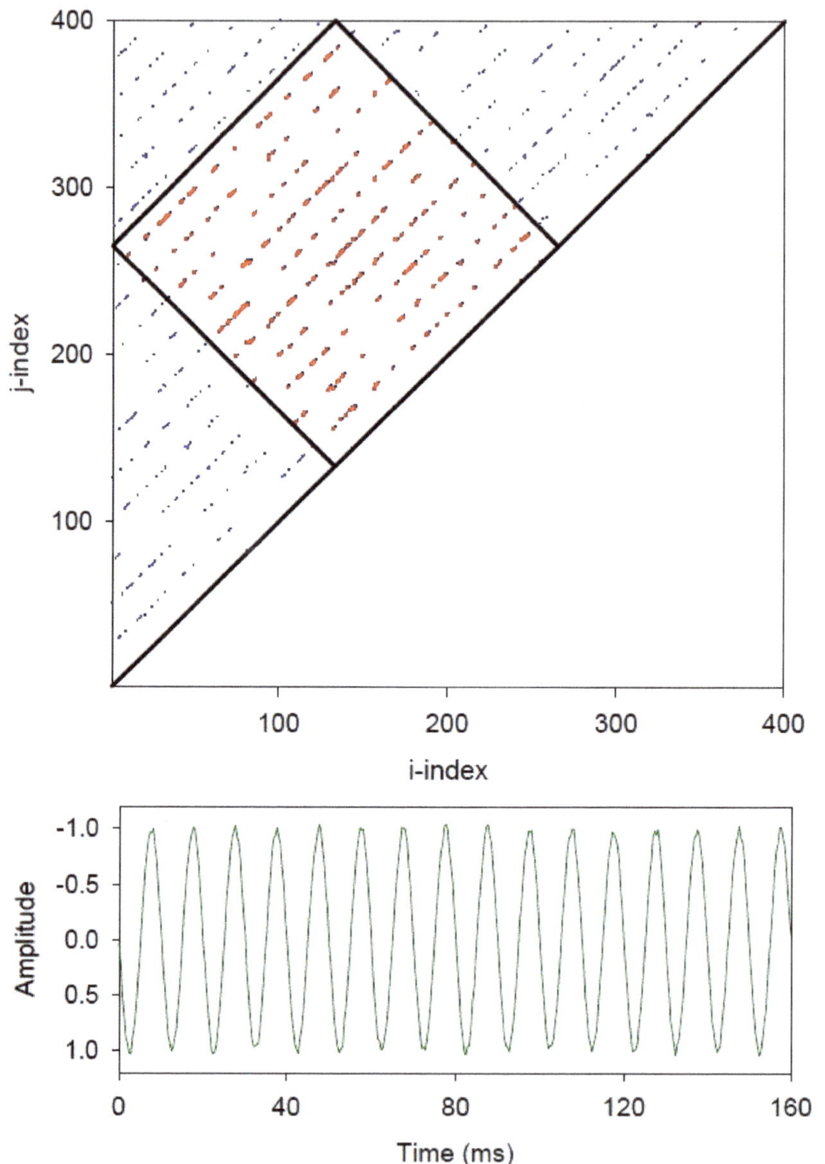

Figure 2. Recurrence plot of a sinewave with 5% added noise. The 16 noise-free sinewave cycles represent a 100 Hz sinewave digitized at 2500 Hz (25 points/cycle). The boxed recurrences (red lines) are constrained within the tilted box. The triangular recurrences (blue lines + red lines) are constrained with the upper half of the recurrence plot (above the LOI). DELAY = 1; EMBED = 2; NORM = Euclid; WINDOW = 400; RESCALE = max distance; RADIUS = 1%; LINE = 2.

4. Tilted Box Boundaries and Line Entropy Modifications

Figure 3 is key to understanding the differences of the boxed entropies over the triangular entropies computed from the histogram distribution of diagonal lines within the recurrence plot. The upper histogram is the distribution of diagonal lines within the noise-free sinewave (Figure 1) and the lower histogram is the distribution of diagonal lines within the sinewave with 5% added noise (Figure 2). Note that both horizontal and vertical scales are base-10 logarithmic. In the first case (blue lines), all 15 line lengths are of differing lengths (375, 350, 325, 300, ..., 100, 75, 50 25 points) and there is but one unique line length of each. However, for the second case (red line), the 10 diagonal lines are all of identical length (133 points) and fall into a single histogram bin. Consequently, the triangular entropy computes as 3.907 bits/bin, whereas the box entropy computes as 0.000 bits/bin (see below).

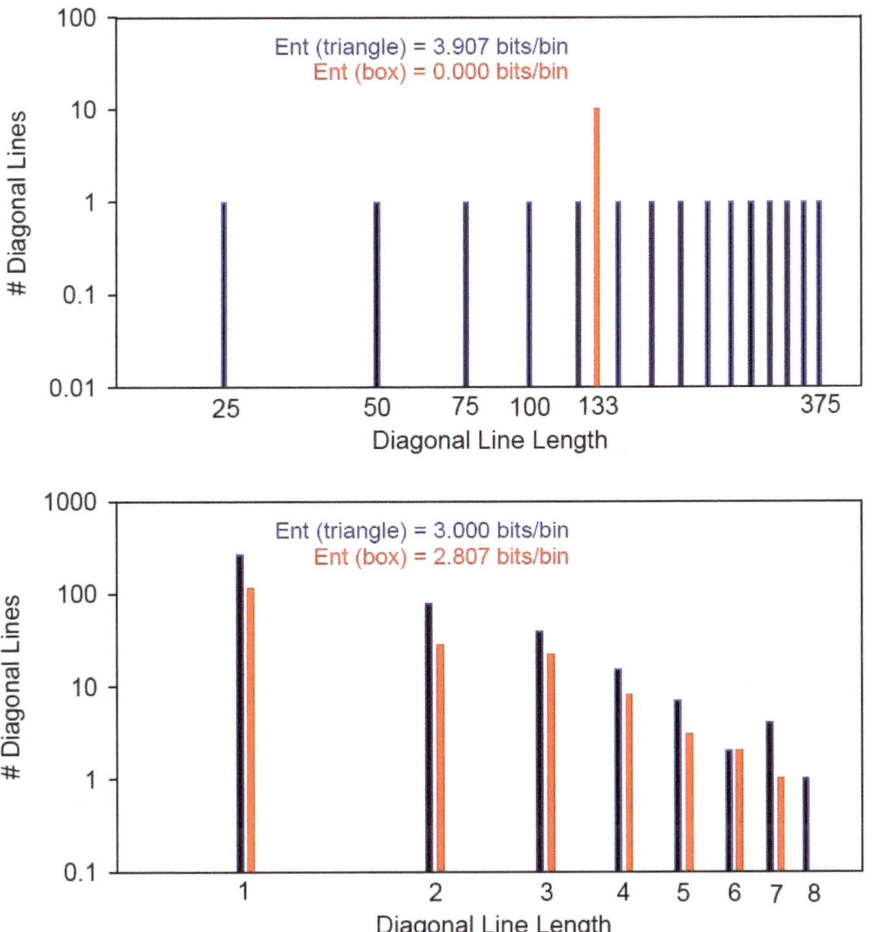

Figure 3. Histograms of diagonal line length distributions for noise-free sinewave (upper panel) and sinewave with 5% added noise (lower panel). Diagonal line lengths and their counts are shown for triangular recurrences (blue) versus boxed recurrences (red). Corresponding, color-coded entropy values are also given. Note that entropy values in the top panel are very different, but entropy values in the lower panel are very similar.

Computation of line entropy follows the generalized formula used for Shannon information entropy [8] as shown in Equation (1). The entropy value is maximized when each non-zero bin has the same number of counts (identical probabilities) as computed by Equation (2). Conversely, the entropy value is minimized if bin counts are restricted to a single bin. In this case, the entropy falls out as 0.000 bits/bin as given by Equation (3). So from Figure 3, where 15 bins are each filled with the count of one (blue) the entropy maximized ($-\log_2(1/15) = 3.907$ bits/bin). However, for the 10 boxed lines (red), each line is 133 units long and the entropy is minimized ($-\log_2(1/1) = 0.000$ bits/bin).

$$\text{ENT}_{\text{gen}} = -\Sigma(\text{P}_{\text{bin}})\log_2(\text{P}_{\text{bin}}) \tag{1}$$

$$\text{ENT}_{\text{max}} = -\log_2(1/\text{N}_{\text{bin}}) \tag{2}$$

$$\text{ENT}_{\text{min}} = -\log_2(1/1) = 0.000 \text{ bits/bin} \tag{3}$$

So for noise-free sinewaves, at least, the computed values for line entropy range from maximal entropy (triangular recurrence area) to minimal entropy (tiled box recurrence area). It is the tilted box masking that makes all the difference. Both entropy values are mathematically correct, but the boxed entropy makes more sense with respect to the complexity of the signal. This is not globally true for all situations and signals.

With the jostling of the pure sinewave with 5% random noise, the long diagonal lines both within the boxed area and within the triangular area get parceled (chopped up) into shorter segments (Figure 2). Quantitatively, the distributions of these shortened line segments are very similar (Figure 3), which is why the information entropy values for both the triangular recurrences and boxed recurrences are likewise very similar (3.000 versus 2.807 bits/bin, respectively). Since noise is ubiquitous in real-world systems, possibly the two methods of entropy computations are not that much different after all. However, such a (good) conclusion has yet to be verified using experimental data.

5. Boundary Details of the Tilted Recurrence Box within the Recurrence Triangle

The edges of the recurrence box are not smooth, and the shape of the box is not perfectly square as implied by Figures 1 and 2. For example, let us construct the largest 5×5 tilted square within a 14×14 recurrence matrix, half of which is shown in the upper panel of Figure 4. Here, the edges of the tilted square box are marked with black dots. As can be seen, each side of the box is 5 units in length. However, the area of this square is not 25 square units (dark pink pixels) but must include all gap spaces as well (light pink pixels). Secondly, the length of the diagonal must all be the same unit length (5 pixels here) increasing the box area to 45 square units (Equation (4)). Thirdly, the area of the triangle includes the box area plus all other empty pixels (white pixels) excluding the LOI (green pixels) (Equation (5)). Taking the ratio of box to rectangle areas computes the percentage of the recurrence plot occupied by the tilted box (Equation (6)).

$$\text{Area}_{\text{box}} = \text{Side}_{\text{box}} \cdot \text{Side}_{\text{box}} + \text{Side}_{\text{box}} \cdot (\text{Side}_{\text{box}} - 1) \tag{4}$$

$$\text{Area}_{\text{triangle}} = (\text{Side}_{\text{triangle}} \cdot \text{Side}_{\text{triangle}} - \text{Side}_{\text{triangle}})/2 \tag{5}$$

$$\%\text{Box} = 100 \cdot \text{Area}_{\text{box}}/\text{Area}_{\text{triangle}} \tag{6}$$

Now, if the 14×14 recurrence matrix size is increased to 15×15 square units or even to 16×16 square units, the shape and area of the tilted recurrence box as defined above remain the same. This is proven graphically in panels 2 and 3, respectively, in Figure 4. At the same time, however, the area of the triangles increases as more empty pixels are added. Consequently, the ratios of box area to triangle area (%Box) must necessarily decrease. For this trio of pairs (Figure 4) the ratios decrease from 49.5% to 42.9% to 37.5% as indicated. However, by incrementing the size of the recurrence matrix up by one step to 17×17 square units, the legal tilted-box size likewise increases from 5×5 to 6×6 (not shown).

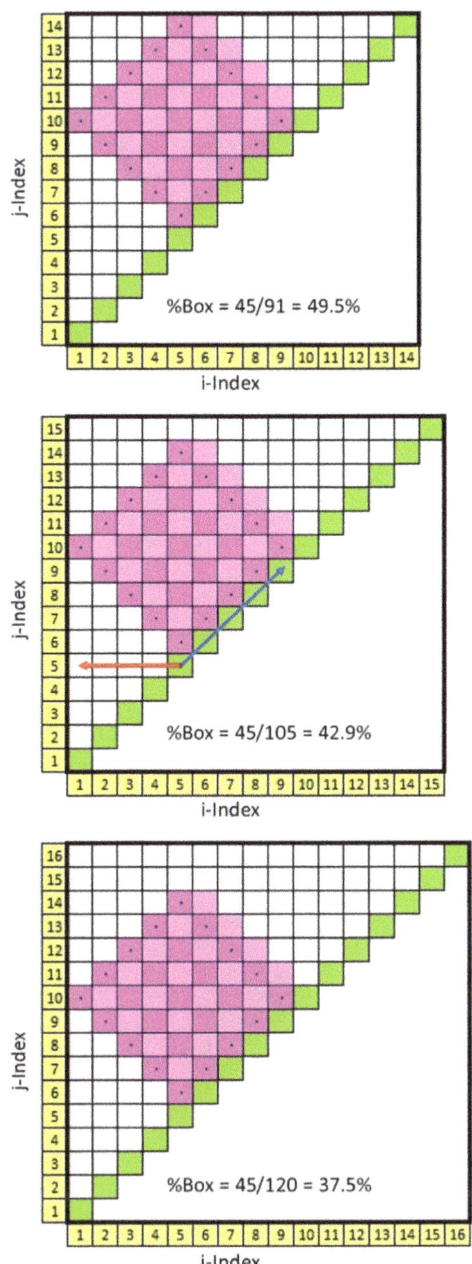

Figure 4. Tilted recurrence box of 5 × 5 units (dark and light pink pixels) fits within 3 recurrence matrices of 14 × 14 units (top), 15 × 15 units (middle), and 16 × 16 units (bottom). %Box is the ratio of the box areas to the triangle areas (excluding the green LOI) which progressively decreases as the size of the recurrence matrix increases. The blue arrow designates the middle third of five input points (e.g., P5–P9) from which the mean and standard deviation are computed. The red arrow indicates the left shift in the boxed recurrence variables by 4 points (P5–P1) for alignment with the triangular recurrence variables.

Clearly, the "square" tilted boxes can fit within 3threeincremental sizes of recurrence matrices. To compute the integer length of the box side one needs only to divide the number of points in the triangular recurrence window by 3. Thus, for our trio example: 14/3 = 4.667 (round up to 5); 15/3 = 5.000 (retain as 5); 16/3 = 5.333 (round down to 5). Now the question arises, what happens to %Box (area ratios) as the number of points in the recurrence window increases? This answer is shown graphically in Figure 5 for multiple sets of triplets. The 5 × 5 example is indicated in red. The slope of each trio decreases as the number of points increases. For example, for set 1001-1002-1003, the area ratios are 45.51%, 44.42% and 44.33%; for set 2000-2001-2002, the area ratios are 44.48%, 44.43% and 44.39%; for set 2999-3000-3001, the area ratios are 44.47%, 44.44% and 44.41%. In the limit for large set numbers, the area ratios converge on 44.44% for all trios. All this means is that the area of the new tilted box is approximately 44% of the area of the traditional recurrence triangle.

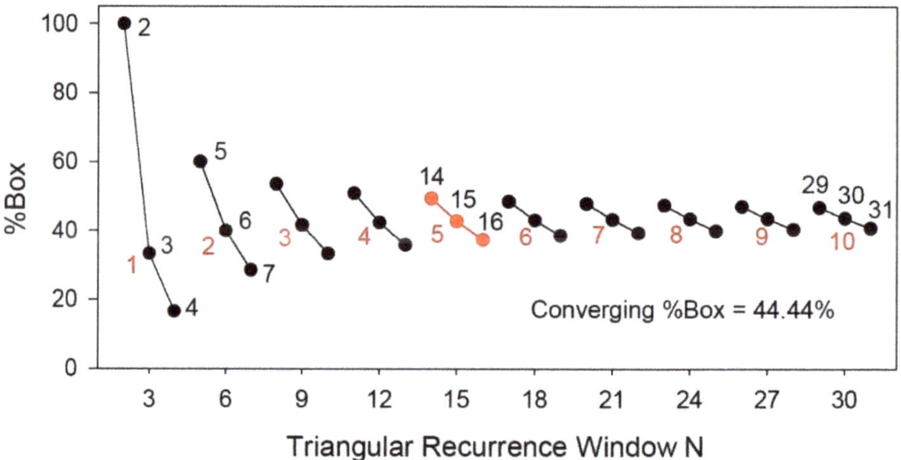

Figure 5. Ratio of tilted-box area to recurrence triangular area (%Box) in sets of 3 for increasing number of points in triangular recurrence window. Box sizes are shown in red integers. The triplet points (14-15-16) for the 5 × 5 box size as described in the text are also shown in red.

6. Logistic Equation

At least three studies have examined the logistic map in terms of entropy over a range of control parameter a values [10,12,16] (Equation (7)). Each takes a different perspective on entropy calculations with pros and cons. The present study followed the procedures of the earlier study [10] in which parameter a was incremented on each iterated cycle in steps of 0.00001 from a = 2.8 to 4.0 yielding 120,001 points. This series of points was subjected to 800-point moving-window recurrence analyses using the traditional recurrence and boxed recurrence programs (program3 RQE and RQEB). The results for all eight recurrence variables are superimposed in Figure 6 with blue representing the traditional values and red indicating the boxed values. As shown by the tilted blue arrow in Figure 4, the mean and standard deviation values were computed from the middle third of time series points. Additionally, as shown by the horizontal red arrow in Figure 4, the boxed recurrence variables were shifted left by one-third of the window size (267 points = 800 points/3) to better align with the triangular recurrence variables.

$$X_{n+1} = a \cdot X_n \cdot (1 - X_n) \quad (7)$$

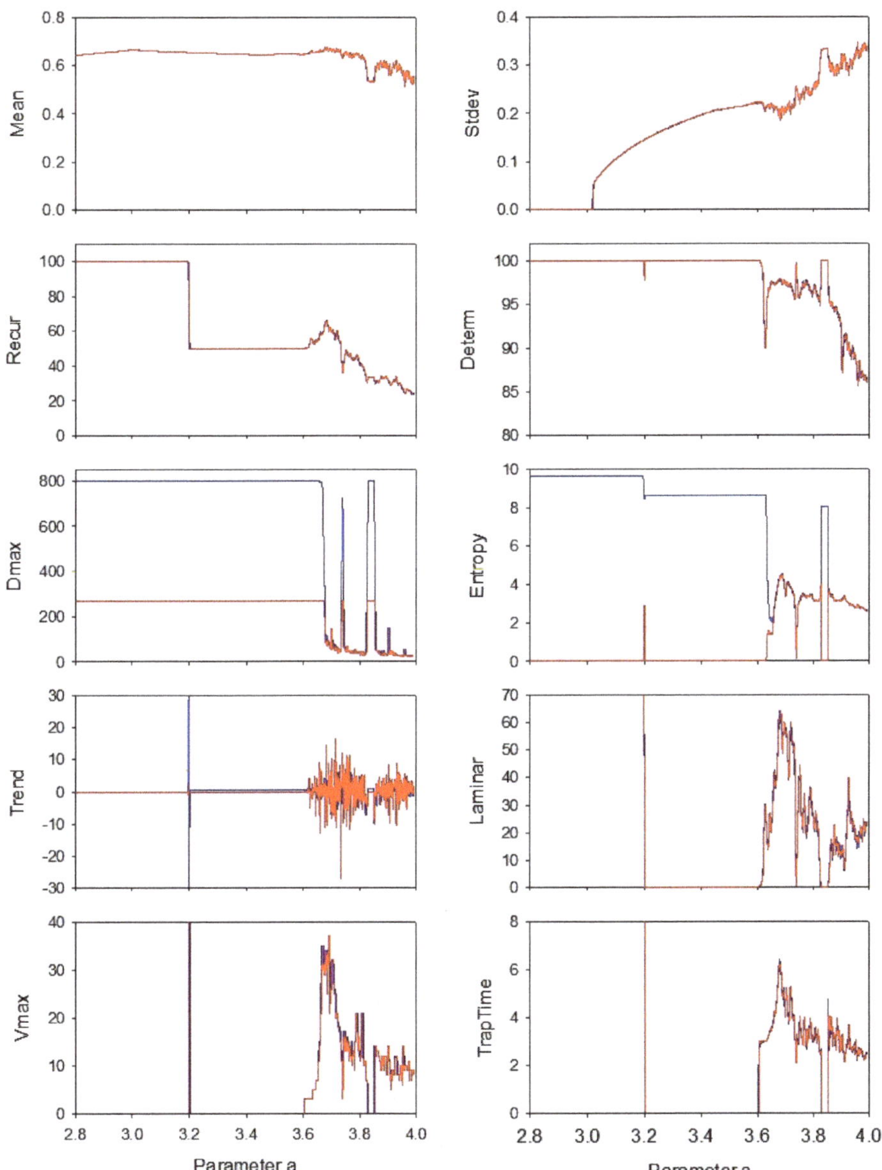

Figure 6. Recurrence variables of the logistic map computed by traditional RQA (blue lines) and boxed RQA (red lines). The logistic time series consisted of 120,001 points for a values ranging from 2.8 to 4.0. Each sliding window epoch consisted of 800 points which were shifted by 10 points (98.75% overlap) to yield 11,920 points per variable. Superimposed values are shown in red; divergent values are shown wherever blue appears. Parameter settings: DELAY = 1; EMBED = 3; NORM = Euclid; WINDOW = 800; SHIFT = 10; RESCALE = absolute; RADIUS = 1; LINE = 2.

In most cases, the red traces overwrite the blue traces, proving excellent agreement with both approaches. However, there are three notable exceptions. First, the ENT values are widely divergent when the logistic system was in its fully periodic modes (period 1, period 2, period 4, period 8, etc.) over the approximate range of a = 2.8 to 3.6 with spikes just over a = 3.2. Additionally, during the period 3 window (a = 3.83) the 2 entropy values diverge.

Second, the DMAX values superimposed nicely, save during the sliding periodic windows. Because the window size was selected as 800 points, the traditional DMAX peaks at 799 diagonal points (just next to the LOI). However, the boxed window was only one-third the size of 800 which is why the boxed DMAX is clipped at 267 diagonal points.

7. Radius versus Embedding

In RQA, one of the most difficult parameters to select is the radius. If the radius is too low, the number of recurrent points will be too sparse; if radius is too high, the number of recurrence points will be too dense. Indeed, when the radius equals or exceeds the maximum distance in the distance matrix, the recurrence plot will completely saturate (RECUR = near 100% and DETER = near 100% for LINE = 2). Taking advantage of this principle, it is possible to increment the radius and see at which point the entropy values for the triangular RQA and boxed RQA diverge. This was accomplished by using the new boxed RQA program RQSB. Secondly, these divergent points must be a function of the embedding dimension. To quantify these ideas, a single 800-point window was selected from the logistic map (a = 3.9074 to 3.9873) as shown in Figure 7. This window represents a purely chaotic window with no regular periodicities present. The signal looks stochastic, but because it is derived by iterating a nonlinear equation it is actually fully deterministic. This does not mean that the DETERM is 100%, however, because the various orbits are not equally close to one another. Increasing the radius, includes more and more orbits within the recurrence zone.

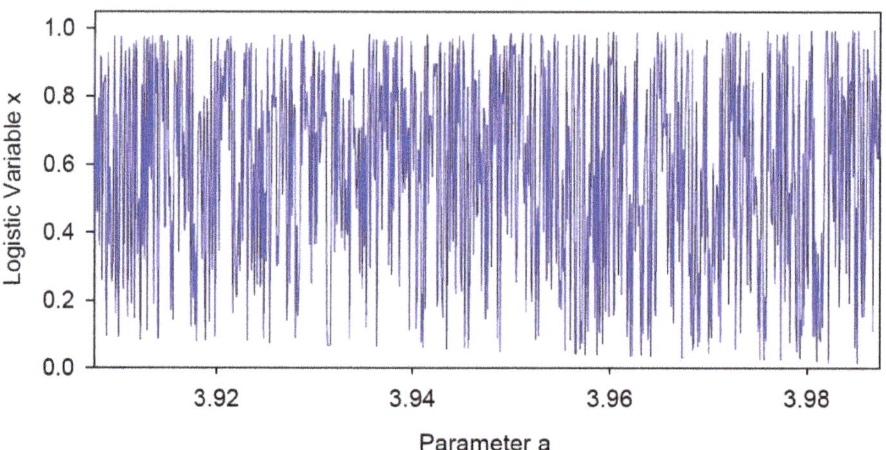

Figure 7. Fully chaotic mode of the logistic equation within an 800-point window (a = 3.9074 to 3.9873). The logistic x variable fluctuates wildly within the bounds of 0.0 (minimum) to 1.0 (maximum).

The computational results are shown in Figure 8, where traditional entropy calculations are displayed by the blue lines and the boxed entropy calculations are displayed by the red lines (programs RQS and RQSB). As can be seen, both entropy values track very closely until the density of the recurrence plots becomes too great at very high radius values. At these turning points the two entropy values begin to diverge as identified by the black dots in each trace. As the embedding dimension increases the turning point shifts to the left. For the five increasing embedding dimensions, the five decreasing radius values at the turning points are, respectively: RADIUS = 94.94%, 85.44%, 80.95%, 79.04% and 77.86%. At the various turning points, all entropy values, triangular and boxed, are bunched within a rather tight or narrow range (ENT = 7.438 to 7.913 bits/bin) albeit at high values. Additionally, as expected, when the radius value reaches 100% of the maximum distance, the two entropy values depart to their separate corners: regular entropies to their maxima of 9.64 bits/bin; boxed entropies to their minima of 0.00 bits/bin. What can be concluded from these results is that the triangular and boxed entropy values are so similar that the triangular RQA computations for entropy remain valid just as long as the systems under study are not fully periodic in the absence of noise.

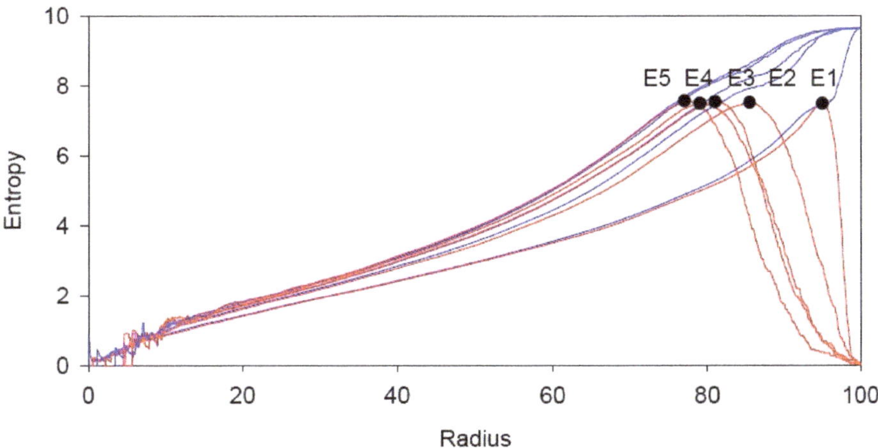

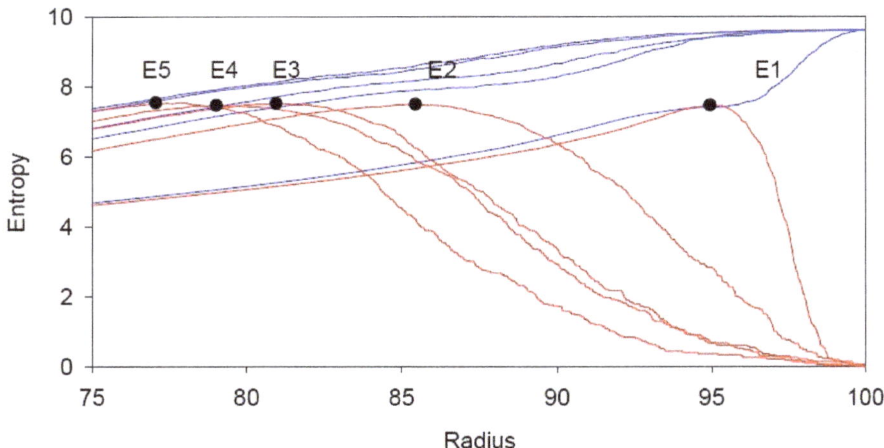

Figure 8. Line entropy values as a function of increasing radius values for five different embedding dimensions (E1, E2, E3, E4 and E5) for regular RQA (blue lines) and boxed RQA (red lines) computations. The black points indicate the radius value when the two entropy values diverge (turning points). The full range of radius values (top traces) are magnified for the upper quartile of radius values (lower traces) to show better discrimination of the parameter settings. Parameter settings: DELAY = 1; EMBED = 5; NORM = Euclid; WINDOW = 500; RESCALE = maximum distance; RADIUS = 10%; LINE = 2.

8. Dow Jones Industrial Averages

As shown above, the noise-free sinewave signal presented above represents an extreme example of divergent entropy computations for verses triangular recurrences verses boxed recurrences (ENT = 3.907 vs. 0.000 bits/bin). Adding noise to the sinewave converges the two entropy values (ENT = 3.000 vs. 2.807 bits/bin). However, what happens to all eight recurrence variables when the recurrence area is constrained (masked) to the tilted box as compared against the standard recurrence triangular recurrence area?

To answer this question, real-world data (noisy) were studied by both recurrence methods and the results were compared. The input time series consisted of downloaded scores of the Dow Jones Industrial Average for over a decade [17]. Recurrence quantifications were computed within a sliding window (programs RQE and RQEB) and the results are shown in Figure 9. Since the window size was 500 points, the length of the box side was 167 points (500/3) and %Box was 44.58% (see Equations (4)–(6)). Again the mean and standard deviation were computed for the middle third of the time series within each window, and the recurrence variables were left time shifted as explained in Figure 4.

1. RECUR: Recurrence rates for triangular and boxed recurrence areas are very comparable, not with exact value matching, but by their directional shifts while moving through the Dow Jones scores.
2. DETERM: Determinism scores are very comparable, quantitatively and qualitatively for the most part, save for dips (red) surrounding trading days 1000 and 2000.
3. DMAX: the longest line lengths with the moving window have minimal agreement. Both series are flat-topped around days 300 to 500 and days 1400 to 1500. The triangular measurement peaks at 499 points, but the boxed measurement peaks at 167 points because its area is lower by 44%.
4. ENTROPY: Despite some variability the two entropy traces track very nicely with each other. In one sense this is proof of concept that the presence of noise in the time series actually has entropy values converging.
5. TREND: This variable quantitates the paling of the recurrence plot away from the LOI. The two traces are almost superimposed save points surrounding trading day 1500.
6. LAMINAR: This variable reports the percentage of recurrent points that form vertical line structures. Qualitatively, the two curves match nicely. Quantitatively, the two curves also match nicely except for dips (red) near days 1000 and 1800.
7. VMAX: This variable is analogous to DMAX, but it measures the length of the longest vertical line. There is some qualitative agreement between both traces, but these results are more difficult to interpret.
8. TRAPTIME: This variable defines the average vertical line lengths which are rather identical for triangular and boxed recurrences.

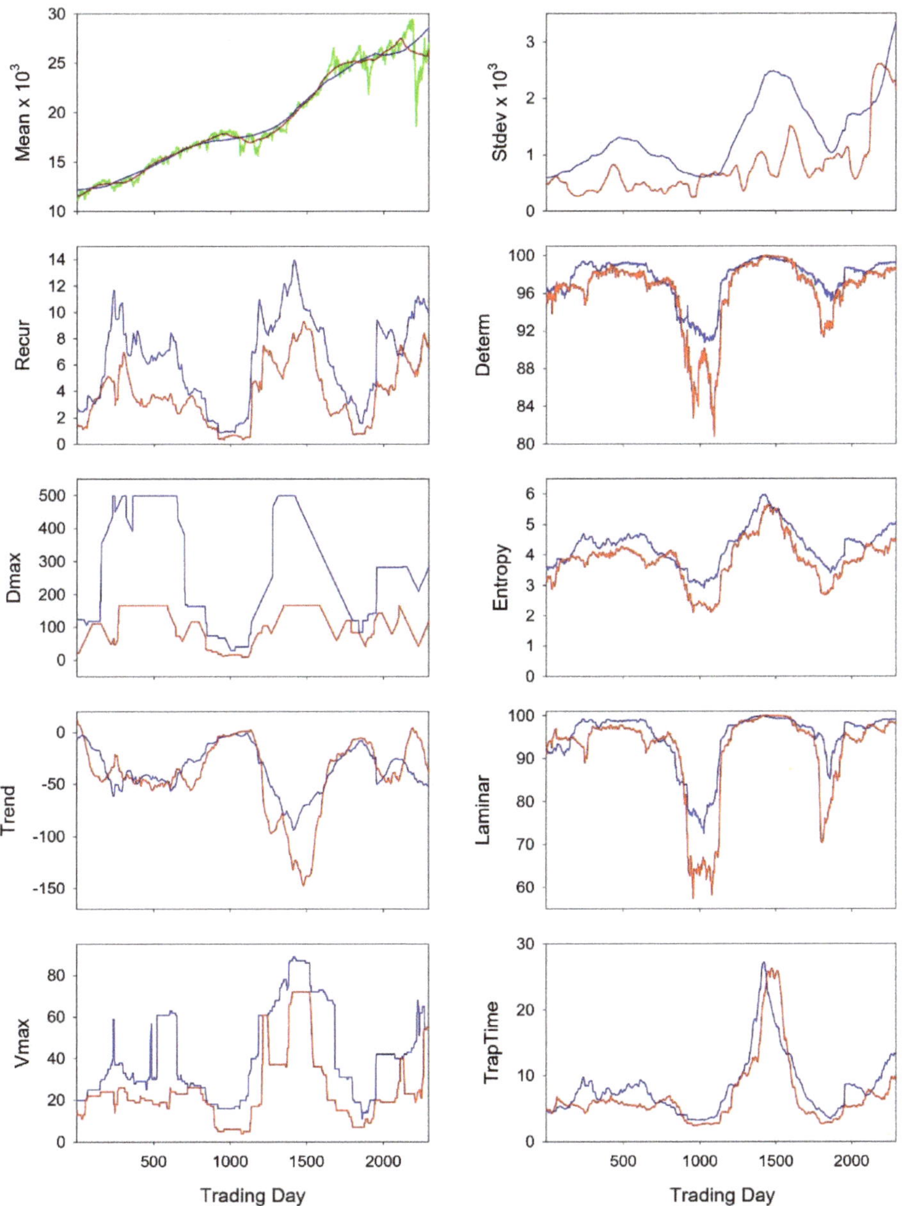

Figure 9. The mean values (left) and standard deviation values (right) are plotted for both the triangular area (blue) and tilted-box area (red). The means line up nicely, but there is less agreement for the standard deviation values. The raw Dow Jones scores are superimposed in the graph of means (green). Second, nonlinear descriptors of the Dow Jones scores are shown in the remaining eight panels (Figure 9). Each recurrence variable will be reviewed one by one, remembering that blue curves refer to triangular recurrences while red curves refer to tilted-box recurrences.

9. Square Wave Pulse

To better understand timing and amplitude relationships among all eight recurrence variables computed by the standard RQA versus the boxed RQA, the two methods were compared using a simple square wave input signal. This signal consisted of 7500 points in which the first 1500 points and last 4000 points were set to 0, but the middle 2000 points were set to 10. The first and second transitions were abrupt (pulse up: 0 to 10; pulse down: 10 to 0, respectively). Figure 10 presents the results of this square wave as processed by both the traditional RAQ and boxed RQA computations for all eight recurrence variables (programs RQE and RQEB). The sliding window approach was taken at the highest sensitivity (only 1 point between epochs) to allow for clear timing delineations. Again It should be noted that the boxed recurrences were aligned with the regular recurrences, plotting both time series from point 1. That is the boxed recurrences were plotted from point 1, not point 167 (or 500/3). Additionally, the double vertical green lines in each panel show the exact onset and offset of the square wave pulse.

The first observation is that width of the means and standard deviations are broader for the standard RQA than for the boxed RQA. This is because the tilted box spans fewer pulse points than the triangular space. Both means start from the same base (0.0) and rise to the same height (10.0), but the rise in the boxed mean begins later and has a greater slope. It can also be noted that the two standard deviations both reach identical maxima.

Second, prior to the onset and offset of the square pulse, the paired recurrence variables have peaks or nadirs that line up in time. Additionally, as is necessary, the interval between peaks or nadirs is exactly 2000 points which is the width of the square wave pulse.

Third, the contours and amplitudes of the paired recurrence variables are not the same. This is explained by the fact that the boxed recurrence area consists of only 55,611 pixels which is but 44.58% of the triangular recurrence area (124,750 pixels) (see Equations (4)–(6)).

Fourth, during the steady-state phases of the input signal (string of 0 s pre- and post-pulse; string of 10 s during the pulse), the regular entropy is pegged at 8.963 bits/bin (incorrect) whereas the boxed entropy is minimized to 0.000 bits/bin (correct). Only during the two transition phases do the two entropy values approach each other, but never meet. With the exception of these two entropy values and possibly the LAMINAR and TRAPTIME variables, the other five RQA variables exhibit similar directional changes during the two transition phases.

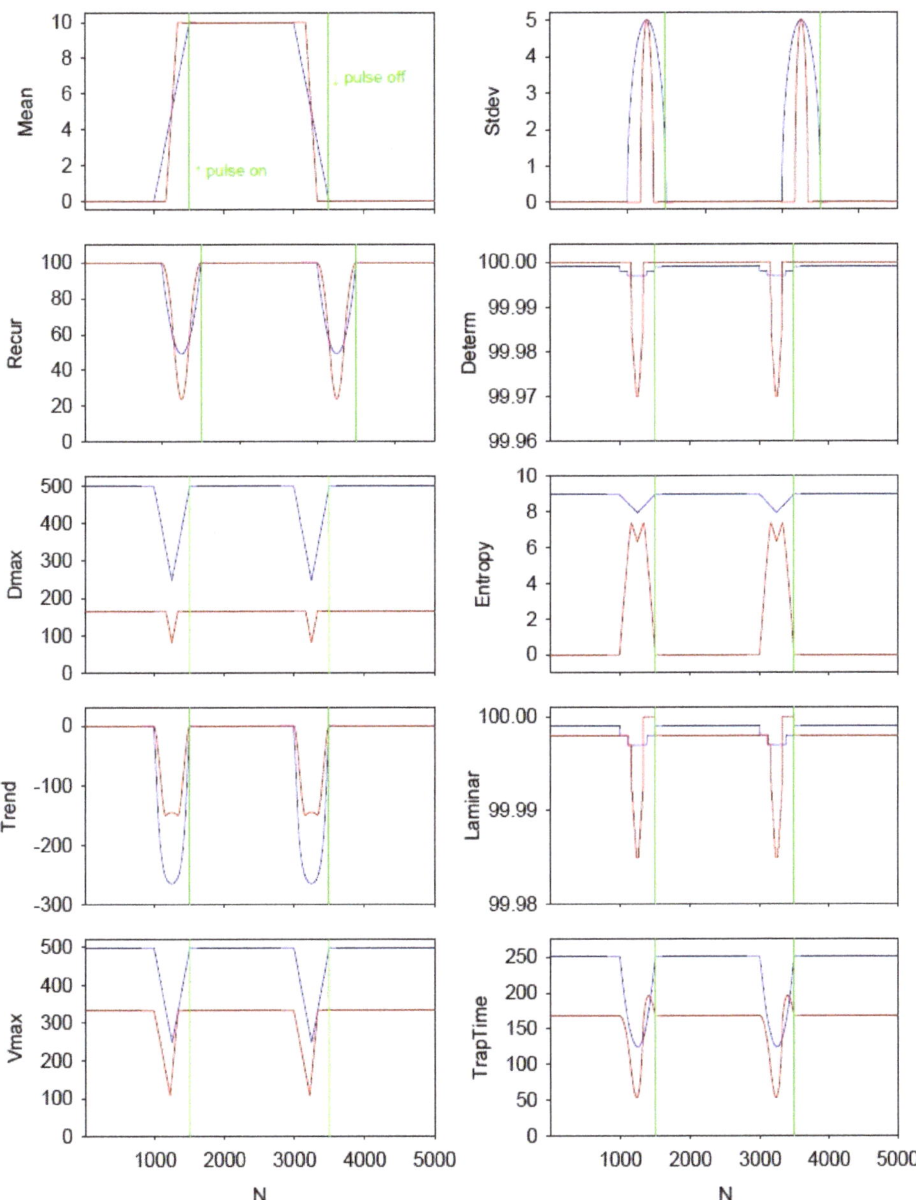

Figure 10. Comparison of traditional and boxed recurrences on a single square wave of 7500 points for 2 linear and 8 nonlinear recurrence variables. Triangular RQA computations (blue traces) and boxed RQA computations (red traces) are performed within a sliding window of 500 points offset by a single point between epochs for a total of 6977 epochs. However, only the first 5000 epochs are plotted. The onset of the rising square wave is indicated by the first vertical green line (pulse on); the end of the falling square wave is indicated by the second vertical green line (pulse off). Both time series were aligned to point 1 (see text). Parameter settings: DELAY = 1; EMBED = 5; NORM = Euclid; WINDOW = 500; SHIFT = 1; RESCALE = maximum distance; RADIUS = 10%; LINE = 2.

10. Discussion and Conclusions

Back in 2015, Marwan and Webber [18] discussed in detail the mathematical and computational foundations of recurrence quantifications. However, no mention was made about masking the recurrence matrix to modify the triangular border truncations of diagonal lines. Later, both of these authors independently posited a tilted rectangle [9] or tilted box (present paper) masking of the recurrence matrix such that long diagonal lines are clipped to identical lengths thereby minimizing the line entropy computations. It should be emphasized that both entropy calculations, triangular entropies and boxed entropies, are accurate, but have different meanings. In the traditional case, line-entropy values really describe the actual probability distributions of truncated lines. However, in the tilted box case, line-entropy values better correspond to the underlying dynamic itself. That is, low entropy values are related to simple (or random) systems, whereas high entropy values are usually correlated with more complex systems, but not necessarily always. As mentioned previously, Censi et al. [11], Eroglu et al. [12], Corso et al. [13] and Leonardi [14] all have thoughts about revised entropy computations. The question is not which method is the best, but rather, based upon the assumptions of each approach, how do the various entropy computations perform on systems of all sorts. This becomes an invitation for performing comparative studies applying these differing methodologies to representative systems of interest (beyond the scope of this paper).

For example, Leonardi [14] has a very nice summary description of the meaning of entropy, especially the information entropy of Shannon [8]. Information entropy values approaching maximum entropy (such as for random systems) contain high levels of information, are complex in nature, and unpredictable in principle. However, information entropy values approaching minimum entropy (such as for highly periodic systems) contain low levels of information, are simple in construct, and can be very predictable. This last statement is qualified for white-noise processes which present with very sort line structures which translate to very low entropy values as well. In short, the higher the entropy value, the more interesting is the system. Conversely, the lower the entropy value, the less interesting, even boring, is the system (excepting white noise). So for the sinewave, which is definitely periodic, very linear, and very repetitive (if you have seen one, you have seen them all), the entropy minimum of 0.000 bits/bin best describes the diagonal-line entropy. The tentative conclusion is that boxed recurrence entropies may have an advantage over triangular recurrence entropies. However, confirmation of such a claim must be verified by other correction schemes [9].

Many observations in this paper suggest that entropy values, if not most other recurrence variables, computed from triangular and boxed recurrences may not necessarily be that different after all. First, the addition of subtle noise to a pure sine wave collapses both entropies to nearly the same values insofar that long diagonal lines are disallowed (Figures 1–3). Second, the logistic equation in its non-periodic and chaotic modes assigns very similar quantitative (overlapping) values to all recurrence variables (Figure 6). Third, both entropies track together for the chaotic state of the Logistic system until breaking points of separation occur only at very high radius values (Figure 8). Additionally, fourth, moving-window recurrence analyses of the Dow Jones Industrial Average data reveal correlations between triangular and boxed recurrence variables that have similar qualitative (REC, DET and LAM) or quantitative (ENT, TREND and TRAPTIME) profiles. With the exception of two poorly correlated variables (DMAX and VMAX), such patterns would lead to singular, not diverse, interpretations of the underlying dynamic (Figure 9).

As an aside, it will be appreciated that the two tilted maskings of the recurrence matrix are qualitatively similar for both Kramer and Marwan [9] and the present paper. What distinguishes the two approaches quantitatively, however, is that the area of the masked rectangular recurrence plot is a full 50% in the limit of the triangular matrix whereas the masking of the tilted box is only 44%. This 8% difference may constitute an advantage for the former insofar that it captures a greater proportion of recurrent points. This slight edge may be a fine tune difference between masked versus unmasked recurrence plots and quantifications. This hypothesis could be studied using various signals.

Finally, the moving-window recurrence analyses of the square wave pulse are provided basically for heuristic purposes alone (Figure 10). The input signal is very clean (noise-free) and two instantaneous jumps occur in the time series (pulse on and pulse off). One can visualize the unmasked triangular window and the masked boxed window both moving through the vertical lines. The responses in all eight recurrence variables reflect the two transitional state changes encountered. In addition, the effect of the left shift of the boxed recurrence variables can also be studied clearly. These plots can assist in learning how to read recurrence plot variables when using masked versus unmasked methodologies.

Perhaps the most surprising conclusion of this paper is that box masking of the recurrence matrix is not necessarily called for in most practical situations. Nevertheless, this study reveals that border truncation effects are real when dealing with long (even infinite) time series. If this observation can be verified for different systems it would mean that all previous studies relying on the traditional entropy calculations for systems that are even near periodic, the entropy values and other recurrence variables computed and reported are valid. A good example is one recent master's thesis on which the conclusions relied heavily on traditional triangular entropy values to come to consistent and meaningful conclusions [19].

Funding: This research received no external funding.

Data Availability Statement: The data reported in this paper can be requested directly from the author.

Conflicts of Interest: The author declares no conflict of interest.

Appendix A

The new software programs used in this paper, all written by the author, are bundled within the suite of RQA 2021 programs. These new programs include RQDB (recurrence display) RQEB (recurrence epoch), and RQES (recurrence scale) which were run to compute the boxed recurrences in this paper. They correspond to traditional triangular recurrence programs RQD, RQE and RQS. All RQA programs are embedded within a single ZIP file [20] which is easily downloaded at no cost. C-source codes will be made available to anyone for the asking.

References

1. Eckmann, J.-P.; Kamphorst, S.O.; Ruelle, D. Recurrence plots of dynamical systems. *Europhys. Lett.* **1987**, *4*, 973–977. [CrossRef]
2. Zbilut, J.P.; Webber, C.L., Jr. Embeddings and delays as derived from quantification of recurrence plots. *Phys. Lett. A* **1992**, *171*, 199–203. [CrossRef]
3. Webber, C.L., Jr.; Zbilut, J.P. Dynamical assessment of physiological systems and states using recurrence plot strategies. *J. Appl. Physiol.* **1994**, *76*, 965–973. [CrossRef] [PubMed]
4. Marwan, N.; Wessel, N.; Meyerfeldt, U.; Schirdewan, A.; Kurths, J. Recurrence-plot-based measures of complexity and their application to heart rate variability data. *Phys. Rev. E* **2002**, *66*, 026702. [CrossRef] [PubMed]
5. Webber, C.L., Jr.; Zbilut, J.P. Recurrence quantifications: Feature extractions from recurrence plots. *Int. J. Bifurc. Chaos* **2007**, *17*, 3467–3475. [CrossRef]
6. Webber, C.L., Jr.; Marwan, N.; Facchiani, A.; Giuliani, A. Simpler methods do it better: Success of recurrence quantification analysis as a general-purpose data analysis tool. *Phys. Lett. A* **2009**, *373*, 3753–3756. [CrossRef]
7. Marwan, N.; Romano, M.C.; Thiel, M.; Kurths, J. Recurrence plots for the analysis of complex systems. *Phys. Rep.* **2007**, *438*, 237–329. [CrossRef]
8. Shannon, C.E. A mathematical theory of communication. *Bell Syst. Tech. J.* **1948**, *27*, 379–423, 623–656. [CrossRef]
9. Kraemer, H.; Marwan, N. Border effect corrections for diagonal line based recurrence quantification analysis measures. *Phys. Lett. A* **2019**, *383*, 125977. [CrossRef]
10. Trulla, L.L.; Giuliani, A.; Zbilut, J.P.; Webber, C.L., Jr. Recurrence quantification analysis of the logistic equation with transients. *Phys. Lett. A* **1996**, *223*, 255–260. [CrossRef]
11. Censi, F.; Calcagnini, G.; Cerutti, S. Proposed corrections for the quantification of coupling patterns by recurrence plots. *IEEE Trans. Biomed. Eng.* **2004**, *51*, 856–859. [CrossRef] [PubMed]

12. Eroglu, S.D.; Peron, T.K.D.M.; Marwan, N.; Rodrigues, F.A.; Costa, L.d.F.; Sebek, M.; Kiss, I.Z.; Kurths, J. Entropy of weighted recurrence plots. *Phys. Rev. E* **2014**, *90*, 042919. [CrossRef] [PubMed]
13. Corso, G.; Prado, T.d.L.; Lima, G.Z.d.S.; Kurths, J.; Lopes, S.R. Quantifying entropy using recurrence matrix microstates. *Chaos* **2018**, *28*, 083108. [CrossRef] [PubMed]
14. Leonardi, G. A Method for the computation of entropy in the Recurrence Quantification Analysis of categorical time series. *Phys. A Stat. Mech. Its Appl.* **2018**, *512*, 824–836. [CrossRef]
15. Thiel, M.; Romano, M.C.; Kurths, J.; Meucci, R.; Allaria, E.; Arecchi, F.T. Influence of observational noise on the recurrence quantification analysis. *Physica D* **2002**, *171*, 138–152. [CrossRef]
16. Letellier, C. Estimating the Shannon Entropy: Recurrence plots versus symbolic dynamics. *Phys. Rev. Lett.* **2006**, *96*, 254102. [CrossRef] [PubMed]
17. Dow Jones Industrial Average® Scores. S & P Dow Jones Indices, a Division of S & P Global. 2021. Available online: https://www.spglobal.com/spdji/en/indices/equity/dow-jones-industrial-average/#overview (accessed on 20 August 2021).
18. Marwan, N.; Webber, C.L., Jr. Mathematical and Computational Foundations of Recurrence Quantifications. In *Recurrence Quantification Analysis: Theory and Best Practices;* Springer Series in Complexity; Webber, C., Jr., Marwan, N., Eds.; Springer: Berlin, Germany, 2015; pp. 1–41.
19. Yan, R.; Qian, Y.; Huang, Z.; Gao, R.X. Recurrence plot entropy for machine defect severity assessment. *Smart Struct. Syst.* **2013**, *11*, 299–314. [CrossRef]
20. Webber, C.L., Jr. Introduction to Recurrence Quantification Analysis. RQA Software Version 16.1 README.PDF. 2021. Available online: http://cwebber.sites.luc.edu/ (accessed on 20 August 2021).

MDPI

Article

Affect-Logic, Embodiment, Synergetics, and the Free Energy Principle: New Approaches to the Understanding and Treatment of Schizophrenia

Luc Ciompi [1] and Wolfgang Tschacher [2,*]

1 Formerly Medical Director of the Social-Psychiatric University Hospital, CH-3010 Bern, Switzerland; lucciompi57@gmail.com

2 Department of Experimental Psychology, University Hospital of Psychiatry and Psychotherapy, University of Bern, CH-3060 Bern, Switzerland

* Correspondence: wolfgang.tschacher@upd.unibe.ch

Abstract: This theoretical paper explores the affect-logic approach to schizophrenia in light of the general complexity theories of cognition: embodied cognition, Haken's synergetics, and Friston's free energy principle. According to affect-logic, the mental apparatus is an embodied system open to its environment, driven by bioenergetic inputs of emotions. Emotions are rooted in goal-directed embodied states selected by evolutionary pressure for coping with specific situations such as fight, flight, attachment, and others. According to synergetics, nonlinear bifurcations and the emergence of new global patterns occur in open systems when control parameters reach a critical level. Applied to the emergence of psychotic states, synergetics and the proposed energetic understanding of emotions lead to the hypothesis that critical levels of emotional tension may be responsible for the transition from normal to psychotic modes of functioning in vulnerable individuals. In addition, the free energy principle through learning suggests that psychotic symptoms correspond to alternative modes of minimizing free energy, which then entails distorted perceptions of the body, self, and reality. This synthetic formulation has implications for novel therapeutic and preventive strategies in the treatment of psychoses, among these are milieu-therapeutic approaches of the Soteria type that focus on a sustained reduction of emotional tension and phenomenologically oriented methods for improving the perception of body, self, and reality.

Keywords: schizophrenia; phase transition; emotions; embodiment; self-organization

Citation: Ciompi, L.; Tschacher, W. Affect-Logic, Embodiment, Synergetics, and the Free Energy Principle: New Approaches to the Understanding and Treatment of Schizophrenia. *Entropy* **2021**, *23*, 1619. https://doi.org/10.3390/e23121619

Academic Editor: Andrei Khrennikov

Received: 12 November 2021
Accepted: 29 November 2021
Published: 1 December 2021

Publisher's Note: MDPI stays neutral with regard to jurisdictional claims in published maps and institutional affiliations.

1. Introduction

During recent decades, several comprehensive theoretical accounts have emerged that are potentially relevant for a better understanding of schizophrenia, among them are the concepts of affect-logic [1–3], embodiment [4], synergetics [5,6], and the free energy principle [7]. These theories were developed largely independent of each other and remained only loosely connected in spite of interesting commonalities. The goal of the present paper is to explore shared features of these four concepts by focusing on common or complementary elements with the hope of identifying novel therapeutic approaches to schizophrenia.

Affect-logic is a comprehensive metatheory of cognition that focuses on the interactions between emotions (affectivity) and cognition (logic). Affect-logic is rooted in psychology, psychiatry, neurobiology, and evolutionary theory, adopting a systems-theoretical perspective, including the theory of nonlinear dynamics in complex open systems. The original German term, Affektlogik, points to omnipresent circular interactions between emotion and cognition, where affect (Latin afficere, i.e., "arousing", "attuning") is understood as a global bioenergetic and psychophysical state of varying quality, duration, and degree of awareness. Here, affect is an umbrella term that comprises all variants of overlapping emotion-near phenomena such as feelings, sensations, or moods. Affective and cognitive elements interact in all mental processes, where processes seemingly characterized by neutrality or indifference are likewise affective states in the sense mentioned. Simultaneously active emotions, cognitions, and behaviors are memorized as integrated "programs" for feeling, thinking, and behaving (FTB-programs). FTB-programs are learned schemas that form the essential building blocks of the psyche and are reactivated in similar situations. Conscious and unconscious emotions related to past or present experiences guide and connect all cognitive functions, such as perception, attention, memory, thought, and decision-making. In this, affect-logic concurs with Damasio's somatic marker hypothesis that cognitive decisions are guided by emotions [8]. Emotions have focusing, selecting, and filtering "operator effects" on cognition. They tend to focus on cognitions with similar emotional tuning and to ignore cognitions with dissimilar tuning. Initially intense conscious emotions related to new, exciting, difficult, or potentially dangerous situations gradually become automated and largely unconscious yet may still continue to exert their operator effects on cognition, including in seemingly non-emotional situations. This is true even in scientific or mathematical activities, where initially intense "eureka-feelings", which have accompanied a new discovery or solution, may gradually turn into easy-going "highways" for semi-automatized mental operations.

One aspect of emotion is, as we will see below, of particular importance for the dynamics of psychosis: their energetic aspect. Emotional phenomena are rooted in embodied bioenergetic states, which drive and motivate, sometimes also block and freeze, all motor and cognitive behaviors. These states were selected by evolution to cope adequately with situations relevant to survival, such as exploration, fight, flight, attachment, or loss, which are eventually experienced as curiosity, rage, fear, pleasure, or mourning [9]. The notion of "energy" is not just a metaphor but corresponds to the measurable allocation of energy in the form of sympathetic/parasympathetic activations of the organism in specific situations. In terms of systems theory, emotional energies provide the dynamics (the "fuel"), whereas cognitive distinctions provide the structure (the "channels") for all kinds of mental and social systems and activities.

The concept of affect-logic postulates, furthermore, that affective-cognitive interactions have a scale-free (so-called fractal or self-similar) architecture, as they are formally similar on individual, microsocial, and macrosocial levels [10]. The levels interact through structural coupling [11]. Self-similar selecting and filtering operator effects of emotions on attention, perception, memory, and thought act on various individual or collective levels (Figure 1), in local and short-term as well as in extended and long-lasting mental and social processes [12,13].

The affect-logic approach was developed and elaborated to serve as a metatheoretical background, especially for a better understanding of the outbreak of psychosis in the context of schizophrenia spectrum disorder. In the present theoretical overview, we wish to compare the affect-logic approach with further theoretical 'approaches' in the sense of encompassing theories and paradigms of cognition. We focused on systems approaches that aim at modelling the temporal dynamics of mental processes as a complex system. Three general dynamical-systems approaches were identified: embodied cognition (recently extended to 4E cognition: [14]), Haken's synergetics, and Friston's free energy principle. These theories, described in the following, have remained only loosely connected with each other in spite of apparent commonalities. In the following, we will explore their commonalities to derive implications and novel perspectives for the modelling of psychosis in schizophrenia-spectrum disorder and arrive at inspirations for treatment.

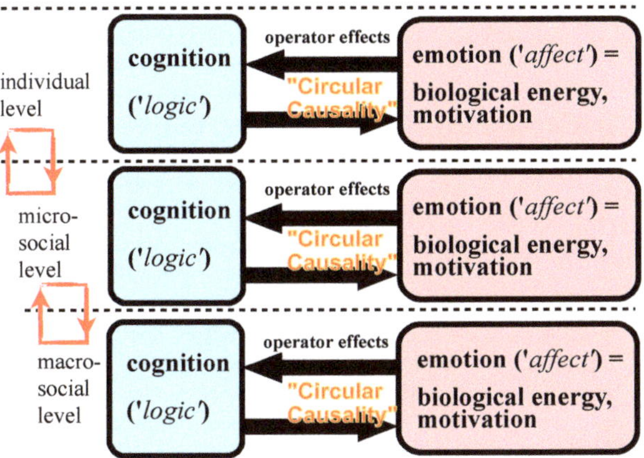

Figure 1. Affect-logic: cognition and emotion interact in a circular way on individual, microsocial, and macrosocial levels. Levels are linked by structural coupling. Cognitions provide the structure, whereas emotions provide the dynamics (energy, motivation) of complex mental and social systems.

2. Embodied Cognition

The core tenet of embodiment is that mental processes, including feelings and thoughts, are anchored in the body, whether through the release of hormones, neural activity, or through behaviors ("body language"). In the process of communication between two (or more) persons, non-verbal synchronies occur regularly, as for example in the unconscious mimicry of gestures, postures, tone of voice, and facial expressions of people we interact with. Interacting individuals also tend to synchronize their physiological processes, such as skin conductance, respiration, or heart rate [15]. A characteristic of embodiment is bidirectionality, which emphasizes that the connections between mind and body go in both directions (Figure 2). In terms of the broad concept of 4E cognition, we focus here on just 2 E's, *E*mbodied and *E*nactive. Enactivism [16] encompasses mind, body and the environment; cognition is to be understood as the active continuous interplay between sensory perception ("environment"), the contingent adaptation of motor behavior ("body"), and cognitive models of the environment ("mind"). This is symbolized by the upper loop in Figure 2. Such enactive sensory-motor loops are the basis of phenomena of nonverbal synchrony in social interaction and psychotherapy, specifically [17].

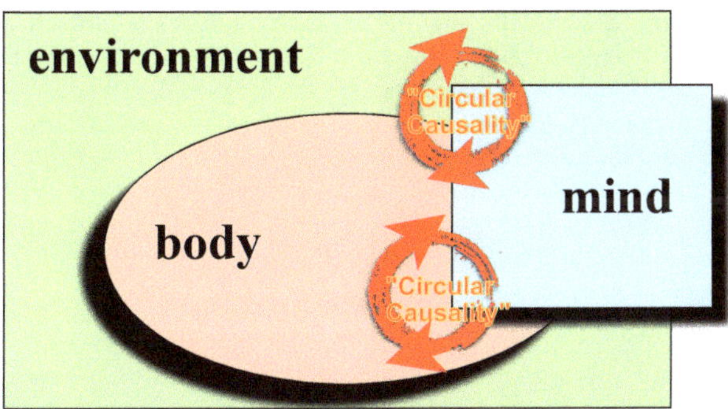

Figure 2. Embodiment: the mind is anchored in the body (lower loop: embodied). Body and mind are embedded in the environment (upper loop: enactive) and interact circularly in all mental activities.

An example for the Embodied loop is as follows: We smile when we feel joy, but joy can also be induced by the activation of smile-related facial muscles under a pretext. In the first case, the emotion is expressed in the body, while in the second the body expresses itself, so to speak, in the perceived emotion. For instance, the manipulation of the motor system by a "depressive" or "happy" gait on a conveyor belt correspondingly changes mental processes and influences the memory of previously learned words. Depressive walkers were found to show a bias in favor of recalling words with negative emotional content [18].

In schizophrenia, sensorimotor and physiological synchronizations are often deficient or completely absent [19]. This phenomenon, called disembodiment, may also be the source of a number of disordered bodily sensations commonly observed in schizophrenia. Such symptoms have been extensively described as so-called basic schizophrenic disorders by Süllwold [20] and Huber [21]. According to current phenomenological research, many of these disturbing body sensations may be at the origin of the distorted perception of reality and self [22–24]. Other mainly somatic manifestations of psychosis, such as catatonia, psychomotor agitation, or mutism-negativism, are closely related to the phenomenon of disembodiment.

The concepts and fields of embodiment/disembodiment and affect-logic are largely overlapping. The first and main reason for their correspondence is that, according to affect-logic, all emotional phenomena are somatically rooted and thus embodied. Secondly, the synchronies studied by embodiment research are crucial for the emergence of emotional contagion [25] and, hence, for the mentioned self-similar effects of emotions at the microsocial and macrosocial levels as proposed by affect-logic.

3. Synergetics

The concepts that underwrite synergetics were developed in the context of the dynamics of complex open systems (or complexity theory) by the physicist Hermann Haken [5,6,26]. Synergetics is an interdisciplinary theory that models processes and mechanisms of pattern formation in physical and chemical, but also biological, mental, and social systems. Spontaneous pattern formation through self-organization depends on a system being driven by external energy sources (hence "open" system). The pattern that arises has dynamical stability and thus can be described as an "attractor". A core phenomenon of synergetics is that sudden nonlinear bifurcations from one global pattern of functioning into another, corresponding to a transition from one basin of attraction to another, occur in complex systems of different kinds when the input of energy to the system reaches a critical level [27,28]; for visual illustrations [29]. In the terminology of synergetics, the energy input acts as a control parameter that determines the moment of bifurcation. The new pattern of functioning is shaped by a new so-called order parameter (or "nucleus of crystallization") around which the new functional dynamic is organized. This order parameter is often a formerly peripheral structural element that suddenly becomes dominant and "enslaves" the dynamics of the whole system (Figure 3). An example from physics is the sudden transition of a random mixture of light waves into a highly ordered laser beam at a certain critical threshold of the energy supply.

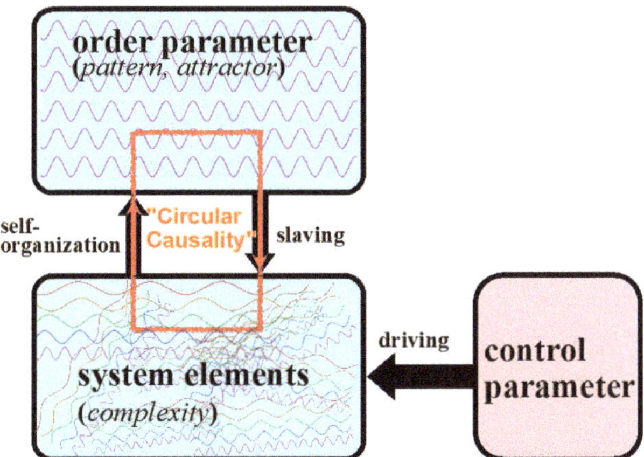

Figure 3. Synergetics: the complex patterns of self-organizing systems, symbolized by disordered wave lines, are synchronized, and coordinated by structural elements, the order parameters. These patterns can be disturbed and globally altered by critically increasing energetic tensions, when the input of energy (=control parameter) reaches a critical level.

Examples from the biological domain include the change of a horse's gait from trot to gallop (both are attractors of movement coordination) under increasing energetic stimulation. In the psychosocial context, it is a common observation that critically increasing emotional tensions can provoke sudden shifts between one global pattern of feeling, thinking, and behaving into another. Thus, an initially merely verbal argument may turn into a raging brawl, a diffuse fear into a collective panic, a festering conflict into open warfare. Tschacher and Haken [30] have also applied synergetic modelling approaches to psychotherapy, distinguishing between different types of intervention in the framework of the Fokker–Planck equation (FPE). The FPE defines the change of the probability of a state variable x depending on time t. In the simplest case, $x(t)$ is represented by Gaussian normal distribution. The probability of this distribution can be changed by a deterministic "drift" term of the equation and a stochastic "diffusion" term, hence by causation and/or chance: "change = causation + chance". Causation can shift the location of the Gaussian distribution, whereas the stochastic diffusion term of the FPE (chance) can be expressed by changing the variance of the Gaussian. The FPE combines both dynamics and thus integrates causation with chance in a change model.

Considering the energetic properties of emotions under affect theory, we hypothesize that such bifurcations are also at work during the emergence of psychotic symptomatology. Emotional tensions can overburden a vulnerable coping system and enforce a transition from normal patterns of feeling, thinking, and behaving into psychotic patterns. The level of emotional tension is the relevant control parameter here. This tension can turn a formerly marginal structural element, such as a vague suspicion or odd behavior, into the new order parameter around which a psychotic pattern (e.g., a structured system of persecutory delusions) emerges. This hypothesis is supported both by classical clinical observations during psychotic breaks [31,32] and by the research on so-called high-expressed emotions, which has shown significant associations between the outbreak of psychosis and excessive emotional tensions in and around individuals at risk [33,34]. Emotional tensions related to traumatic experiences such as sexual abuse, migration, painful separations, or other unfavorable life events, can contribute to the progressive destabilization of a genetically vulnerable "premorbid terrain" [1,3]. In addition, they also play a crucial role in the occurrence of acute relapses during chronic long-term evolution dominated by negative symptoms [35,36].

4. Free Energy Principle

The free-energy principle was proposed by the neuroscientist Karl Friston. He postulated that self-organizing biological systems have the fundamental tendency of continually striving for an equilibrated energy flow with minimal losses of free energy in the service of autopoiesis [7]. In terms of systems theory, this corresponds to the minimum of the potential of an attractor, and in everyday language to the smoothest possible functioning without unpleasant surprises (where surprises can be read as states that are not part of the attracting set). Not only the single neuron but also the brain and the organism as a whole are understood as devices for improving predictions regarding the behavior of their respective environment. Mathematically, the free energy in question here stands in as an upper boundary on the likelihood of states a particular system encounters and plays the role of a potential function. Or in everyday language: free energy arises when the predictions made by the system are wrong. However, unlike thermodynamic free energy that is related to heat, the free energy in question here is a function of sensations and (Bayesian) beliefs about the causes of those sensations. Free energy minimization in biological systems operates by improving the prediction of environmental reactions on the basis of experience (Figure 4). It can then be construed as "active inference", in which both action and perception try to minimize "surprise" (also known as a prediction error). This is equivalent to maximizing the marginal likelihood (i.e., the goodness of fit) of sensory inputs (also known as model evidence) under a generative model that is embodied by the system in question. In this setting, a prediction (i.e., generative model) generates predicted sensory consequences from inferred causes (i.e., model evidence).

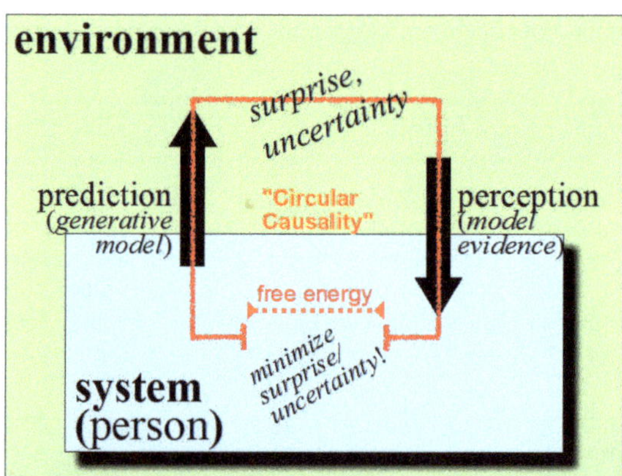

Figure 4. Free energy principle: biological systems continuously work on the discrepancies between predictions and perceptions, corresponding to free energy. Thus, minimizing free energy optimizes their fit with the environment, reducing surprise and uncertainty.

Improved predictions are achieved by actively adapting the implicit model of the real world through structural and functional changes—from fast changes in neuronal activity through to the slow growth of new connections in the neuronal network—and through acting on the environment itself (e.g., by simply moving one's eyes, or by moving one's entire body to a more appropriate environment). Any movement changes both what is perceived and what can be predicted, thus influencing the generative models and the model evidence with the goal of improving adaptation to the environment. Expressed in statistical terms, the organism strives to maximize the likelihood of its continuous modelling loops so as to minimize free energy. According to Friston, the minimization of free energy underlies many biological phenomena. A striking example is the emergence of rhythmic oscillations in the brain known as alpha, beta, and gamma rhythms in the EEG [37,38]. All learning processes, too, can be cast as the minimization of free energy.

Friston's ideas have predecessors in the reafference principle [39] and are also reminiscent of the interplay between assimilation and accommodation during cognitive development described by Jean Piaget [40]. Konrad Lorenz similarly postulated that every development of life is in itself equivalent to an accumulation of knowledge about the surrounding world [41]. The free energy principle shares the fundaments of affect-logic, embodiment, and synergetic formulations: For example, all rest upon a circular causality that lies at the heart of synergetics. In the case of the free energy principle, it is minimizing surprise or prediction errors to attune an individual's implicit model of the world (the generative model) with the embodied world that supplies the sensory evidence for that model. The free energy principle is quintessentially "embodied" in the sense that the generative model is embodied or entailed by both the brain and body. Mathematically, the free energy formulation rests upon the solution to the Fokker–Planck equation and on the very existence of a random dynamical attractor, where "random" pertains to the stochastic term of the equation and "attractor" to its deterministic term. This attractor is constituted by states which the organism is actively striving to attain in the context of exogenous (deterministic) forces and (random) fluctuations that endow nonequilibrium steady states with the dynamical itinerancy evinced by bifurcations and phase transitions. Finally, the existential imperative of minimizing surprise and uncertainty speaks directly to the affective nature of self-organized, autopoietic behavior and the dynamic role of excessive emotional tensions featured in affect-logic theory.

Friston's concepts complete and deepen the approaches of affect-logic and synergetics in an interesting way. They suggest, in particular, that the anxiety, insecurity, and critical emotional tension, which usually precede the outbreak of psychosis, may represent a clinical manifestation of increased free energy (i.e., ego-dystonic unpredictability and uncertainty). The production and perception of psychotic symptoms such as delusions, hallucinations, catatonia etc. may thus correspond to attempts of the mental apparatus to control and minimize free energy through the creation of an "alternative reality" in the service of autopoiesis. This interpretation is supported by clinical observations, which show that the emergence of a coherent system of delusions, or of negative psychotic symptoms such as indifference and social isolation, is often accompanied by a decrease of overt emotional tension and anxiety [31,32]. The inference aspect of active inference manifests clearly in this reading of the free energy principle in terms of false inference: namely, inferring things are there when they are not (e.g., hallucinations and delusions) or inferring things are not there when they are (e.g., dissociative symptoms, derealization, depersonalization).

Tschacher, Giersch, and Friston [19] assume that psychotic interpretations of reality are based on erroneous predictions possibly related to disturbed perceptions of the environment and, in particular, the embodied self. Erroneous predictions may also be the consequence and, simultaneously, the cause of the "loss of normal self-evidence" and disembodiment emphasized by phenomenologists such as Blankenburg and Fuchs. The blurred boundaries between self and others focused on by psychoanalysts [42] point in the same direction. Similarly, unfamiliar environments and conflicting or contradictory life experiences ("double-binds"), which were found to be related to the outbreak of psychosis [43,44], may contribute to false inferences and ensuing (erroneous) predictions.

5. Discussion: Towards a Translational Understanding of Psychosis

A translational, that is to say, multi-conceptual, understanding of schizophrenia arises on the basis of the above, and this translational view can be integrated into the generally accepted vulnerability–stress model of psychiatry, which postulates that acute psychotic symptoms can arise in both genetically and/or biographically vulnerable individuals when emotional tensions induced by stress—surprise or uncertainty—reach a critical level [1,45–47]. These tensions correspond to a clinical manifestation of unbound free energy, which is eventually minimized by a nonlinear phase transition from normal to psychotic patterns of feeling, thinking, and behaving.

In synergetics and the free energy approach, the dynamics of phase transitions is viewed as the process of moving through bifurcations. In other words, the currently active attractor becomes unstable and increasingly weakened before the critical point of the control parameter or free energy is reached [6]. The signature of this destabilization of the old pattern/attractor is called critical slowing down, which can be observed empirically as the system needs more time to relax into its attractor after some perturbation. At the exact point of bifurcation, the system "chooses" between two (new) quasi-attractors, which choice often occurs through a chance event. This means, however, that exactly this period in the development of the system is most accessible for (even small) deterministic interventions that can guide the trajectory into the preferred specific attractor.

This innovative view has the advantage of not only integrating the four theoretical approaches in question. It also better explains the enormous variability of long-term outcomes revealed by follow-up studies over several decades [48–51]. Furthermore, psychotic symptoms such as delusions, hallucinations, emotional indifference, and social withdrawal do not appear, in this light, merely as deficits in the sense of the standard medical model but also as active and productive coping strategies. Symptoms are thus understood as instrumental in the service of free energy minimization and autopoiesis, as symptoms decrease the critical emotional tensions. Over time, these defensive mechanisms may become ingrained, habitual, and hard-wired by neural plasticity—thus partly explaining certain functional and structural modifications found in the brains of persons with schizophrenia [52,53].

6. Therapeutic Implications

The proposed understanding of schizophrenia in the context of complexity science has both therapeutic and preventive implications. We consider the following of particular interest:

We noted above that, according to complexity theory, the critical time of phase transitions are particularly sensitive for interventions. The critical slowing down of a system, the signature of being close to imminent bifurcation, is in principle clinically observable and can thus inform about the optimal time for deterministic interventions towards desired new cognition or behavior.

If excessive emotional tensions do indeed play the postulated key role at the onset of psychosis, then systematically reducing emotional tensions in and around acute patients should be a main therapeutic focus. Many elements of conventional medical interventions, however, rather increase than decrease emotional tensions, among them the often chaotic circumstances of hospitalization, the unfamiliar and opaque atmosphere of psychiatric hospitals, size of wards, architecture, lack of staff continuity, overstimulation by excessive noise, and sometimes also violence. Small, family-like, stimulus-protecting environments should therefore be much more appropriate. In terms of the Fokker–Planck Equation (FPE), such interventions relate to the stochastic term of the FPE. Stochastic interventions can be implemented to protect existing patterns from random fluctuations by boundary regulation [30].

This is realized in therapeutic environments of the Soteria type, which were first created in the 1970s in California [54] and have been functioning successfully since 1984, with some conceptual modifications, in the therapeutic community Soteria Bern in Switzerland [55,56]. An increasing number of similar institutions were recently founded across Europe, especially in Germany. In this approach, sustained emotional relaxation is not mainly obtained by neuroleptic medication, but rather by continual personal support, by "being with" and "doing with" the psychotic patient, creating a trusting therapeutic alliance between specially selected and trained caregivers and the patient. The Soteria milieu is designed to ward off overtaxing external inputs when the patient initially lives in the "soft room", which corresponds to boundary regulation. Further ingredients of the treatment philosophy are the systematic inclusion of the patient's family or other important persons of reference into the therapeutic process and participation in appropriately dosed everyday activities such as cooking, shopping, and housekeeping. According to comparative empirical research, two-year results were obtained that are equivalent to, and subjectively rather better than, standard treatment. Soteria-type environments use significantly fewer neuroleptics and can partly also be run with lower costs [56–59].

A stable, transparent, and securing therapeutic environment of the Soteria type may also improve patients' predictions about the behavior of their environment and, thus, reduce the need for "crazy" alternative explanations. According to a recent pilot study conducted by Soteria Konstanz (Germany), disturbed feelings of reality and self can be significantly improved by flexibly integrating specific behavior–therapeutic elements into everyday contacts and activities [60].

Similar goals are pursued by body-centered therapeutic methods recently proposed in connection with the disembodiment in schizophrenia, such as interventions from dance and movement therapy. These approaches emphasize the significance of sensorimotor experience and body motion for cognition, affect, and social interaction and strive to enhance emotional processing and self-regulation [61,62]. Such methods appear especially effective for reducing negative symptoms such as passivity and social withdrawal [63].

Innovative therapeutic strategies may also be based on the previously mentioned productivity of psychotic symptoms, e.g., by valorizing the creative aspects of the involved emotional energies and eventually steering them towards more constructive directions. An example is Milton Erickson's imitating the incomprehensible artificial language of a chronically psychotic patient and regularly "speaking" with him in this language until the patient suddenly started to speak normally, gradually accepted a friendly relationship and progressively normalized his behavior [64].

Finally, the multifocal understanding of psychosis proposed here strongly argues for restructuring long-term psychotherapy. Conventional long-term neuroleptic medication is not only unable to consolidate the underlying structural vulnerability of the patient but is also burdened by severe long-term side effects. Among the many currently proposed psychotherapeutic methods, the following two are particularly close to the proposed understanding of psychosis: The approach practiced by the Swiss family therapist Ursula Davatz, who describes psychosis as the result of an "emotional tsunami" in the family system [65], and the integrative approach developed by the Italian therapist Giovanni Ariano [66,67]. Since, as we believe, emotions do play a decisive role in the outbreak of psychoses, it may also be worth considering interventions developed in psychotherapies for other disorders, such as depression. For instance, it was found that emotion regulation plays a role in depressive disorders, where some patients are emotionally over-regulated whereas others are under-regulated. It is thus important to take the type of emotion regulation into account in the choice of therapeutic strategies [68], which appears reasonable also in psychotherapy with schizophrenia patients.

7. Open Questions

The synopsis of the concepts of affect-logic, embodiment, synergetics, and free energy minimization appears capable of opening up new ways of understanding and also treating schizophrenia. In this view, embodied emotions and emotional energies play a more important role in the genesis and long-term evolution of schizophrenia than hitherto admitted. Unbounded ("free") emotional energies related to traumatic life experiences not only contribute to further destabilizing a vulnerable "premorbid terrain" but are probably also responsible both for the outbreak of acute psychotic decompensations, as well as for acute relapses during long-term clinical progression characterized by negative symptoms. This same view also stimulates a number of lingering open questions:

- Why do certain people react to critically rising emotional tensions with psychotic symptoms, whereas others respond with violence, fall into panic, or become depressed?
- Is there a specific schizophrenogenic vulnerability, and what does it consist of? It was long suspected by the affect-logic perspective, with special reference to Eugen Bleuler's concept of "loose associations" (cf. disembodiment!), that there may be a

partly genetically and partly biographically determined instability of the links between feeling and thinking. Of particular importance are those links that regulate basic interpersonal relations (in psychoanalytic terminology, the object representations). This hypothesis is at least partly supported by recent neurobiological findings revealing disturbed neuronal connections between frontal cortical and subcortical areas, the amygdalae [69].
- Does the labilization of the psyche that is physically manifest in terms of functional disconnections in the brain result as a direct consequence of aberrant (i.e., false) learning and inference [70,71]?
- There is yet no direct quantitative evidence for the hypothesis that emotional tensions can lead to a phase transition into psychosis, as is proposed by affect-logic. It would be desirable to have objective physiological markers for emotional tension at hand.
- Why does the psychotic phenomenology differ so much from case to case that Eugen Bleuler, the creator of the term schizophrenia in 1911, used to speak of "the group of schizophrenias" [72]? Is this related to the changing influence of a great number of environmental variables or to some genetic or other biological variables?
- As a final and perhaps most important, but to our knowledge astonishingly neglected, question: How do both "spontaneous" and therapy-induced improvements and recoveries arise? According to long-term follow-up studies over several decades [48,50,73], full and lasting recoveries comprise, in the long run, at least one fourth of cases and may even amount to roughly two thirds under especially favorable conditions [51,74].

It is obvious that all these questions and hypotheses, while quite plausible in our eyes and also consistent with a number of clinical observations and empirical findings, need much more specification and confirmation—or rejection—by further research. The goal of the present explorative (and perhaps provocative) overview was to stimulate such research.

8. Endnotes

1. The Fokker-Planck equation plays a central role in nearly all of physics. It describes the evolution of the probability density of a system's states when they are subject to random fluctuations and deterministic impacts. Common variants of the Fokker-Planck equation include the master equation for discrete systems and the Schrödinger wave equation in quantum electrodynamics. It also manifests in theoretical biology as a model of population density dynamics (e.g., the Wright-Fisher model). The solution of the Fokker-Planck equation for any random dynamical system—in a nonequilibrium steady state—also forms the basis of the free energy principle.

2. Mathematically on the background of the Fokker-Planck equation, uncertainty can be read as an expected surprise. In information theory, surprise (or surprisal) corresponds to self-information, while expected surprise or uncertainty is known as entropy. Minimizing expected free energy by choosing appropriate actions can then be read as reducing uncertainty in anticipation of familiar, unsurprising outcomes. From a physicist's perspective, this would look like self-organization to a nonequilibrium steady state. From a psychiatrist's perspective, this would look like an active search for synchrony and predictability and the active avoidance of existential fear.

3. We use 'unbound' here deliberately to conflate Freudian notions of unbound energy with the role of variational free energy as a bound on surprise or marginal likelihood [75]. Indeed, in machine learning, variational free energy is known as an evidence bound [76].

Author Contributions: Conceptualization, L.C.; writing—original draft preparation, L.C., W.T.; writing—review and editing, L.C., W.T.; visualization, W.T. All authors have read and agreed to the published version of the manuscript.

Funding: This research received no external funding.

Institutional Review Board Statement: Not applicable.

Informed Consent Statement: Not applicable.

Acknowledgments: The article is based on a lecture given at the Annual Congress of the German Society for Psychiatry, Psychotherapy and Neurology (DGPPN), 28 November 2020.

Conflicts of Interest: The authors declare no conflict of interest.

References

1. Ciompi, L. *The Psyche and Schizophrenia. The Bond between Affect and Logic*; Harvard University Press: Cambridge, MA, USA, 1988.
2. Ciompi, L. Ein blinder Fleck bei Niklas Luhmann. Soziodynamische Wirkungen von Emotionen nach dem Konzept der fraktalen Affektlogik. *Soz. Syst.* **2004**, *10*, 21–49.
3. Ciompi, L. *Die Emotionalen Grundlagen des Denkens: Entwurf Einer Fraktalen Affektlogik*; Vandenhoeck & Ruprecht: Göttingen, Germany, 1997.
4. Tschacher, W.; Bergomi, C. (Eds.) *The Implications of Embodiment: Cognition and Communication*; Imprint Academic: Exeter, UK, 2011.
5. Haken, H. *Synergetics: An Introduction*; Springer: Berlin/Heidelberg, Germany, 1990.
6. Haken, H.; Portugali, J. Information and Selforganization: A Unifying Approach and Applications. *Entropy* **2016**, *18*, 197. [CrossRef]
7. Friston, K.J. The free-energy principle: A unified brain theory? *Nat. Rev. Neurosci.* **2010**, *11*, 127–138. [CrossRef] [PubMed]
8. Damasio, A.R. *Descartes' Error: Emotion, Reason, and the Human Brain*; Putnam: New York, NY, USA, 1994.
9. Ciompi, L.; Baatz, M. The energetic dimension of emotions—An evolution-based computer simulation with general implications. *Theor. Biol.* **2008**, *3*, 42–50. [CrossRef]
10. Lewis, M.D. Cognition-emotion feedback and the self-organization of developmental paths. *Hum. Dev.* **1995**, *38*, 71–102. [CrossRef]
11. Maturana, H. Autopoiesis, structural coupling and cognition: A history of these and other notions in the biology of cognition. *Cybern. Hum. Knowing* **2002**, *9*, 5–34.
12. Ciompi, L.; Baatz, M. Do mental processes have a fractal structure? The hypothesis of affect-logic. In *Fractals in Biology and Medicine*; Losa, A.L., Merline, D., Nonnenmacher, T.F., Weibel, E.R., Eds.; Birkhäuser: Basel, Switzerland, 2005; Volume IV, pp. 107–119.
13. Ciompi, L.; Endert, E. *Gefühle Machen Geschichte: Die Wirkung Kollektiver Emotionen von Hitler bis Obama*; Vandenhoeck & Ruprecht: Göttingen, Germany, 2011.
14. Newen, A.; De Bruin, L.; Gallagher, S. (Eds.) *The Oxford Handbook of 4E Cognition*; Oxford University Press: Oxford, UK, 2018.
15. Tschacher, W.; Meier, D. Physiological synchrony in psychotherapy sessions. *Psychother. Res.* **2020**, *30*, 558–573. [CrossRef] [PubMed]
16. Varela, F.; Thompson, E.; Rosch, E. *The Embodied Mind. Cognitive Science and Human Experience*; MIT Press: Cambridge, UK, 1991.
17. Wiltshire, T.J.; Philipsen, J.S.; Trasmundi, S.B.; Jensen, T.W.; Steffensen, S.V. Interpersonal coordination dynamics in psychotherapy: A systematic review. *Cogn. Ther. Res.* **2020**, *44*, 752–773. [CrossRef]
18. Michalak, J.; Rohde, K.; Troje, N.F. How we walk affects what we remember: Gait modifications through biofeedback change negative affective memory bias. *J. Behav. Ther. Exp. Psychiatry* **2015**, *46*, 121–125. [CrossRef]
19. Tschacher, W.; Giersch, A.; Friston, K.J. Embodiment and schizophrenia: A review of implications and applications. *Schizophr. Bull.* **2017**, *43*, 745–755. [CrossRef]
20. Süllwold, L. Basis-Störungen. Ergebnisse und offene Fragen. In *Schizophrenie, Stand und Entwicklungstendenzen der Forschung*; Huber, G., Ed.; Schattauer: Stuttgart, Germany, 1981.
21. Huber, G. Das Konzept substratnaher Basisstörungen und seine Bedeutung für Theorie und Therapie schizophrener Erkrankungen. *Nervenarzt* **1983**, *54*, 23–32.
22. Fuchs, T. Corporealized and disembodied minds: A phenomenological view of the body in melancholia and schizophrenia. *Philos. Psychiatry Psychol.* **2005**, *12*, 95–107.
23. Fuchs, T.; Schlimme, J.E. Embodiment and psychopathology: A phenomenological perspective. *Curr. Opin. Psychiatry* **2009**, *22*, 570–575. [CrossRef]
24. Parnas, J.; Henriksen, M.G. Disordered self in the schizophrenia spectrum. *Harv. Rev. Psychiatry* **2014**, *22*, 251–265. [CrossRef] [PubMed]
25. Hatfield, E.; Cacioppo, J.T.; Rapson, R.L. *Emotional Contagion*; Cambridge University Press: Paris, France, 1994.
26. Haken, H.; Haken-Krell, M. *Erfolgsgeheimnisse der Wahrnehmung. Synergetik als Schlüssel zum Gehirn*; Deutsche Verlags-Anstalt: Stuttgart, Germany, 1991.
27. Haken, H.; Portugali, J. Information and Self-Organization II: Steady State and Phase Transition. *Entropy* **2021**, *23*, 707. [CrossRef] [PubMed]
28. Friedenberg, J. *Dynamical Psychology: Complexity, Self-Organization and Mind*; Emergent Publishing: Litchfield Park, AZ, USA, 2009.
29. Abraham, R.H.; Shaw, C.D. *Dynamics—The Geometry of Behavior*; Addison-Wesley: Redwood City, CA, USA, 1992.
30. Tschacher, W.; Haken, H. *The Process of Psychotherapy: Causation and Chance*; Springer Nature: Cham, Switzerland, 2019.
31. Mishara, A.L. Klaus Conrad (1905–1961): Delusional mood, psychosis, and beginning schizophrenia. *Schizophr. Bull.* **2010**, *36*, 9–13. [CrossRef]
32. Bowers, M. *Retreat from Sanity: The Structure of Emerging Psychosis*; Human Sciences Press: New York, NY, USA, 1974.

33. Vaughn, C.; Leff, J. The influence of family and social factors on the course of psychiatric illness: A comparison of schizophrenic and depressed neurotic patients. *Br. J. Psychiatry* **1976**, *129*, 125–137. [CrossRef] [PubMed]
34. Kavanagh, D.J. Recent developments in expressed emotion and schizophrenia. *Br. J. Psychiatry* **1992**, *160*, 601–620. [CrossRef]
35. Dohrenwendt, P.; Egri, G. Recent stressful life events and episodes of schizophrenia. *Schizophr. Bull.* **1981**, *7*, 12–23. [CrossRef]
36. Abrahamyan Empson, L.; Baumann, P.S.; Söderström, O.; Codeluppi, Z.; Söderström, D.; Conus, P. Urbanicity: The need for new avenues to explore the link between urban living and psychosis. *Early Interv. Psychiatry* **2019**, *14*, 398–409. [CrossRef]
37. Friston, K.J.; Stephan, K.E. Free-energy and the brain. *Synthese* **2007**, *159*, 417–458. [CrossRef] [PubMed]
38. Palacios, E.F.; Isemura, T.; Friston, K.J. The emergence of synchrony in networks of mutually inferring neurons. *Sci. Rep.* **2019**, *9*, 6412. [CrossRef] [PubMed]
39. von Holst, E.; Mittelstaedt, H. Das Reafferenzprinzip. Wechselwirkung zwischen Zentralnervensystem und Peripherie. *Naturwissenschaften* **1950**, *37*, 464–476. [CrossRef]
40. Piaget, J. *Die Äquilibration der Kognitiven Strukturen. [L'équilibration des Structures Cognitives]*; Klett: Stuttgart, Germany, 1976.
41. Lorenz, K. Die Rückseite des Spiegels. In *Behind the Mirror: A Search for a Natural History of Human Knowledge*; Pieper: München, Germany, 1973.
42. Silver, A.S. Psychoanalysis and Psychosis: Trends and Developments. *J. Contemp. Psychother.* **2001**, *31*, 21–30. [CrossRef]
43. Wynne, L.C.; Ryckoff, J.; Dave, S.; Hirsch, I. Pseudomutuality in the family relations of schizophrenics. *Psychiatry* **1958**, *21*, 205–220. [CrossRef]
44. Tienari, P.; Sorri, A.; Lathi, I.; Naurala, M.; Wahlberg, K.E.; Pohojola, J.; Moring, J. Interaction of genetic and psychosocial factors in schizophrenia. *Acta Psychiatr. Scand.* **1985**, *71*, 19–30. [CrossRef] [PubMed]
45. Zubin, J.; Spring, B. Vulnerability—A new view on schizophrenia. *J. Abnorm. Psychol.* **1977**, *86*, 103–126. [CrossRef]
46. Nuechterlein, K.H.; Dawson, M.E. A heuristic vulnerability/stress model of schizophrenic episodes. *Schizophr. Bull.* **1984**, *10*, 300–312. [CrossRef] [PubMed]
47. Peters, A.; McEwen, B.S.; Friston, K.J. Uncertainty and stress: Why it causes diseases and how it is mastered by the brain. *Prog. Neurobiol.* **2017**, *156*, 164–188. [CrossRef] [PubMed]
48. Bleuler, M. *The Schizophrenic Disorders: Long Term Patient and Family Studies*; Yale University Press: New Haven, CT, USA, 1978.
49. Ciompi, L.; Müller, C. *Lebensweg und Alter der Schizophrenen: Eine katamnestische Langzeitstudie*; Springer: Berlin, Germany, 1976.
50. Huber, G.; Gross, G.; Schüttler, R.; Linz, M. Longitudinal studies of schizophrenic patients. *Schizophr. Bull.* **1992**, *6*, 592–605. [CrossRef]
51. Harding, C.M.; Brooks, G.W.; Ashikaga, T.; Strauss, J.S.; Breier, A. The Vermont longitudinal study: II. Long-term outcome of subjects who retrospectively met DSM-III criteria for schizophrenia. *Am. J. Psychiatry* **1987**, *144*, 727–735. [PubMed]
52. Shepherd, A.M.; Laurens, K.R.; Matheson, S.L.; Vaughan, J.C.; Green, M.J. Systematic meta-review and quality assessment of the structural brain alterations in schizophrenia. *Neurosci. Biobehav. Rev.* **2012**, *36*, 1342–1356. [CrossRef] [PubMed]
53. Ciompi, L. The key role of emotions in the schizophrenia puzzle. *Schizophr. Bull.* **2015**, *41*, 318–322. [CrossRef] [PubMed]
54. Mosher, L.R.; Menn, A.Z. Community residential treatment for schizophrenia: Two-year follow-up data. *Hosp. Community Psychiatry* **1978**, *29*, 715–723. [CrossRef]
55. Ciompi, L.; Hoffmann, H. Soteria Berne. An innovative milieu therapeutic approach to acute schizophrenia based on the concept of affect-logic. *World Psychiatry* **2004**, *3*, 140–146.
56. Ciompi, L. Soteria Berne: 32 years of experience. An alternative approach to acute schizophrenia. *Swiss Arch. Neurol. Psychiatry Psychother.* **2017**, *168*, 10–13.
57. Ciompi, L.; Dauwalder, H.P.; Maier, C.; Aebi, E.; Trütsch, K.; Kupper, Z.; Rutishauser, C. The pilot project "Soteria Berne". Clinical experiences and results. *Br. J. Psychiatry* **1992**, *161*, 145–153. [CrossRef]
58. Bola, J.R.; Mosher, L. Treatment of acute psychosis without neuroleptics: Two-year outcomes from the Soteria Project. *J. Nerv. Ment. Dis.* **2003**, *191*, 219–229. [CrossRef]
59. Carlton, T.; Ferriter, M.; Hubard, N.; Spandler, H. A systematic review of the Soteria paradigm for the treatment of people diagnosed with schizophrenia. *Schizophr. Bull.* **2008**, *34*, 181–192. [CrossRef]
60. Nischk, D.; Rusch, J. What makes Soteria work? On the effect of therapeutic milieu on self-disturbances in the schizophrenia syndrome. *Psychopathology* **2019**, *52*, 213–220. [CrossRef] [PubMed]
61. Röhricht, F.; Priebe, S. Effect of body-oriented psychological therapy on negative symptoms in schizophrenia: A randomized controlled trial. *Psychol. Med.* **2006**, *36*, 669–678. [CrossRef] [PubMed]
62. Martin, L.A.; Koch, S.C.; Hirjak, D.; Fuchs, T. Overcoming disembodiment: The effect of movement therapy on negative symptoms in schizophrenia—A multicenter randomized controlled trial. *Front. Psychol.* **2016**, *7*, 483. [CrossRef]
63. Pfammatter, M.; Junghan, U.; Brenner, H.D. Efficacy of psychological therapy in schizophrenia: Conclusions from meta-analyses. *Schizophr. Bull.* **2006**, *32*, 64–80. [CrossRef] [PubMed]
64. Bandler, R.; Grinder, J. *Patterns of the Hypnotic Techniques of Milton H. Erickson*; Meta Publications: Cupertino, CA, USA, 1975.
65. Davatz, U. *ADHS und Schizophrenie: Wie Emotionale Monsterwellen Entstehen und Wie Sie Behandelt Werden*; Verlag Rüegger: Chur, Switzerland, 2019.
66. Ariano, G. *La Psicoterapia D'integrazione Strutturale*; Armando: Roma, Italy, 1997.
67. Ariano, G. *Dolore per la Crescita*; Armando: Roma, Italy, 2005.
68. Greenberg, L.S.; Watson, J.C. *Emotion-Focused Therapy for Depression*; American Psychological Association: Washington, DC, USA, 2006.

69. Comte, M.; Zendjidjian, X.Y.; Coull, J.T.; Cancel, A.; Boutet, C.; Schneider, F.C.; Sage, T.; Lazerges, P.-E.; Jaafari, N.; Ibrahim, E.C.; et al. Impaired cortico-limbic functional connectivity in schizophrenia patients during emotion processing. *Soc. Cogn. Affect. Neurosci.* **2018**, *13*, 381–390. [CrossRef]
70. Friston, K.; Brown, H.R.; Siemerkus, J.; Stephan, K.E. The dysconnection hypothesis. *Schizophr. Res.* **2016**, *176*, 83–94. [CrossRef]
71. Preller, K.H.; Razi, A.; Zeidman, P.; Stämpfli, P.; Friston, K.J.; Vollenweider, F.X. Effective connectivity changes in LSD-induced altered states of consciousness in humans. *Proc. Natl. Acad. Sci. USA* **2019**, *116*, 2743–2748. [CrossRef] [PubMed]
72. Bleuler, E. *Dementia Praecox or the Group of Schizophrenias*; International University Press: New York, NY, USA, 1950.
73. Ciompi, L. The natural history of schizophrenia in the long term. *Br. J. Psychiatry* **1980**, *136*, 413–420. [CrossRef]
74. Harding, C.M.; Brooks, G.W.; Ashikaga, T.; Strauss, J.S.; Landerl, P.D. Aging and social functioning in once-chronic schizophrenic patients 21–58 years after first admission: The Vermont Story. In *Schizophrenia, Paranoia, and Schizophreniform Disorders in Later Life*; Hudgins, G., Miller, N., Eds.; Guilford Press: New York, NY, USA, 1985.
75. Solms, M. The Conscious Id. *Neuropsychoanalysis* **2014**, *15*, 5–19. [CrossRef]
76. Winn, J.; Bishop, C.M. Variational message passing. *J. Mach. Learn. Res.* **2005**, *6*, 661–694.

MDPI

Article

Clients' Emotional Experiences Tied to Therapist-Led (but Not Client-Led) Physiological Synchrony during Imagery Rescripting

Jessica Prinz [1,*], Eshkol Rafaeli [2], Jana Wasserheß [1] and Wolfgang Lutz [1]

1 Department of Clinical Psychology and Psychotherapy, University of Trier, 54296 Trier, Germany; s1jawass@uni-trier.de (J.W.); lutzw@uni-trier.de (W.L.)
2 Department of Psychology, Bar-Ilan University, Ramat Gan 5290002, Israel; eshkol.rafaeli@gmail.com
* Correspondence: prinzj@uni-trier.de; Tel.: +49-(0)651-2013743

Abstract: Imagery rescripting (IR), an effective intervention technique, may achieve its benefits through various change mechanisms. Previous work has indicated that client–therapist physiological synchrony during IR may serve as one such mechanism. The present work explores the possibility that therapist-led vs. client-led synchrony may be differentially tied to clients' emotional experiences in therapy. The analyses were conducted with data taken from an open trial of a brief protocol for treating test anxiety (86 IR sessions from 50 client–therapist dyads). Physiological synchrony in electrodermal activity was indexed using two cross-correlation functions per session: once for client leading and again for therapist leading (in both cases, with lags up to 10 s). The clients' and therapists' in-session emotions were assessed with the Profile of Mood States. Actor–partner interdependence models showed that certain client (but not therapist) in-session emotions, namely higher contentment and lower anxiety and depression, were tied to therapist-led (but not client-led) physiological synchrony. The results suggest that therapist-led synchrony (i.e., clients' arousal tracking therapists' earlier arousal) is tied to more positive and less negative emotional experiences for clients.

Keywords: imagery rescripting; physiological synchrony; electrodermal activity; actor–partner interdependence models

Citation: Prinz, J.; Rafaeli, E.; Wasserheß, J.; Lutz, W. Clients' Emotional Experiences Tied to Therapist-Led (but Not Client-Led) Physiological Synchrony during Imagery Rescripting. *Entropy* **2021**, *23*, 1556. https://doi.org/10.3390/e23121556

Academic Editors: Franco Orsucci and Wolfgang Tschacher

Received: 20 October 2021
Accepted: 22 November 2021
Published: 23 November 2021

1. Introduction

Over the past two decades, psychotherapy researchers have demonstrated that imagery-based techniques are a very effective means of intervention for various disorders [1]. Because emotions are more strongly associated with images than with verbal thoughts, imagery-based techniques appear to activate emotions more strongly than simple conversation [1,2]. Much of the work on imagery-based techniques has centered on imagery with rescripting (i.e., imagery rescripting (IR)), an approach which was originally developed for work with clients who had undergone traumatic experiences (e.g., [3,4]). In IR, imagery is used to activate distressing memories replete with vivid sensory and emotional and cognitive content, an activation which also helps clarify unmet needs that still plague the client (e.g., [5]). The reactivated experience is then "rescripted" (i.e., changed in the imagination in a positive, desired direction) so that the unmet needs of the vulnerable or traumatized self are satisfied, at least in part. To accomplish this, the client is asked to imagine the scene from the perspective of their present self and step into the image to do whatever is necessary to satisfy the needs of their vulnerable selves [6–8].

Mental images simulate perceptual processes and elicit reactions that are quite similar to real experiences [9,10]. Consequently, as numerous laboratory studies have shown, imagery can activate strong physiological responses (e.g., [11–13]). To date, however, the role of physiological arousal vis-à-vis emotional activation in IR has received little attention.

This appears to be an important lacuna. After all, physiological data can serve as an objective measure of the arousal component of participants' emotional responses, particularly those of anxiety and stress [14,15]. They can be collected with minimal client burden and disruption to the treatment itself [14]. Unlike self-reports, physiological measures can be recorded continuously with a high temporal resolution and are therefore able to capture important nuanced responses. Consequently, physiological measures may open a window into identifying additional mechanisms of change in IR that go beyond cognitive accounts (e.g., [6,16,17]) and may allow us to detect beneficial emotional processes.

The evocative potency of IR often leads to emotional activation (and to its attendant physiological reactions) in the therapist alongside the client. It is quite possible that this synchronous activation could serve as a mechanism of change by increasing the sense of a shared experience within the dyad and by catalyzing the client's intrapersonal emotional processes.

To date, interpersonal processes in IR have received mostly theoretical attention. Rafaeli et al. [18] hypothesized that therapists' emotional activation during IR may serve as a mechanism of change having to do with shared emotions, shared focus and greater empathy or attunement. More generally, Koole and Tschacher's In-Sync model [19] postulated that synchronous (i.e., shared) emotions and experiences from clients and therapists lead to a shared experience and better client emotion regulation.

Several studies have investigated the dynamics and clinical meaning of client–therapist physiological synchrony (for a review, see [20]). As a group, these studies generally demonstrate positive associations between synchrony on the one hand and adaptive processes (e.g., empathy or attachment) on the other. To our knowledge, only two studies have examined physiological synchrony specifically during imagery interventions (with or without rescripting). Both studies involved multiple sessions and used electrodermal activity (EDA) as an index of arousal (based on extensive work demonstrating the sensitivity of EDA to emotional and cognitive processing, such as in [21,22]). Additionally, both studies compared client–therapist EDA synchrony during imagery vs. cognitive behavioral (CB) segments of sessions and found the former (but not the latter) to be tied to therapeutic bond ratings [23] and to the next session's (as well as overall treatment's) outcome [24].

Notably, both studies utilized an overall synchrony index (namely cross-correlation functions computed using $\pm$10-s lags on the dyads' residualized EDA time series). As such, they do not allow us to distinguish between synchronous experiences that are led by one party (e.g., the client) or by the other (e.g., the therapist). In other words, these studies' results could not be used to identify which party typically drives the physiological synchrony, nor could they tell us whether both therapist-driven and client-driven synchrony—or only one of the two—were tied to adaptive therapeutic processes.

To begin answering these questions, we may draw on the study of synchrony processes within other dyadic contexts. In developmental psychology, for example, studies on caregiver–infant interactions have demonstrated that both parties co-regulate their physiology and emotions in a dynamically responsive way to each other [25]. Feldman (e.g., [26,27]; see also [28]) described caregiver–infant synchrony as reciprocal processes (on the sensory, hormonal and physiological levels) and suggested that it serves as an important part in infant's development, such as in self-regulation capacity.

Several authors (e.g., [29]) have argued that synchrony between therapists and clients is likely to lead to similar self-regulation benefits. This may be particularly true for synchrony experienced during emotionally charged therapeutic moments, such as IR segments.

Specifically, in IR, the client and the therapist deliberately activate a memory or an experience which elicits an emotional reaction. Sharing and processing this experience, as well as communicating the emotions that accompany it, often generates emotional reactions in both parties, setting off dynamic dyadic affective processes [29]. Such processes may be characterized by synchrony or asynchrony, as well as by leading or following. Synchrony may exert its benefits best when a therapist tracks their client's arousal levels—which may suggest empathic accuracy or a shared experience (e.g., [30])—or when the client tracks their therapist's arousal level, which may suggest that co-regulation is occurring (e.g., [31]).

To examine these two (not necessarily incompatible) possibilities, the present study used EDA data from 50 client–therapist dyads who participated in a 6-session imagery-based treatment addressing test anxiety. Sessions 3 and 4 of the protocol use the traditional IR of past situations. We used the EDA data from these two sessions to investigate the associations between a client's emotional experience during these sessions and both therapist-led and client-led physiological synchrony. Specifically, we calculated two synchrony indices. Therapist-led synchrony was defined as the cross-correlation within a lag of 0–10 s with the therapist preceding the client, while client-led synchrony was defined as the cross-correlation within a lag of 0–10 s with the client preceding the therapist. We expected our data to provide a conceptual replication of the positive associations between physiological synchrony and adaptive therapy processes found previously [23,24]. Moreover, our central (though exploratory) goal was to distinguish between therapist-led and client-led physiological synchrony and determine whether either or both are associated with adaptive emotional experiences (i.e., lower negative emotions and higher positive ones) in IR sessions.

2. Materials and Methods

2.1. Clients

A total of 90 potential participants were recruited using flyers and a campus newsletter. The following inclusion criteria were applied: (1) a score higher than 54 on the Test Anxiety Inventory (TAI [32]); (2) absence of an imminent risk for suicide; and (3) currently no other psychological treatment for test anxiety. Based on these criteria, three participants were excluded. Twelve additional participants dropped out after the intake examination because of timing or setting concerns. Seventy-five clients met the inclusion criteria and began treatment. Of these, 11 clients dropped out during the treatment period. Thus, 64 clients completed the entire 6-session protocol. The present study utilized data solely from Sessions 3 and 4, in which traditional IR techniques were used. Physiological data from 14 clients (or their therapists) were lost due to poor signals or technical problems; thus, the final sample consisted of 50 clients (44 female, MAge = 25.3, SDAge = 6.17). The clients differed in terms of their academic fields, with psychology, law, education, business and computer science being the most frequent ones. Table 1 provides additional client information.

Table 1. Client characteristics.

	M	SD
Academic Year	4.82	3.82
TAI	65.24	7.78
Marital Status	N	%
Single	35	70
In a Relationship	12	24
Married	2	4
Divorced	1	2
Degree Being Pursued		
Bachelor	36	72
Master	6	12
Other	8	16

Written informed consent was obtained from all clients. Other than receiving the treatment at no cost, participants were not compensated in any way. All procedures performed in this study involving human participants were in accordance with the ethical standards of the institutional or national research committee and with the 1964 Declaration of Helsinki and its later amendments or comparable ethical standards. This study was approved by the local research ethics committee (Nr. 01/2020, ethics committee of Trier University).

2.2. Therapists, Treatment, Training and Supervision

Twenty-two therapists treated between one and six clients each (M = 2.27 clients, SD = 1.09). The therapists were either psychotherapy trainees with at least one year of experience as clinicians (N = 5) or masters students in clinical psychology with no prior clinical experience (N = 17). All therapists received intensive training in using the treatment protocol and were supervised by a senior therapist in weekly video-based group supervision.

The treatment used was based on a six-session protocol integrating different cognitive behavioral (CB) as well as imagery-based techniques (for the full protocol, see www.osf.io/hraqd (accessed on 21 November 2021)). This protocol's effectiveness for the treatment of test anxiety has been previously reported [33]. The protocol's CB components include psychoeducation at the beginning of treatment, discussion of present and elaboration of alternative perceptions, feelings, behaviors and cognitions in past and future learning and exam situations and the optimization of learning strategies and procedures in exams. The protocol's imagery-based components differ from session to session. The present study uses data from Sessions 3 and 4, in which the traditional IR of past situations was used.

2.3. Measures

2.3.1. POMS

To assess the clients' and therapists' in-session emotions, we used a shortened version of the Profile of Mood States [34]. Each of the seven mood variables (contentment, vigor, calmness, anxiety, depression, anger and fatigue) was assessed by three items (e.g., for depression: hopeless, discouraged and sad). Clients and therapists were asked to rate the extent to which they had felt these feelings during the session on a five-point Likert scale ("1" = "not at all" and "5" = "extremely"). The POMS has been validated and applied in several previous studies (e.g., [34,35]), and it demonstrated excellent internal consistency in present study (α ranged from 0.80 (Session 3) to 0.81 (Session 4)).

2.3.2. Client–Therapist EDA Synchrony

EDA was monitored simultaneously for the client and therapist. Two Ag/AgCl electrodes were attached to the third and fourth digits of the non-dominant hand. The signal was recorded at a sampling rate of 500 Hz using a Becker Meditec EDA module amplifier (Karlsruhe, Germany; with 0–100 μS, sensitivity: 25 mV/μS) connected to the acquisition computer via a Cesys C028149 USB-ISOLATOR and downsampled to 25 Hz. Several preprocessing steps were conducted before the cross-correlation functions (CCFs) were applied. First, the raw data were screened, and any artifacts were removed. Nonresponsive signals (EDA > 1 μs in at least 10% of the time series) were excluded from the analyses. Second, the signal was recorded in 1-s intervals and averaged across 2-min segments. Third, the auto.arima function (forecast package for R [36]) was applied to remove the autocorrelated component for each EDA time series. (For a similar approach, see [23,24])

Two CCFs were computed per session, one in which the client led (with a maximum lag of +10 s) and one in which the therapist led (with a maximum lag of −10 s). The maximal positive (in-phase) correlations (one for each CCF) were used as the two synchrony indices.

A total of 86 sessions (out of $50 \times 2 = 100$) were analyzed. Three sessions were not recorded due to technical problems. An additional 11 sessions were excluded because of non-responsive signals from either the clients or the therapists.

2.4. Data Analysis

Actor–Partner Interdependence Models

Given the dyadic nature of our data, we used a series of actor–partner interdependence models (APIMs; see Figure 1), with one for each POMS scale. The two dependent variables (client's POMS ratings and therapist's POMS ratings) are modeled on four independent variables (client-led and therapist-led physiological synchrony, as well as lagged client and therapist POMS ratings from the previous session). Paths marked as "a" represent actor effects (i.e., the degree to which the client-led or therapist-led synchrony predicts their own post-session POMS ratings). Paths marked as "p" represent partner effects (i.e., the degree to which the client-led or therapist-led synchrony predicts their partner's post-session POMS ratings). The actor and partner paths are estimated simultaneously while adjusting for each actor's lagged POMS ratings, and U and U′ denote the residual error terms for the two dependent variables.

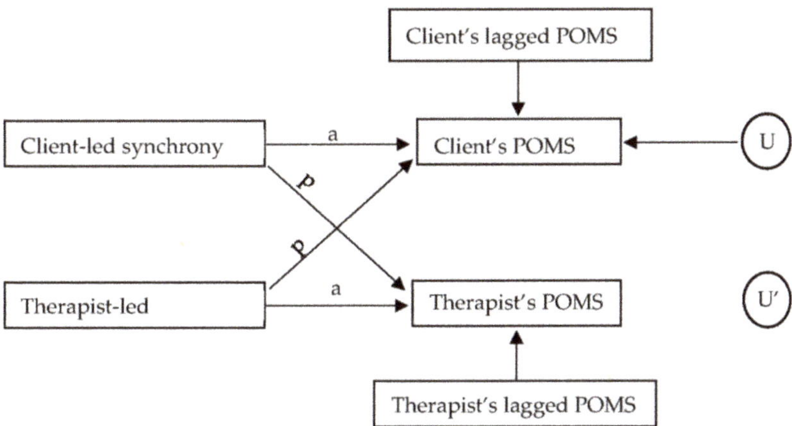

Figure 1. The Actor–Partner Interdependence Model. Note: U = residual error for client; U′ = residual error for therapist; a = actor effect; p = partner effect.

The models were estimated using the two-intercept approach to multilevel modeling [37] with the following equation:

$$\begin{aligned} POMS_{tp/cd} &= b0_{cd} + b1_{cd} \times POMS_{(t-1)cd} \\ &+ b2_{cd} \times \text{Client-led Synchrony}_{tcd} + b3_{cd} \times \text{Therapist-led Synchrony}_{tcd} + e_{tcd} \\ &+ b4_{pd} + b5_{pd} \times POMS_{(t-1)pd} \\ &+ b6_{pd} \times \text{Therapist-led Synchrony}_{tpd} + b7_{pd} \times \text{Client-led Synchrony}_{tpd} + e_{tpd} \end{aligned}$$

Here, the POMS score in each session (t) for the client (c) or psychotherapist (p) in each dyad (d) is modeled using their own lagged (t − 1) POMS score, as well as the two (client-led and therapist-led) synchrony scores (which serve as actor and partner effects, interchangeably) and a within-person residual error score. This model was run seven times: once for each POMS emotion (contentment, vigor, calmness, anxiety, depression, anger and fatigue).

3. Results

The results from the three APIM analyses predicting positive emotions showed that only therapist-led synchrony was associated with clients' emotional experience, significantly so for contentment and marginally for vigor and calmness. Client-led synchrony was not significantly associated with therapists' or clients' emotional experiences. All results are presented in Table 2.

Table 2. APIMs of therapist- and client-led synchrony as predictors of both actor and partner positive POMS ratings.

POMS Emotion	Synchrony Predictors	Estimate	Std. Error	*p*
Contentment				
	Therapist-Led Actor	6.048	3.891	0.122
	Client-Led Actor	−3.428	3.859	0.376
	Therapist-Led Partner	10.083	3.893	0.010
	Client-Led Partner	1.993	3.859	0.606
Vigor				
	Therapist-Led Actor	3.169	3.122	0.313
	Client-Led Actor	1.420	3.087	0.646
	Therapist-Led Partner	*5.557*	*3.113*	*0.076*
	Client-Led Partner	3.928	3.085	0.205
Calmness				
	Therapist-Led Actor	4.747	3.093	0.143
	Client-Led Actor	−2.400	3.057	0.441
	Therapist-Led Partner	*5.840*	*3.087*	*0.075*
	Client-Led Partner	−0.009	3.057	0.997

Note: Actor effects involve therapist-led or client-led synchrony predicting the actor's own emotional experience as the outcome. Partner effects involve therapist-led or client-led synchrony predicting the partner's emotional experience as the outcome.

Negative Emotions as Outcomes

The results from the four APIM analyses predicting negative emotions showed that higher therapist-led synchrony was significantly associated with a lower client emotional experience of anxiety and depression, as well as marginally lower fatigue. No associations were found for the emotional experience of anger. In addition, client-led synchrony was significantly associated with the higher client emotional experience of anxiety. All results are presented in Table 3.

Table 3. APIMs of therapist- and client-led synchrony as predictors of both actor and partner negative POMS ratings.

POMS Emotion	Synchrony Predictors	Estimate	Std. Error	*p*
Anxiety				
	Therapist-Led Actor	−2.924	3.172	0.358
	Client-Led Actor	9.383	3.120	0.003
	Therapist-Led Partner	−8.439	3.147	0.008
	Client-Led Partner	−0.299	3.142	0.924
Depression				
	Therapist-Led Actor	−1.872	3.618	0.606
	Client-Led Actor	3.790	3.580	0.292
	Therapist-Led Partner	−10.499	3.616	0.004
	Client-Led Partner	1.976	3.564	0.580
Anger				
	Therapist-Led Actor	−0.792	3.109	0.799
	Client-Led Actor	2.298	3.058	0.454
	Therapist-Led Partner	−0.999	3.084	0.746
	Client-Led Partner	0.151	3.079	0.961
Fatigue				
	Therapist-Led Actor	−3.586	3.276	0.275
	Client-Led Actor	0.192	3.244	0.953
	Therapist-Led Partner	*−5.908*	*3.272*	*0.073*
	Client-Led Partner	−3.012	3.268	0.358

Note: Actor effects involve therapist-led or client-led synchrony predicting the actor's own emotional experience as the outcome. Partner effects involve therapist-led or client-led synchrony predicting the partner's emotional experience as the outcome.

4. Discussion

The present study aimed to identify specific EDA dynamics in client–therapist dyads during IR that are associated with in-session emotional experiences. To our knowledge, this is the first study examining physiological synchrony in regard to a leading or following partner in IR. Therapist-led synchrony was significantly associated with clients' in-session emotional experiences of greater contentment, lower anxiety and lower depression and marginally associated with the experience of greater vigor and calmness and lower fatigue. In addition, client-led synchrony was significantly associated with their own greater feelings of anxiety. No association was found between either synchrony score and the emotional experience of anger.

As expected, the results highlight the importance of distinguishing between therapist-led and client-led synchrony. Therapist-led synchrony was associated with more positive and less negative in-session client emotional experiences. In contrast, with one exception (i.e., greater anxiety), client-led synchrony was unrelated to the clients' emotional experiences.

These results are in line with the previous literature conferring a mood-regulatory role on the therapist (e.g., [29]). Specifically, the results of the present study indicate that beneficial emotion regulation occurs not only because clients share their emotions with their therapists (as has been predicted, for example, by the social baseline theory [38]) but also because they synchronize their arousal levels with those of their therapists (but only when the temporal sequence has the therapist in the lead and is client-following). This is particularly interesting because this process typically happens outside of awareness. During the emotionally intensive IR segments, therapists often empathize with their clients' narratives [23] and, presumably, the more this occurs, the better they are able to help the client process and regulate their emotions. The present results suggest that more effective therapists may actually be "guiding" their clients through the emotions that arise during IR, with the therapists being slightly ahead of their clients' emotional responses.

In the present study, only certain emotions were significantly associated with therapist-led synchrony. This finding may have to do with the specific distress with which clients in this study were contending, namely test anxiety. Test anxiety is characterized by negative thoughts about consequences or failure in exams or other evaluation-related situations. Its symptoms include both emotional ones (e.g., fear and anxiety) and behavioral ones (e.g., sleep disturbance, procrastination, impaired motivation and rumination). Notably, these behavioral symptoms are quite similar to depressive reactions. This may explain why therapist-led physiological synchrony was associated most strongly with feelings of anxiety and depression.

Methodologically, the marginal findings (with fatigue, vigor and calm) and the non-significant finding (with anger) may be attributable to floor effects, namely ratings that were low and lacked much variability. Such ratings make the detection of significant effects nearly impossible.

Strengths, Limitations and Future Directions

The current study is novel in several respects. To our knowledge, it is the first study to examine leading and following with regard to physiological synchrony in psychotherapy, rather than simply examining the overall cross-correlations. Its use of multiple (in this case, two) sessions per dyad is an additional strength, as it its focus on IR segments (in which emotional activation tends to be most consequential). By focusing on IR, a technique without eye contact, our results demonstrate that the observation of others' emotional responses is not needed for synchronization or its benefits.

These strengths notwithstanding, several limitations of this study are noteworthy. The POMS items were used to assess emotions experienced during the entire session and not only during the IR segments. Even though the CB segments, which tended to be more conversational and less experiential, are likely generate lower emotional activation than the IR segments (see also [1,2]), these segments may have also influenced the post-session POMS ratings.

Additionally, prior research on IR mechanisms has mostly relied on laboratory studies, which induce memories or feelings in a controlled setting (e.g., [39,40]). To better understand how IR can address the emotional beliefs and dysfunctional schemas that underlie emotional disorders, we chose to examine IR mechanisms in a more ecologically valid treatment setting. However, our choice robs us of the possibility (available in lab studies) to control both the content and the temporal aspects of the experience.

Another shortcoming of the present study is its focus on sympathetic arousal. Our use of EDA, an indicator of sympathetic activation, is considered to be positively associated with emotional arousal and specifically with emotions such as anger, anxiety and fear [41]. The use of alternative measures which were unavailable to us (e.g., heart rate variability (HRV), a measure of parasympathetic functioning) may have painted a different picture, as such measures are more strongly associated with self-regulation [42]. Indeed, our non-significant results (with most positive emotions) may stem from this limitation, as our EDA responses did not capture parasympathetic functioning. Furthermore, the indexation of synchrony as a maximum positive (in-phase) correlation only reflected co-regulation to a limited extent. Future studies could benefit from the collection of both the EDA and HRV channels as well as the separate investigation of in-phase and anti-phase correlations.

5. Conclusions

The present study adds to the growing body of research investigating IR mechanisms, and the findings highlight the importance of examining physiological processes. The results provide the first evidence that therapists' physiological reactions—and their clients' subsequent synchronization with these reactions—may be tied to better session-level outcomes, at least with respect to client anxiety and depression. Methodologically, this study provides further evidence for the importance of assessing leading vs. following in physiological synchrony.

Author Contributions: Conceptualization, J.P.; methodology, J.P., E.R. and W.L.; formal analysis, J.P. and J.W.; writing—original draft preparation, J.P.; writing—review and editing, E.R. and W.L. All authors have read and agreed to the published version of the manuscript.

Funding: This research received no external funding.

Institutional Review Board Statement: The study was conducted according to the guidelines of the Declaration of Helsinki, and approved by the Institutional Review Board (or Ethics Committee) of the University of Trier (protocol code Nr. 01/2020 and date of approval 27 July 2020).

Informed Consent Statement: Informed consent was obtained from all subjects involved in the study.

Data Availability Statement: The datasets analyzed in the current study are available from the corresponding author on reasonable request.

Conflicts of Interest: The authors declare no conflict of interest.

References

1. Wheatley, J.; Hackmann, A.; Brewin, C. Imagery rescripting for intrusive sensory memories in major depression following traumatic experiences. In *A Casebook of Cognitive Therapy for Traumatic Stress Reactions*; Grey, N., Ed.; Routledge: Oxfordshire, UK, 2009; pp. 78–92.
2. Rafaeli, E.; Bernstein, D.P.; Young, J. *Schema Therapy: Distinctive Features*; Routledge: Oxfordshire, UK, 2010.
3. Arntz, A.; Tiesema, M.; Kindt, M. Treatment of ptsd: A comparison of imaginal exposure with and without imagery rescripting. *J. Behav. Ther. Exp. Psychiatry* **2007**, *38*, 345–370. [CrossRef]
4. Smucker, M.R.; Dancu, C.; Foa, E.B.; Niederee, J.L. Imagery rescripting: A new treatment for survivors of childhood sexual abuse suffering from posttraumatic stress. *J. Cogn. Psychother.* **1995**, *9*, 3–17. [CrossRef]
5. Arntz, A.; Thoma, T.; McKay, D. Imagery rescripting for posttraumatic stress disorder. In *Working with Emotion in Cognitive-Behavioral Therapy: Techniques for Clinical Practice*; Thoma, N.C., McKay, D., Eds.; Guilford Publications: New York, NY, USA, 2014; pp. 203–2015.

6. Arntz, A. Imagery rescripting as a therapeutic technique: Review of clinical trials, basic studies, and research agenda. *J. Exp. Psychopathol.* **2012**, *3*, 189–208. [CrossRef]
7. Arntz, A.; Weertman, A. Treatment of childhood memories: Theory and practice. *Behav. Res. Ther.* **1999**, *37*, 715–740. [CrossRef]
8. Holmes, E.A.; Arntz, A.; Smucker, M.R. Imagery rescripting in cognitive behaviour therapy: Images, treatment techniques and outcomes. *J. Behav. Ther. Exp. Psychiatry* **2007**, *38*, 297–305. [CrossRef]
9. Holmes, E.A.; Mathews, A. Mental imagery in emotion and emotional disorders. *Clin. Psychol. Rev.* **2010**, *30*, 349–362. [CrossRef]
10. Ji, J.L.; Heyes, S.B.; MacLeod, C.; Holmes, E.A. Emotional mental imagery as simulation of reality: Fear and beyond—A tribute to peter lang. *Behav. Ther.* **2016**, *47*, 702–719. [CrossRef]
11. Cuthbert, B.N.; Lang, P.J.; Strauss, C.; Drobes, D.; Patrick, C.J.; Bradley, M.M. The psychophysiology of anxiety disorder: Fear memory imagery. *Psychophysiology* **2003**, *40*, 407–422. [CrossRef]
12. Lang, P.J.; Levin, D.N.; Miller, G.A.; Kozak, M.J. Fear behavior, fear imagery, and the psychophysiology of emotion: The problem of affective response integration. *J. Abnorm. Psychol.* **1983**, *92*, 276–306. [CrossRef]
13. Miller, G.A.; Levin, D.N.; Kozak, M.J.; Cook, E.W.; McLean, A.; Lang, P.J. Individual differences in imagery and the psychophysiology of emotion. *Cogn. Emot.* **1987**, *1*, 367–390. [CrossRef]
14. Deits-Lebehn, C.; Baucom, K.J.W.; Crenshaw, A.O.; Smith, T.W.; Baucom, B.R.W. Incorporating physiology into the study of psychotherapy process. *J. Couns. Psychol.* **2020**, *67*, 488–499. [CrossRef]
15. Hill-Soderlund, A.L.; Mills-Koonce, W.R.; Propper, C.; Calkins, S.D.; Granger, D.A.; Moore, G.A.; Gariepy, J.-L.; Cox, M.J. Parasympathetic and sympathetic responses to the strange situation in infants and mothers from avoidant and securely attached dyads. *Dev. Psychobiol.* **2008**, *50*, 361–376. [CrossRef]
16. Brewin, C.R.; Wheatley, J.; Patel, T.; Fearon, P.; Hackmann, A.; Wells, A.; Fisher, P.; Myers, S. Imagery rescripting as a brief stand-alone treatment for depressed patients with intrusive memories. *Behav. Res. Ther.* **2009**, *47*, 569–576. [CrossRef]
17. Dibbets, P.; Poort, H.; Arntz, A. Adding imagery rescripting during extinction leads to less aba renewal. *J. Behav. Ther. Exp. Psychiatry* **2012**, *43*, 614–624. [CrossRef]
18. Rafaeli, E.; Maurer, O.; Thoma, N.C. Working with modes in schema therapy. In *Working with Emotion in Cognitive Behavioral Therapy: Techniques for Clinical Practice*; Guilford Publications: New York, NY, USA, 2014; pp. 263–287.
19. Koole, S.L.; Tschacher, W. Synchrony in psychotherapy: A review and an integrative framework for the therapeutic alliance. *Front. Psychol.* **2016**, *7*, 862. [CrossRef]
20. Kleinbub, J.R. State of the art of interpersonal physiology in psychotherapy: A systematic review. *Front. Psychol.* **2017**, *8*, 2053. [CrossRef]
21. Dawson, M.E.; Schell, A.M.; Filion, D.L. The electrodermal system. In *Handbook of Psychophysiology*; Cacioppo, J.T., Tassinary, L.G., Berntson, G.G., Eds.; Cambridge University Press: Cambridge, UK, 2017; pp. 217–243.
22. Sequeira, H.; Hot, P.; Silvert, L.; Delplanque, S. Electrical autonomic correlates of emotion. *Int. J. Psychophysiol.* **2019**, *71*, 50–56. [CrossRef]
23. Bar-Kalifa, E.; Prinz, J.; Atzil-Slonim, D.; Rubel, J.A.; Lutz, W.; Rafaeli, E. Physiological synchrony and therapeutic alliance in an imagery-based treatment. *J. Couns. Psychol.* **2019**, *66*, 508–517. [CrossRef]
24. Prinz, J.; Rafaeli, E.; Reuter, J.K.; Bar-Kalifa, E.; Lutz, W. Physiological activation and co-activation in an imagery-based treatment for test anxiety. *Psychother. Res. J. Soc. Psychother. Res.* **2021**, 1–11. [CrossRef]
25. Buhler-Wassmann, A.C.; Hibel, L.C. Studying caregiver-infant co-regulation in dynamic, diverse cultural contexts: A call to action. *Infant Behav. Dev.* **2021**, *64*, 101586. [CrossRef]
26. Feldman, R. Parent-infant synchrony and the construction of shared timing; physiological precursors, developmental outcomes, and risk conditions. *J. Child Psychol. Psychiatry Allied Discip.* **2007**, *48*, 329–354. [CrossRef]
27. Feldman, R. Bio-behavioral synchrony: A model for integrating biological and microsocial behavioral processes in the study of parenting. *Parenting* **2012**, *12*, 154–164. [CrossRef]
28. Hofer, M.A. Psychobiological roots of early attachment. *Curr. Dir. Psychol. Sci.* **2006**, *15*, 84–88. [CrossRef]
29. Fosha, D. The dyadic regulation of affect. *J. Clin. Psychol.* **2001**, *57*, 227–242. [CrossRef]
30. Ramseyer, F.; Tschacher, W. Nonverbal synchrony in psychotherapy: Coordinated body movement reflects relationship quality and outcome. *J. Consult. Clin. Psychol.* **2011**, *79*, 284–295. [CrossRef]
31. Paz, A.; Rafaeli, E.; Bar-Kalifa, E.; Gilboa-Schectman, E.; Gannot, S.; Laufer-Goldshtein, B.; Narayanan, S.; Keshet, J.; Atzil-Slonim, D. Intrapersonal and interpersonal vocal affect dynamics during psychotherapy. *J. Consult. Clin. Psychol.* **2021**, *89*, 227–239. [CrossRef]
32. Spielberger, C.D. *Test Anxiety Inventory: Preliminary Professional Manual*; Consulting Psychologist Press: Palo Alto, CA, USA, 1980.
33. Prinz, J.; Bar-Kalifa, E.; Rafaeli, E.; Sened, H.; Lutz, W. Imagery-based treatment for test anxiety: A multiple-baseline open trial. *J. Affect. Disord.* **2019**, *244*, 187–195. [CrossRef]
34. Cranford, J.A.; Shrout, P.E.; Iida, M.; Rafaeli, E.; Yip, T.; Bolger, N. A procedure for evaluating sensitivity to within-person change: Can mood measures in diary studies detect change reliably? *Personal. Soc. Psychol. Bull.* **2006**, *32*, 917–929. [CrossRef]
35. Bar-Kalifa, E.; Abba-Daleski, M.; Pshedetzky-Shochat, R.; Gleason, M.E.J.; Rafaeli, E. Respiratory sinus arrhythmia as a dyadic protective factor in the transition to parenthood. *Psychophysiology* **2021**, *58*, e13736. [CrossRef]
36. Hyndman, R.J.; Athanasopoulos, G. *Forecasting: Principles and Practice*; OTexts: Melbourne, Australia, 2018.

37. Cook, W.L.; Kenny, D.A. The actor–partner interdependence model: A model of bidirectional effects in developmental studies. *Int. J. Behav. Dev.* **2005**, *29*, 101–109. [CrossRef]
38. Coan, J.A.; Sbarra, D.A. Social baseline theory: The social regulation of risk and effort. *Curr. Opin. Psychol.* **2015**, *1*, 87–91. [CrossRef]
39. Dibbets, P.; Arntz, A. Imagery rescripting: Is incorporation of the most aversive scenes necessary? *Memory* **2016**, *24*, 683–695. [CrossRef]
40. Kunze, A.E.; Arntz, A.; Kindt, M. Investigating the effects of imagery rescripting on emotional memory: A series of analogue studies. *J. Exp. Psychopathol.* **2019**, *10*, 204380871985073. [CrossRef]
41. Kreibig, S.D. Autonomic nervous system activity in emotion: A review. *Biol. Psychol.* **2010**, *84*, 394–421. [CrossRef]
42. Thayer, J.F.; Hansen, A.L.; Saus-Rose, E.; Johnsen, B.H. Heart rate variability, prefrontal neural function, and cognitive performance: The neurovisceral integration perspective on self-regulation, adaptation, and health. *Ann. Behav. Med.* **2009**, *37*, 141–153. [CrossRef]

MDPI

Article

Affective Saturation Index: A Lexical Measure of Affect

Alessandro Gennaro [1], Valeria Carola [1,2], Cristina Ottaviani [2,3], Chiara Pesca [1], Arianna Palmieri [4] and Sergio Salvatore [1,*]

1 Department of Dynamic and Clinical Psychology and Health Studies, Sapienza University of Rome, 00185 Rome, Italy; a.gennaro@uniroma1.it (A.G.); valeria.carola@uniroma1.it (V.C.); chiara.pesca@uniroma1.it (C.P.)
2 IRCCS Santa Lucia Foundation, 00179 Rome, Italy; cristina.ottaviani@uniroma1.it
3 Department of Psychology, Sapienza University of Rome, 00185 Rome, Italy
4 Padua Neuroscience Center, Department of Philosophy, Sociology, Education and Applied Psychology, University of Padua, 35122 Padua, Italy; arianna.palmieri@unipd.it
* Correspondence: sergio.salvatore@uniroma1.it

Abstract: Affect plays a major role in the individual's daily life, driving the sensemaking of experience, psychopathological conditions, social representations of phenomena, and ways of coping with others. The characteristics of affect have been traditionally investigated through physiological, self-report, and behavioral measures. The present article proposes a text-based measure to detect affect intensity: the Affective Saturation Index (ASI). The ASI rationale and the conceptualization of affect are overviewed, and an initial validation study on the ASI's convergent and concurrent validity is presented. Forty individuals completed a non-clinical semi-structured interview. For each interview transcript, the ASI was esteemed and compared to the individual's physiological index of propensity to affective arousal (measured by heart rate variability (HRV)); transcript semantic complexity (measured through the Semantic Entropy Index (SEI)); and lexical syntactic complexity (measured through the Flesch–Vacca Index (FVI)). ANOVAs and bi-variate correlations estimated the size of the relationships between indexes and sample characteristics (age, gender), then a set of multiple linear regressions tested the ASI's association with HRV, the SEI, and the FVI. Results support the ASI construct and criteria validity. The ASI proved able to detect affective saturation in interview transcripts (SEI and FVI, adjusted $R^2 = 0.428$ and adjusted $R^2 = 0.241$, respectively) and the way the text's affective saturation reflected the intensity of the individual's affective state (HRV, adjusted $R^2 = 0.428$). In conclusion, although the specificity of the sample (psychology students) limits the findings' generalizability, the ASI provides the chance to use written texts to measure affect in accordance with a dynamic approach, independent of the spatio-temporal setting in which they were produced. In doing so, the ASI provides a way to empower the empirical analysis of fields such as psychotherapy and social group dynamics.

Keywords: affect; affective saturation index; meaning; text analysis; physiology; heart rate variability

Citation: Gennaro, A.; Carola, V.; Ottaviani, C.; Pesca, C.; Palmieri, A.; Salvatore, S. Affective Saturation Index: A Lexical Measure of Affect. *Entropy* **2021**, *23*, 1421. https://doi.org/10.3390/e23111421

Academic Editors: Franco Orsucci and Wolfgang Tschacher

Received: 6 August 2021
Accepted: 24 October 2021
Published: 28 October 2021

1. Introduction

The measurement of the level of affective intensity is a relevant issue for both theoretical and practical reasons. Affect plays a central role in how individuals make sense of experience [1–3] as well as in how executive [4] and higher-order cognitive processes [5] operate. A high level of affect intensity is associated with psychopathological conditions [6–8] as well as maladjusted forms of behaviors, such as gambling [9–12] or academic drop-out [13]; again, the elaboration of affect and the mode of variation of its intensity within and throughout sessions is a core focus of the analysis of the psychotherapeutic process and clinical change [14–18]. Affect is also involved in psychosocial and socio-political phenomena. Social representations of forms of alterity have proven to frame their objects in terms of affect-laden meanings—e.g., the securitization frame, i.e., the view of alterity as an incumbent radical threat, is the most common way migration is conveyed by

Western media [19]. In/out-group polarization, populism, xenophobia, hate speech, and conspiracy theories are phenomena that, though different in content and determinants, are all characterized by the role played by affect, and, more specifically, by the affect-laden friend–foe schema [20–22]. More generally, affective activation has been considered a major response by which individuals and social groups cope with uncertainty [23–26]; accordingly, the analysis of the impact and design of social communication (in contexts such as politics, media, but also health, education, marketing, security, urban planning, and civic engagement) benefit from understanding the capacity of the message to pander to and/or oppose the target's affective response [27].

Due to the relevance of the topic to so many areas of psychological investigation, it is not surprising that numerous attempts have been made to measure the main characteristics of affect. Measures can be collected in three broad clusters: physiological, self-report, and, behavioral measures.

Physiological measures focus on both central (e.g., electroencephalography) and peripheral (e.g., electrodermal conductance, heart rate variability (HRV)) signals to estimate bodily affective activation (for a review, see [28]). The validity of these indexes is generally robust, given that they can be considered direct measures of the intensity of the body's physiological activation. However, physiological measures are generally not easy to use, given that, in most cases, their application and computation require technical devices and skilled researchers. Moreover, they often require implementation in controlled conditions, and this limits their ecological validity.

Self-report measures skip some of these limitations. Yet, although these kinds of measures are widely used in many domains of investigation, their validity is considerably jeopardized by the subjects' inherently low capability to reliably detect their own inner state [29,30]. This limitation is reflected in the weak association with both psychophysiological [31] and behavioral measures [32].

Behavioral measures offer an alternative to self-report. Many aspects of overt behavior —such as vocal fundamental frequency [33], speech rate [34], facial expressions [35], and whole-body posture [36]—have been proposed, conceived as markers of one or more features of the intensity of the affective state. However, these measures have been criticized because the association between behavior and affective states is not invariant but depends on contextual conditions [37–43]. Above all, it must be considered that it is not always possible to involve participants in individual tasks, or even that certain analyses are not based on participants as direct sources of data. For instance, in many areas of investigation—e.g., psychotherapy process, media representations of social objects (migration, Islam, COVID-19, etc.), organizational dynamics, and marketing communication—many studies cannot but be based on textual data such as verbatim transcripts of sessions, interviews, focus groups, newspaper articles, etc. Thus, the estimation of affect in research contexts of this kind requires a measure based on textual data, rather than on behavioral responses.

The present paper intends to address this need. It presents the Affective Saturation Index (ASI), a textual-based measure of the intensity of affect. To this end, we first present the ASI, the definition of affect it is grounded on, and its rationale; then, an initial validation study is reported. Three aspects of the ASI make it relevant from a dynamic systems theory standpoint. First, the ASI is based on a semiotic interpretation of affect, which is in turn grounded on a field-dynamic conception of meaning, which models it as an emergent property of sensemaking [44]. Second, dynamic systems theory informs the methodological framework on which the ASI's computation is based (in particular, the dimensional model of meaning). Third, the ASI is designed to detect the ongoing flow of meaning-making, enabling the application of strategies for data analysis (e.g., Time Series Analysis) aimed at modeling the dynamic evolution of communication and cognitive processes.

2. The Affective Saturation Index

2.1. The Semiotic Definition of Affect at the Basis of the ASI: Affect as Meanings

The ASI is based on the view of affect as global embodied meanings [1–3,44–46] —namely, patterns of activation of the whole body that provide the individual with a global experience of the world as a totality. For instance, when a person is happy—though they may be so because of something—their sense of pleasantness radiates over and fills their whole current experience of the world. As proposed by Feldman Barrett [29] (p. 30), affect is the "neurophysiological barometer" thanks to which the body maps the ongoing, immediate coupling with the world and, in so doing, prepares itself to address its variations.

It is worth underscoring that it is this function of dynamic mapping that makes affect a *meaning*—albeit of an embodied kind. Indeed, insofar as one assumes the pragmatist view of meaning—i.e., a response triggered in the interpreter by something in order to interpret that something [47] (p. 228)—then affect is meaning because it consists precisely of a response activated in the (body of the) interpreter by something, in order to interpret that something. Thus, as Peirce claims explicitly [48], affect can be conceived as a specific type of sign—namely, signs that make sense of the world in terms of basic patterns of bodily activation [49].

This semiotic interpretation of affect is framed in, and fully consistent with, the bodily nature of the cognitive process recognized by embodied cognition theories over the last quarter-century [50–53]. Moreover, it finds further support in psychoanalytic theory [54,55] and in semiotic-oriented cultural psychology [3,55,56]. Incidentally, this view enables affect to be distinguished from emotions, where the latter are discrete inner states (e.g., anger, joy) combining the state of body activation (i.e., the affect) with its categorization, occurring in accordance with contextual cues [29].

For the current discussion, it is worth highlighting the specificity of affective meaning. Given its embodiment and globality, affect is a (a) hypergeneralized, (b) homogenizing, (c) bipolar, (d) asemantic, and (e) basic class of meaning. It refers to the relationship with the whole environment, rather than with specific objects (*hypergeneralization* [29]; from a psychoanalytic perspective, [57]). In so doing, the discrete aspects of the context are assimilated to the whole affective meaning, as in all cows being black at night (*homogenization* [58]). Moreover, the fact that affect is a hypergeneralized global pattern implies that it works in dichotomous mode, namely as the juxtaposition of opposite states—e.g., pleasantness/unpleasantness (*bipolarism* [21,59–61]). Again, it has to be underlined that these characteristics make affective meaning-making work differently from the rule-based rational mode of thought. Indeed, the fact that affect works as a single, global, hypergeneralized, homogenizing class of meaning entails it establishing relationships between elements in spite of semantic, logical, and functional differences—e.g., from the standpoint of affective meaning-making, that which is beautiful is good and trustworthy (*asemanticity* [62]). Finally, affect is an embodied form of the organization of experience that emerges from early ontogenetic stages, and therefore comes before, paves the way for, frames, and channels the subsequent rule-based processing of the semantic content (*basicity* [29,45,63,64]).

Several streams of thought converge in providing support to the characterization of affective meaning proposed above. First, one can refer to classical studies [65,66] that have shown that the prime effect works also when the association between the prime and the stimulus consists of the sharing of an affective value (e.g., pleasantness) in the absence of any semantic linkage. Similarly, the Emotional Categorization Theory (ECT [67]) states that the meaning-maker tends to assimilate objects that have the same emotional valence for him/her into the same class, regardless of their semantic relationships. The Homogenization of Classification Function Measure (HOCFUN [63]) is a method of measuring affective intensity, based on a generalization of the ECT; it assumes that affective meaning is not limited to assimilating objects, but also properties and qualities—i.e., propositional functions (e.g., an object which is seen as pleasant tends to also be seen as important). Accordingly, HOCFUN estimates the affective intensity in terms of the degree to which the individual in an evaluation task uses two semantically independent evaluation criteria (pleasantness and relevance) in a homogeneous way. Second, the psychoanalytic theory of the unconscious, more specifically the tradition triggered by Freud's *Interpretation of Dreams* [68], has conceptualized the unconscious as a mode of thinking (rather than a place within the mind) characterized by a specific logic, that of the primary process ([45,59]; for a review of this tradition, see [55]). This tradition emphasizes that affective meaning-making is a mode of interpreting reality (*affective semiosis*, see [44]) that has its own specific logic (the primary process), which is different from rational, rule-based logic, but still systematic and endowed with inner consistency. Third, the copious literature on the semantic differential [69,70] has shown that semantic representation is grounded in three basic dimensions of meaning—evaluation, power, and activity. With few exceptions, these dimensions emerge systematically from hundreds of studies focused on a great many objects, adopting many different semantic scales, carried out in many cultural contexts, over more than half a century. Due to their generality and transversality, the three dimensions have been interpreted as basic affective meanings—for instance, the evaluation dimension channels the meaning of semantic scales as pleasant/unpleasant, beautiful/ugly, good/bad, making people use these scales in quite a similar way, despite their semantic differences. Fourth, the transversality and basicity of affect are further supported by the fact that affective meaning appears to be shared among cultures [71] and is active within all human languages [72]. Finally, framed by the Semiotic Cultural Psychology Theory, Salvatore and colleagues [5,21] have recently identified five global beliefs about life (symbolic universes, as defined by the authors) that are active in the cultural milieu of various European societies. As underscored by the authors, each symbolic universe is an implicit generalized meaning providing a global vision of what the world is/should be, which channels the way of feeling, thinking, and acting of those who identify with it. Like the semantic differential framework, symbolic universes have been considered forms of affective meaning, because each of them comprises a set of beliefs lacking semantic linkage, while associated with each other by reason of the same basic affective values (e.g., the connotation of the world as an enemy one must protect oneself from) [21].

2.2. ASI Rationale

2.2.1. Saturation and Intensity

The ASI is designed to detect the contribution of affective meanings—as defined above: hypergeneralized, homogenizing, bipolar, asemantic, and basic classes of meaning—to the whole semantic content of the text under investigation (e.g., individual interviews, focus groups, newspaper articles, or verbatim transcripts of psychotherapy sessions). The ASI calls the extent of this contribution *affective saturation*—the more the affective meaning contributes to the whole text's meaning, the greater the affective saturation of the text. Thus, affective saturation is like the chromatic saturation of an image—the more a given color contributes to the image, the more that image is saturated by that color.

The ASI assumes that the saturation reflects, at the level of textual output, the intensity of the meaning-maker's affective state, associated with and influencing the cognitive process underlying the production of the text. The higher the intensity of the affective state, the more power it has to influence the meaning-making, and therefore the greater the affective saturation of the meaning-making's textual output (see [44,64,73] for a discussion of this tenet in the context of the analysis of the psychotherapy process). In terms of the previous analogy, if an image is saturated by a given color, this is expected to be because that color plays a major role in the painter's aesthetic desire and taste.

In summary, the ASI assumes that the affective state promotes affective saturation as a function of its intensity. On this basis, the ASI considers affective saturation the textual marker of the level of affective intensity that characterizes the meaning-maker during the production of the text.

2.2.2. A Geometric Model of Affective Saturation

The ASI's estimation of affective saturation is based on a geometrical model of meaning [45]. According to this view, meaning (e.g., an attribute, a belief, or a representation) can be modeled as a point in a semiotic space—the *Phase Space of Meaning [PSM]*, as modeled by Venuleo and colleagues [74]. The PSM is made up of dimensions, each of which maps a relevant component of environmental variability. Accordingly, the meaning is represented by the coordinates of the corresponding point in the space. For instance, the meaning of "orange" could be represented as the point in the semiotic space composed of the dimensions: form, color, and function, and with the coordinates: almost spherical, orange, food. To give a topical example, Salvatore and colleagues [20] found that the five symbolic universes they identified are framed within a semiotic space composed of three dimensions: (1) connotation of the world (polarities: friend vs. foe); (2) direction of the desire (polarities: passivity vs. engagement); and (3) form of the demand (demand for systemic resources vs. demand for community identity). Incidentally, one can see that the first two of these dimensions overlap, similar to many dimensions emerging from the semantic differential [63], providing further evidence as to the basicity of affective meaning.

Two core points of the PSM are relevant here (for a theoretical discussion, cf. [45]; for a computational model, see [6]). First, PSM dimensionality requires modulation as a function of the contextual conditions. Indeed, in any circumstance, most of the environmental sources of variation—therefore of the PSM dimensions—are not relevant, and therefore must be backgrounded to enable the cognitive system to make the relevant information pertinent and thus process it. For instance, when driving, one should focus on the aspects of the car and road that are relevant to the regulation of the vehicle, while other non-pertinent aspects should be backgrounded as dangerous sources of control loss.

Second, the PSM dimensions can be divided into two broad classes—primary and secondary dimensions. The former consists of affective meanings, each of which corresponds to a specific semantic component that maps a given quality/property of the environment, to achieve the fine attunement with the environment needed for action to be regulated. As discussed above, the affective dimensions are basic; that is, they tend to be stable, both within cultures and between individuals—namely, they work as the grounds of any meaning-making process, regardless of contextual conditions. This means that the modulation of PSM dimensionality is due to secondary dimensions, that is, to the semantic components that are activated in order to elaborate the pockets of information required to regulate action in contingent circumstances.

The above points give rise to the view of affective saturation as *the relative weight of the primary over the secondary dimensions within the PSM.* A PSM that is highly saturated by primary dimensions models a text whose global meaning is largely defined by basic affective meanings, with little room left for further semantic components. For instance, various media and political discourses provide a representation of the concept of migration almost completely saturated by the securitization frame, an expression of the friend–foe affective meaning [21], with very little room for any other meaning. Thus, framed in the geometric model of meaning, the ASI computes affective saturation in terms of the contribution of the affective meanings to the text compared to the other semantic components.

The geometric view of meaning and the related definition of affective saturation are supported by growing evidence, even if this is mostly indirect. At the level of the individual, low dimensional semiotic spaces have proven to be associated with proxies of affective activation—e.g., higher response speed to evaluation tasks and a tendency to characterize semantically distinct social objects in homogeneous ways [63]; a high need for closure and a highly negative attitude towards foreigners [21]; and a lower tendency to explore the marginal area of the attentional perception field [63]. Finally, the geometrical model of meaning outlined above was recently used by some of the present authors to frame a computational model of the form of meaning-making underpinning psychopathology—the Harmonium Model [6]. In a subsequent work, the Harmonium Model was tested by means of a simulation based on a neural network deep learning procedure. Findings showed that the neural networks simulating psychopathological meaning-making were characterized by a lower dimensional micro-dynamic compared to the neural networks simulating non-psychopathological meaning-making. At a psycho-social level, symbolic universes whose content was characterized by connotations of the social world that were positively (very high trust in people and the future, very high satisfaction, or idealization of interpersonal bonds) or negatively (very high distrust, fatalism, or rejection of rules) polarized, reactive, and generalized—in that sense affect-laden—showed a lower dimensionality than symbolic universes whose content was characterized by moderate, differentiated beliefs [20].

2.3. Relation to Other Text-Based Measures of Affect

The ASI shares the interest of other measures in the measurement of affect in text. The Therapeutic Cycle Model [16] focuses on the interplay of two lexical measures, based on vocabularies estimating the Emotional Tone (ET)—i.e., the affective level of the text, as detected by the use of emotion-laden words—and the Abstraction (AB)—the abstract thought underpinning the text, as estimated by the use of abstract nouns or words—respectively. Similarly, the Discourse Attribute Analysis Program (DAAP [75]) is based on dictionaries and measures the text producer's mental activity of connecting affective and cognitive domains. Furthermore, the Linguistic Inquiry and Word Count (LIWC [76]) detects specific keywords assumed as the marker of relevant characteristics of cognitive processing (e.g., emotions, cognitions, or perceptions). However, in spite of their validity, reliability, and spread, these measures imply an invariant value of the words composing the dictionaries. Thus, they do not take into due account the indexicality of meaning [44], due to the field-dynamic nature of meaning-making.

3. Aims and Hypotheses

A preliminary version of the ASI was used in a recent study analyzing the evolution of meaning characterizing the dreams of a patient through the course of psychotherapy [77]. In the context of that study, the ASI proved successful in estimating the saturation of the affect-laden meanings in the patient's dreams. Analyses showed that the saturation followed a meaningful, though nonlinear, trajectory, globally indicative of the progressive increase in the patient's capacity to elaborate unconscious, affectively relevant areas of her mental landscape. Such findings can be viewed as encouraging, preliminary, indirect evidence in support of the ASI; however, results are based on a single case study and do not provide information on the relationship between the ASI and other measures of affect.

The present study starts from these preliminary findings and is aimed at providing an initial validation of the ASI, with a specific focus on its validity as a measure of affective intensity. More particularly, it intends to test the two core assumptions underlying the ASI: (i) the ASI is able to detect the affective saturation of a text, and (ii) the affective saturation of the text reflects the intensity of the affective state characterizing the meaning-maker involved in producing the text. This gives rise to the following three hypotheses.

First, an association between the ASI and an independent, content-based textual index of the affective saturation of the text is expected. Given that no other direct measure of a text's affective saturation was found in the literature, we developed an ad hoc indirect index of this characteristic, not related to the ASI: *semantic complexity*. Semantic complexity is the degree of variability of textual content, namely the heterogeneity of the spectrum of content active within the text—the more heterogeneous it is, the greater its semantic complexity. We assume that semantic complexity is inversely associated with affective saturation: the lower the former, the higher the latter. This assumption derives from the definition of affective saturation; insofar as one assumes that affective saturation consists of the magnitude of the contribution of affective meaning compared to other meanings, then the greater the affective saturation, the less the contribution provided by other semantic components, and therefore the lower the global variability of the text. Two studies provide support for the interpretation of semantic complexity as a proxy of affective saturation. First, in the context of the analysis of a therapist–patient exchange, the semantic complexity of narratives proved to be associated inversely with an index of the relevance of generalized affective meanings [78]. Second, in the context of a study of European societies' cultural milieus, secondary quali-quantitative analyses highlighted that the semantic richness of the cultural worldviews identified proved to be inversely associated with their affective saturation [20] (see above paragraph, *A geometric model of affective saturation*). An analysis of the association between the ASI and semantic complexity was carried out, paying attention to checking the potential effect of the lengths of the texts under investigation (estimated in terms of the number of words).

Second, the ASI is expected to be associated with an independent measure of the intensity of the meaning-maker's affective state related to text production. To this end, a physiological index of propensity to engage in context-appropriate affective responses or affective arousal was adopted: heart rate variability (HRV), a measure of parasympathetic autonomic nervous system function. The relationship between the ASI and HRV was estimated by parsing out the potential effect of the individual's capacity for affective regulation. This was done because it is plausible to think that the impact of the intensity of the affective state on the meaning-making underpinning the production of the text, namely the affective saturation, is moderated by the meaning-maker's capacity to "filter" her/his affective activation. One can expect that the lower the capacity for affective regulation, the weaker the elaborative filter, and, therefore, the stronger the relationship between affective intensity and saturation.

Third, the ASI is expected to be associated with an independent, content-unrelated impact of the affective intensity on the text. To this end, we implemented lexical-syntactic complexity as a proxy of that impact. This choice is based on the combination of the following ideas. First, the lexical-syntactic complexity reflects the efficiency of the meaning-making underpinning the text production [79,80]. The production of lexically and syntactically complex texts (e.g., texts comprising long sentences, based on networks of multi-level hierarchized statements) requires computational efficiency (e.g., working memory and abstract conceptualization); correspondingly, low-efficiency meaning-making reduces the text's lexical-syntactic complexity. Second, affective intensity prevents the computational and functional efficiency of meaning-making—e.g., it reduces metacognitive processes, the availability of working memory, and access to abstract reasoning [4,81,82]. Thus, affective intensity has a negative impact on syntactic complexity, via the reduction in the efficiency of the underpinning meaning-making. There is already indirect empirical evidence of the inversely proportional association between lexical-syntactic complexity and affective mental state. (A) It has been shown that in patients with schizophrenia—i.e., patients who are characterized by overwhelming affective states—the number of words per sentence in spontaneous speech is significantly lower than in patients with less severe psychiatric diseases and in non-clinical samples. It is worth adding that this is not due to an impairment in the global production of narratives, as shown by the fact that individuals with schizophrenia produce narratives with approximately the same number of words as control groups do [83]. (B) DePaulo and colleagues [84] reviewed empirical studies of deception cues. They moved from the well-supported premise that lie-telling is more emotionally challenging than telling the truth [85]. On this basis, they highlighted that research converges on the finding that the self-presentations of liars, in their free verbal narratives, are characterized by shorter responses and simpler syntactical configurations. (C) In the context of the Terror Management Theory [86], it was found that individuals who received prime activating meanings related to one's death—assumed to trigger deep states of anxiety—generated shorter (in terms of fewer words and fewer letters per word) autobiographical narratives compared to controls. The analysis of the relationship between the ASI and syntactic complexity was carried out while paying attention to controlling the level of affective regulation. This control is expected to address potential bias due to the fact that the affective saturation measured by the ASI and syntactic complexity can be influenced by differing abilities to regulate affect. In summary, the alternative hypotheses tested against the null hypotheses were:

Hypothesis 1 (H1). *ASI is negatively associated with text semantic complexity.*

Hypothesis 2 (H2). *ASI is positively associated with affective intensity.*

Hypothesis 3 (H3). *ASI is negatively associated with lexical-syntactic complexity.*

The first hypotheses concern the ASI's convergent validity; the others concern concurrent validity.

4. Method

4.1. Sample

The study used a convenience sample of 42 academic students with Italian as their mother tongue. Participants were excluded in the case of self-reported (current or past) psychiatric diagnoses or if they reported psychopathological symptoms over the threshold of clinical relevance (the SCL-R's GSI index was adopted to this end, see below Section 4.3.1). As a result, 2 participants were not included in the analysis. Thus, the sample consisted of N = 40 (34 F; age: M = 25.33; SD = 2.77).

4.2. Procedure

Participants, screened as to the exclusion criterion (presence of current or past psychiatric diagnoses), were contacted through a snowballing procedure. Each participant was invited to the laboratory, provided with a description of the study, informed about the procedure, and asked to sign a written informed consent (Ethical Committee of the Department of Dynamic and Clinical Psychology and Health studies, Sapienza, University of Rome; Prot. n. 0000453). Then, ECG electrodes were attached, and a 5 min assessment at rest was obtained. After that, the participant was asked to fill out the questionnaire (SCL-90-R and DERS) and to undergo the semi-structured interview.

The semi-structured interview was carried out by a trained clinical psychologist, unaware of the aims of the study. The interview focused on neutral issues concerning participants' involvement in their academic course as well as on life and lifestyle matters. It ended after 10 min. In Table 1, some examples of the questions asked to participants are reported. Interviews were audio-recorded and then subjected to verbatim transcription, from which the textual-based indexes (ASI, SEI, and FVI) were derived.

Table 1. Examples of questions used in the semi-structured interview.

Which degree program are you enrolled in?
What is the subject that you are most passionate about?
Whom do you live with?
Are you still with your family or do you live alone?
Do you have any brothers or sisters?
What do you like to do together?
What do you like to do in your free time?
Do you have any friends?
Do you have any hobbies?

The choice of basing the interview on neutral issues responded to the aim of focusing on the participants' baseline levels of affective activation. This was done with the twofold purpose of controlling for potential bias due to the between-subject variability in ways of coping with higher levels of affective activation and of avoiding a potential ceiling effect, reducing data variance.

4.3. Measures

The study implemented the following measures:

(a) Symptom CheckList-90-Revised (SCL-90-R; [87]), to assess the presence of psychopathological symptoms, considered the exclusion criterion.

(b) Difficulties in Emotion Regulation Scale (DERS; [88]), used to measure the participant's capability for affective regulation.
(c) Affective Saturation Index (ASI).
(d) Semantic Entropy Index (SEI), used to measure the text's semantic complexity.
(e) Flesch–Vacca Readability Index (FVI; [89]), used to measure the text's syntactic complexity.
(f) Resting Heart Rate Variability, (indexed by the root mean square of the successive differences between normal heartbeats; rMSSD [90]) an index of parasympathetic control of the heart, used to measure the participant's propensity for affective arousal.

4.3.1. Symptom Check List 90-Revised (SCL-90-R)

The SCL-90-R [87] is a widely used self-report multidimensional inventory, measuring a broad range of symptoms. The SCL-90 has been validated over various populations [91–93]. The SCL-90 measures the symptomatic intensity of mental and physical impairment over the last 7 days. It consists of 90 items answered on a five-point Likert scale with a score of 0 (not at all) to 4 (extremely). The inventory fits a 9-factor structure (Cronbach's alphas ranging from 0.77 for psychoticism to 0.90 for depression) and provides a Global Severity Index (GSI), which is considered a reliable indicator of the current level of overall psychological distress [94,95]. In the present study, we adopted a GSI of <60 as the exclusion criterion. This threshold is, in fact, indicative of clinically relevant conditions of psychological distress [93–96].

4.3.2. Difficulties in Emotion Regulation Scale (DERS)

The DERS [88] is a 36-item self-report measure developed to assess multiple facets of emotion regulation, including abilities to identify, differentiate, and accept emotional experiences, engage in goal-directed behavior, inhibit impulsive behavior in the context of negative emotions, and use effective emotion modulation strategies.

The DERS' items are rated on a five-point Likert scale ranging from 1 (almost never) to 5 (almost always). Despite the fact that the original 6-factor structure is debated [97–101], there is agreement on the DERS total score as an index of emotional impairment or dysregulation (Cronbach's alpha = 0.93) [89,101–105]. Higher scores in the DERS indicate more difficulties in emotion regulation [106]. The DERS is proven to have high internal consistency and test–retest reliability and good predictive and construct validity [103]. The DERS is proven to be sensitive to change due to successful clinical intervention [106,107] and to be correlated with behavioral measures of emotion dysregulation [103]. Based on these results, the measure has gained wide acceptance as a reliable and valid measure of emotion dysregulation in adults [107].

In the present study, we adopted the total DERS scores as an independent measure of the individual's capability for affective regulation, introduced to control the effect of this facet on the relationship between the ASI and independent measures of affective activation and as well as text syntactic complexity.

4.3.3. Affective Saturation Index (ASI)

The ASI estimates affective saturation by computing the contribution of the primary dimensions of the semiotic space, which models the meaning of the textual corpus under analysis (see above, Section 2.2). The ASI does this through the automated procedure of textual analysis (ACASM [108,109]) described below.

ACASM Procedure

ACASM is a computer-based procedure of textual analysis. ACASM can be implemented by several kinds of software. The current analysis used T-Lab [110,111]. T-Lab is appliable to any text based on the Latin alphabet. For several languages (e.g., English, Italian, French, Spanish, German) T-Lab is able to perform both text preprocessing (e.g., the disambiguation of words) and all computations in a fully automatized way. However, these computations are inspectable, their parameters (e.g., the number and type of words under analysis) can be modulated, and their outputs adjusted by the researcher in view of the specific aim of the investigation.

Below, the main steps of the ACASM procedure are described, as it is implemented by the T-Lab software.

(A) The text is automatically segmented into Elementary Context Units (ECUs) according to the following criteria: (i) each ECU begins just after the end of the previous ECU; (ii) each ECU ends with the first punctuation mark ("." or "!" or "?") occurring after a threshold of 250 characters from the first character; and (iii) if an ECU is longer than 500 characters, it ends with the last word within this length, even if there is no punctuation mark.

(B) The software builds a list of all lexical forms (i.e., any string of characters comprised between two blank spaces) present in the transcripts.

(C) The lexical forms are subjected to lemmatization, namely any lexical form is tracked back to the lemmas it belongs to (e.g., lexical forms "I go," "you went," and "they are going" are classified as the lemma "to go"). For many languages, T-lab can perform this step in a fully automatized way.

(D) Instrumental lexical forms (e.g., "to," "and," "of," etc.), as well as lexical forms devoid of meaning (e.g., due to typos) are excluded. This operation, too, is carried out automatically.

(E) ACASM procedures set T-Lab to exclude from the analysis the first 5% of the most frequent lemmas. This is because very high-frequency lemmas tend to co-occur in too many different ECUs, thus reducing their ability to discriminate among different patterns of co-occurrence. Then, only 10% of the most frequent lemmas of the remaining list are selected. This choice is aimed at reducing the dispersion of information within the dataset. In sum, the analysis is limited to lemmas comprised between the fifth and fifteenth percentile of frequency distribution.

(F) A digital matrix of the text is generated, with the segments of text produced in step A (i.e., the ECUs) as rows and the selected lemmas as columns. The ij-th cell assumes a value of 1 if the j-th lemma is contained in the i-th ECU; a value of 0 is assigned otherwise. Appendix A reports an example of such a digital matrix.

(G) The ECU $*$ lemma matrix is subjected to a Lexical Correspondence Analysis (LCoA). The output of the LCoA is a factorial space, each factorial dimension of which maps a component of the meaning active within the whole textual corpus (in the current analysis, the whole set of interviews). Moreover, the LCoA output provides the coordinate on the factorial space of each ECU, as well as of any super-ordered categories in terms of which the ECUs are classified (in the current analysis, the interview). Henceforth, the object of the factorial coordinate (i.e., ECU or super-ordered category) is denoted as a text unit. The factorial coordinate is the measure of the degree of association between the text unit and the factorial dimension—the higher the coordinate, the higher the association, therefore the higher the contribution of that factorial dimension to the meaning of that text unit. In the current study, we focus on the factorial coordinates of the interviews as text units. However, depending on the researcher's aim, the ASI can also be computed by taking the ECU as a text unit. In doing so, a more fine-grained map of the ASI trends throughout the text would be obtained, with a value of ASI for each ECU—i.e., for each sentence. For instance, take the case of a text comprising the verbatim transcript of a whole psychotherapy course. The ASI can be computed either by adopting each session as a text unit (i.e., at the level of super-ordered category) or at the level of a single ECU as a text unit. In the former case, the ASI trend would concern the whole psychotherapy course; in the latter case, the focus would be on the moment-by-moment within-session ASI evolution. Incidentally, the fine-grained focus could be used to analyze the interplay between the participants of the exchange (note that the latter kind of analysis requires the researcher's intervention to differentiate the ASI trends of the interlocutors).

(H) Factorial coordinates are used to compute the ASI. More specifically, the ASI is calculated for each unit (in the current analysis, each interview) as its Euclidean distance from the origin of the factorial space. It is worth highlighting that the Euclidean distance is computed by using the coordinates of the first two factorial dimensions only, as in Equation (1). This is so because the ASI assumes that the first two dimensions are the computational equivalent of the PSM primary dimensions (see Section 2.2):

$$\mathrm{ASI_t} = \sqrt{\mathrm{C}_{\mathrm{F1}(t)}{}^2 + \mathrm{C}_{\mathrm{F2}(t)}{}^2} \tag{1}$$

where $\mathrm{ASI_t}$ is the Affective Saturation Index of the textual unit t and $\mathrm{C}_{\mathrm{F1}(t)}$ and $\mathrm{C}_{\mathrm{F2}(t)}$ are the factorial coordinates of the textual unit t on the first and second factorial dimension, respectively.

From Equation (1), it can be seen that the ASI increases when one or both of the factorial coordinates increase. Accordingly, the ASI can be interpreted as a measure of the magnitude of the contribution of the two first dimensions of the factorial space to the meaning of the textual unit—the higher the ASI, the greater the contribution. Therefore, the ASI can be considered an index of the degree of saturation of the affective meanings comprising the textual unit.

4.3.4. Semantic Entropy Index (SEI)

To estimate the semantic complexity, we used an ad hoc index: the *Semantic Entropy Index (SEI)*. The SEI is derived from the Information Entropy formula [112], applied to the text's thematic contents. However, previous studies which applied Informative Entropy to text analysis focused on lexical features [113], language comprehension, meaning representation [114,115], and associative strengths among words [111,115]. Different from these studies, the SEI is aimed at detecting the amount of semantic variability in the text. The higher the SEI, the higher the content variability (i.e., the complexity) of the text under analysis is. In the current study, the SEI was applied to each interview.

The SEI was estimated in accordance with the following procedure. First, the digital matrix of the textual corpus (i.e., the same matrix used in the first stage of the computation of the ASI, i.e., the LCoA) was subjected to a Lexical Cluster Analysis (LClA). The LClA groups sentences (Elementary Context Units, ECUs) that tend to share the same co-occurring lemmas. In this way, each cluster can be considered indicative of a thematic content active in the textual corpus and characterized semantically by the pattern of co-occurring lemmas making those ECUs similar to each other (for details, see [108,109]). The number of clusters in which the text is segmented is established in accordance with an iterative algorithm [110,111]; the procedure of clustering stops when further partitions produce no significant improvement of the inter-/intra-cluster ratio, which means that increasing the number of clusters does not produce an appreciable increment of information. In the current analysis, the LClA generated 5 clusters/thematic contents as optimal partitions. Accordingly, each interview was characterized by the relative frequency of each cluster/thematic content within it—i.e., the proportion of ECUs covered by each cluster within the interview.

The SEI was computed in accordance with the following formula:

$$\mathrm{SEI} = -\sum_{i=1}^{n} \mathrm{p}(x_i)\; In\, \mathrm{p}(x_i) \tag{2}$$

where SEI is the degree of the Semantic Entropy Index of the interview I, n is the number of clusters (i.e., thematic contents) obtained by the LClA, and $\mathrm{p}(x_i)$ stands for the probability that a cluster x_i occurs.

According to Equation (2), the entropy consists of the homogeneity of the probability of the occurrence of clusters/thematic content. The more each cluster has an occurrence probability similar to that of others, the more the thematic variability of the text, thus its complexity. This means that, in the context of the current analysis, the highest SEI is given if each cluster/thematic content has f = 0.2 (i.e., 1/5) of relative frequency. In that case, indeed, the text could be characterized by one or more specific clusters; rather, all clusters need to be considered to represent its content.

4.3.5. Flesch–Vacca Index (FVI)

The FVI is a measure of text readability [89]. Text readability is the extent to which a text can be understood by a reader as a result of its lexical and syntactic characteristics. Specifically, the FVI is an Italian language adjustment of the Flesch [116,117] Reading Ease formula. The FVI assumes that long words are typically used less frequently than short ones and that long sentences are usually more complex, from a syntactical point of view, than short ones.

The FVI was computed for each interview separately, in accordance with the following formula (see Equation (3)):

$$FVI_t = 217 - (1.3 * ASL_t) - 0.6 * AWL_t \quad (3)$$

where ASL_t is the length of sentences of the text unit t, measured as the average number of syllables in 100 words, and AWL_t denotes the length of the sentences, expressed as the average mean of words per sentence.

4.3.6. Heart Rate Variability (HRV)

HRV is the variability of the time between adjacent heartbeats, resulting from the dynamic interplay between the fast-acting parasympathetic nervous system and the relatively slower sympathetic nervous system [118]. A higher resting-state HRV reflects better adaptive and flexible prefrontal inhibition to meet various situational demands [119]. In contrast, a lower resting HRV has been related to hypoactive prefrontal regulation, leading to hyperactive subcortical structures and the release of physiological defensive responses [120,121]. Tonic HRV was assessed through the following procedure. A 5 min baseline was started, during which time the participants browsed a gardening magazine (i.e., "vanilla baseline" [122–124]), while their beat-to-beat intervals were recorded using the Bodyguard 2 (Firstbeat). The device has been shown to provide reliable measures of beat-to-beat intervals [124].

HRV analyses were performed using Kubios HRV software [125]. This software uses an advanced detrending method based on smoothness priors formulation in which the filtering effect is attenuated at the beginning and the end of the data, thus avoiding the distortion of data end-points. Moreover, the frequency response of the method is adjusted with a single smoothing parameter, selected in such a way that the spectral components of interest are not significantly affected by the detrending [126]. Kubios HRV includes two methods for correcting any artifacts and ectopic beats: (1) a threshold-based correction, in which these are simply corrected by comparing every RR interval value against a local average interval; and (2) automatic correction, in which artifacts are detected from a time series consisting of differences between successive RR intervals.

HRV was mapped by means of the rMSSD, which reflects the vagal regulation of HR [126]. A higher resting rMSSD reflects better psychological and emotional flexibility and the capability to engage in context-appropriate responses [127–129]. Accordingly, we used the rMSSD as an index of the individual's disposition for affective arousal—the lower the rMSSD, the higher the propensity to be subjected to affective arousal (for a similar interpretation of the index, see [118,130,131]). As the distribution of the rMSSD was non-normal, the variable was transformed into its natural logarithms.

4.4. Data Analysis

A set of one-way ANOVAs and bivariate correlations were first used to estimate the size of the relationships among indexes and sample characteristics (age, gender) and between them. Then, three multiple linear regressions (standard multiple regression method) were performed, one for each hypothesis:

1. H1 was tested by means of a regression model, with the index of saturation (SEI) as the dependent variable and the ASI and the length of the interview (measured by the number of words) as predictors. The latter was introduced in order to control for its effect on the SEI–ASI relationship.
2. H2 was tested by means of a regression model, with the physiological measure (rMSSD) as the dependent variable and the ASI as the predictor. The DERS was introduced in the model as a further predictor, in order to estimate the rMSSD–ASI association and the net effect of the individual's capacity for emotion regulation.
3. H3 was tested by means of a regression model, with the index of syntactic complexity (FVI) as the dependent variable and the ASI as a predictor. We introduced the length of the interview (in number of words) and the DERS as further predictors, to control for their effects on the SEI–ASI relationship.

5. Results

Table 2 reports descriptive statistics of the variables adopted. As shown by the values of kurtosis and skewness, the indexes (in the case of the rMSSD, after logarithmic transformation) proved to approximate the normal distribution (the distribution of ASI is shown in Figure 1).

Table 2. Descriptive statistics of the measures under analysis.

	N	Mean	Min	Max	Std. Dev.	Skewness	SE	Kurtosis	SE
Age	39	25.107	21	31	2.80784	0.348	0.378	−0.879	0.741
SCL90-R	40	44.025	33	75	11.226	1.252	0.374	0.498	0.733
ASI	40	0.124	0.01	0.34	0.08	1.134	0.374	1.391	0.733
SEI	40	1.456	1.081	1.592	0.133	−1.358	0.374	1.091	0.733
rMSSD	40	3.846	2.25	4.94	0.62	−0.885	0.374	0.242	0.733
FVI	40	70.651	54	80	6.133	−0.79	0.374	0.683	0.733
Words	40	1362.43	988	1746	170.569	0.035	0.374	−0.021	0.733
DERS	40	85.75	49	128	21.584	0.367	0.374	−0.651	0.733

Note. Age N = 39 given that 1 participant had unknown age; SCL90-R = Symptom Checklist-90 Revised; ASI = Affective Saturation Index; SEI = index of saturation; rMSSD = log-transformed root mean square of the successive differences between normal heartbeats; FVI = index of syntactic complexity; DERS = Difficulties in Emotion Regulation Scale.

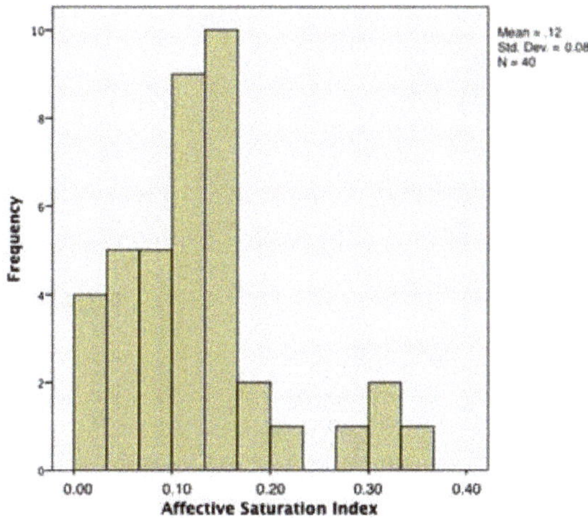

Figure 1. Affective Saturation Index (ASI) distribution in the sample under analysis.

No significant differences resulted between males and females as to the level of the ASI (ANOVA test: F [1,38] = 0.337, $p = 0.565$). Accordingly, and considering that the sample is characterized by a higher proportion of women than men, we did not carry out separate analyses. The length of the interview (in number of words) did not correlate with any of the indexes examined. However, we have used this as a control variable in regression models 1 and 3, given that these models have a text-based index as the dependent variable.

Table 3 reports the correlations between the main variables of the study. The ASI proved to be negatively associated with the SEI (r = −0.657), rMSSD (r = −0.468), and FVI (r = −0.426). No significant correlations emerged for Age, Words, DERS, and SCL-90R. The log-transformed rMSSD and the FVI proved to correlate robustly (r = 0.489).

Table 3. Pearson correlations between the measures under analysis.

	SEI	rMSSD	FVI	Age	Words	DERS	SCL-90R
ASI	−0.657 *	−0.468 *	−0.426 *	0.288	−0.084	0.023	0.047
SEI		0.464 *	0.388 *	−0.247	0.213	−0.012	−0.053
rMSSD			0.489 *	−0.064	−0.012	0.178	0.156
FVI				−0.452 *	0.026	−0.034	−0.59
Age					−0.212	0.155	0.147
Words						0.133	0.084
DERS							0.653 *

* significant correlations at 0.01 level two-tailed. ASI = Affective Saturation Index; SEI = index of saturation; rMSSD = log-transformed root mean square of the successive differences between normal heartbeats; FVI = index of syntactic complexity; DERS = Difficulties in Emotion Regulation Scale; SCL-90 = Symptom Check List-90 Revised.

All three multiple regression models proved to be significant ($p < 0.001$, $p < 0.004$, and $p < 0.005$, respectively; cf. Table 4). Tables 5 and 6 report the main parameters of regression model 1, with SEI as the dependent variable and the ASI and Words (i.e., the number of words in the interview) as predictors. The model did not suffer from problems of collinearity (VIF = 1.007); the adjusted R square was 0.428 (std. err. of estimation = 0.101). The inclusion of Words did not modify the parameters of the model significantly (change of R from model 1 and model 2: $p = 0.199$; cf. Table 5). The ASI beta coefficient was −0.644 (t = −5.296, $p < 0.000$); the Words beta coefficient (0.159) was not significant (cf. Table 6). The distribution of residuals approximated the normal distribution (Figure 2).

Table 4. Regression models. ANOVA test.

	Sum of Squares	df	Mean Square	F	Sig.
Regression Model 1: Dependent Variable: SEI; Predictors: ASI, Words					
Regression	0.315	2	0.158	15.58	0.001
Residual	0.375	37	0.010		
Total	0.691	39			
Regression Model 2: Dependent Variable: rMSSD; Predictors: ASI, DERS					
Regression	3.813	2	1.907	6.313	0.004
Residual	11.175	37	0.302		
Total	14.988	39			
Regression Model 3: Dependent Variable: FVI; Predictors: ASI, DERS, age					
Regression	441.309	3	147.103	5.028	0.005
Residual	1023.922	35	29.255		
Total	1465.231	38			

Note. ASI = Affective Saturation Index; SEI = index of saturation; rMSSD = log-transformed root mean square of the successive differences between normal heartbeats; FVI = index of syntactic complexity.

Table 5. Regression model with the index of saturation (SEI) as the dependent variable. Summary of the model.

Model *	R	R^2	Adjusted R^2	Std. Error	F Change	df1	df2	Sig. F Change
1	0.657	0.432	0.417	0.101				
2	0.676	0.457	0.428	0.101	1.714	1	37	0.199

* Model 1 predictor: Affective Saturation Index (ASI); Model 2 predictors: ASI, Words.

Table 6. Regression model with the index of saturation (SEI) as the dependent variable (model 2, with all predictors included).

	B	Stand. Error	Beta	t	Sig.	VIF
Constant	1.42	0.135		10.55		
ASI	−1.064	0.201	−0.64	−5.3	0	1.007
Words	0	0	0.159	1.309	0.199	1.007

Note. ASI = Affective Saturation Index.

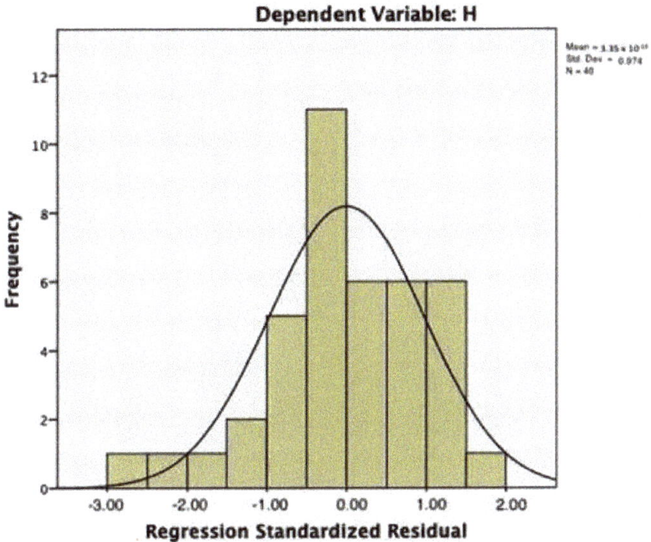

Figure 2. Regression model with the index of saturation (SEI) as the dependent variable. Distribution of residuals.

Tables 7 and 8 report the main parameters of regression model 2, with the rMSSD as the dependent variable and the ASI and DERS as the predictors. The model did not suffer from problems of collinearity (VIF = 1.001); the adjusted R square was 0.214 (std. err. of estimation = 0.549). The inclusion of the DERS did not modify the parameters of the model significantly (change of R from model 1 and model 2: $p = 0.192$; cf. Table 7). The ASI beta coefficient was −0.472 (t = −3.325; $p < 0.002$); the DERS beta coefficient was not significant (0.189; cf. Table 8). The distribution of residuals approximates the normal distribution (Figure 3).

Table 7. Regression model with the log-transformed root mean square of the successive differences between normal heartbeats (rMSSD) as the dependent variable. Summary of the model.

Model *	R	R^2	Adjusted R^2	Std. Error	F Change	df1	df2	Sig. F Change
1	0.468	0.219	0.198	0.55506				
2	0.504	0.254	0.214	0.54956	1.764	1	37	0.192

* Model 1 predictor: Affective Saturation Index (ASI); Model 2 predictors: ASI, Difficulties in Emotion Regulation Scale (DERS).

Table 8. Regression model with the log-transformed root mean square of the successive differences between normal heartbeats (rMSSD) as the dependent variable (model 2, with all predictors included).

	B	Stand. Error	Beta	t	Sig.	VIF
Constant	3.835	0.382		10.03	0.001	
ASI	−3.636	1.093	−0.47	−3.33	0.002	1.001
DERS	0.005	0.004	0.189	1.328	0.192	1.001

Note. ASI = Affective Saturation Index; DERS = Difficulties in Emotion Regulation Scale.

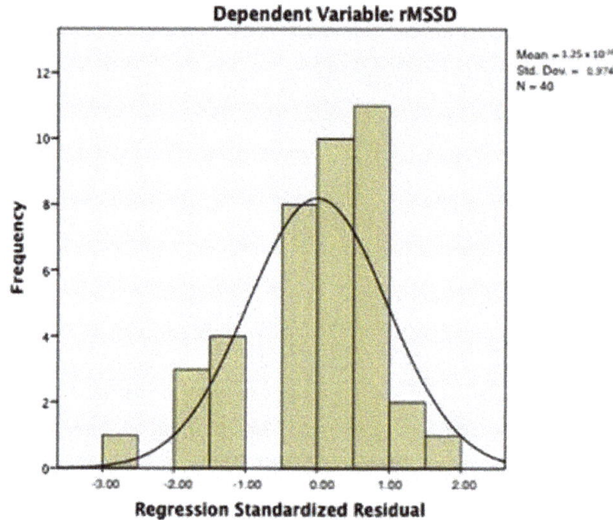

Figure 3. Regression model with the log-transformed root mean square of the successive differences between normal heartbeats (rMSSD) as the dependent variable. Distribution of residuals.

Tables 9 and 10 report the main parameters of regression model 3, with the FVI as the dependent variable and the ASI, DERS, and age as the predictors (we included age as a covariate, due to its high correlation with the FVI; this means that in this case the model was calculated on n = 39 group, given that 1 participant had unknown age). The model did not suffer from problems of collinearity (VIFs close to 1); the adjusted R square was 0.241 (std. err. of estimation = 5.409). The inclusion of the DERS and age did not modify the parameters of the model significantly ($p = 0.064$; cf. Table 9). The ASI beta coefficient is −0.301 (t = −2.187; $p < 0.035$); the age beta coefficient is also significant (−0.364, $p < 0.002$). The DERS beta coefficient (.029) was not significant; (cf. Table 10). The distribution of residuals approximates the normal distribution (Figure 4).

Table 9. Regression model with the index of syntactic complexity (FVI) as the dependent variable. Summary of the model.

Model *	R	R^2	Adjusted R^2	Std. Error	F Change	df1	df2	Sig. F Change
1	0.427	0.182	0.16	5.691				
2	0.549	0.301	0.241	5.409	2.978	2	35	0.064

* Model 1 predictor: Affective Saturation Index (ASI); Model 2 predictors: Affective Saturation Index (ASI), Difficulties in Emotion Regulation Scale (DERS), age.

Table 10. Regression model with the index of syntactic complexity (FVI) as the dependent variable (model 2, with all predictors included).

	B	Stand. Error	Beta	t	Sig.	VIF
Constant	93.251	8.251		11.3	0.001	
ASI	−24.958	11.41	−0.32	−2.19	0.035	1.091
DERS	0.008	0.041	0.029	0.204	0.839	1.025
age	−0.804	0.331	−0.36	−2.43	0.02	1.118

Note. ASI = Affective Saturation Index; DERS = Difficulties in Emotion Regulation Scale.

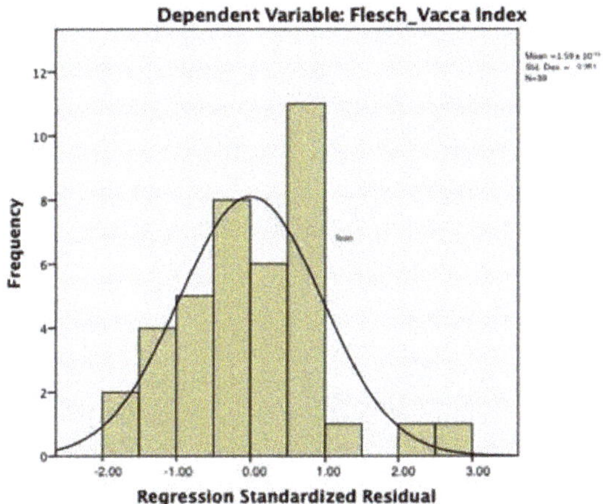

Figure 4. Regression model with the FVI as the dependent variable. Distribution of residuals.

6. Discussion and Conclusions

As hypothesized, the ASI proved to be significantly and inversely associated with the SEI, the independent proxy of affective saturation. The association was robust, in the expected direction (ASI beta = −0.644), and was not weakened by the control of the potential effect of the length of the interviews.

Second, the ASI was significantly and inversely correlated with the rMSSD, the physiological index of the disposition for affective arousal. This relationship was in the expected direction (once one considers that a lower rMSSD indicates higher arousal), and robust, both when estimated directly (r = −0.468) and once the index of affective regulation (DERS) was introduced in the regression model (ASI beta = −0.472). This finding suggests that the ASI is able to detect the intensity of the affective state of the meaning-making involved in the production of the text, in a way that is not influenced by the individual's capability to regulate affective arousal.

Finally, these findings are consistent with the idea that the ASI is able to detect the lexical-syntactic complexity of the text, assumed as an independent correlate of the affective intensity. As to this latter assumption, it is worth highlighting that it was supported by the high correlation between the index of lexical-syntactic complexity (FVI) and the rMSSD (r = 0.489). These results are consistent with and further support lines of thinking that view lexical-syntactic complexity as a property of the textual output that is subject to the influence of the affective state over the meaning-making underpinning its production. The ASI proved to be associated with the FVI (ASI beta = −0.323), in the expected direction. Moreover, in this case, the effect was estimated after the individual's capability for affective regulation was checked.

Taken as a whole, the present findings support both the aspects of the ASI construct and the criterion validity investigated. The ASI proved to be a measure capable of detecting the structural organization of textual meaning—more specifically, of estimating to what extent the text is saturated by affective meaning. Moreover, the ASI's measurement of affective saturation proved to be a valid estimator of the producer's physiological affective state at rest as well as of the impact of affective intensity on meaning-making (as marked by the text's syntactic complexity).

These findings are promising for their theoretical, methodological, and practical implications. From a theoretical standpoint, they enforce the semiotic framework upon which the ASI is based. What needs to be highlighted here is that the intensity of the meaning-maker's affective state proved to be associated not only with the efficiency of the cognitive process underpinning the text production—as signaled by the relationship between affective intensity and lexical-syntactic complexity—but also with the inherent structural organization of the textual meaning—i.e., the relevance of the primary dimensions of meaning over the others, which is the specific property on which the ASI focuses. This legitimates the semiotic view of affect—namely, the idea that affect is an embodied form of meaning that, due to its nature, operates directly on the text's semantic organization. In other words, the findings are consistent with the ASI viewpoint, which does not see affect as an exogenous factor influencing the text from the outside. Rather, it conceives affect as an inherent characteristic of the text; affect does not work on meaning-making by constraining or channeling it but is part and parcel of it [20]. From a methodological standpoint, the ASI opens new opportunities for measuring affect. Two characteristics of the ASI are worth mentioning here. First, the ASI is almost completely insensitive to the size of the data—this means that it can be implemented on a large textual dataset, therefore enabling large-scale studies that can link individual and social levels of analyses as well as studies based on the density of units of observation required for dynamic time series. Second, the affective meaning is frozen, as it were, in the text. Therefore, the ASI's use of texts as a source of the measurement of affective intensity enables the off-line detection of that dimension, namely the possibility for measuring affect in an independent spatio-temporal setting with respect to the setting in which it was activated. The combination of these characteristics envisages thrilling new opportunities—e.g., large-scale retrospective time-series analyses, to model the dynamics of affective activation characterizing the socio-cultural historical evolution of given social groups, and time-series analyses to map the dynamic evolution of meaning-making over the psychotherapy process [77]. The ASI's methodological flexibility has practical implications, too. One can envisage a plurality of applications of the method, in the many fields where the measurement of affect and its impact on meaning-making can be relevant for both interpretative and interventional aims—e.g., fields such as clinical and community interventions, social communication, marketing, and media monitoring (on the role of affective sensemaking in society [27]).

However, the fact that the findings are encouraging must not lead us to underestimate the limitations of the study. There are three main shortcomings to highlight. First, the study was based on a convenience sample of Italian students in a degree course in psychology characterized by homogeneous age and a higher proportion of women. This made it impossible to test the role of language, age, and gender on the relationship between the ASI and other indexes and therefore generalize the findings beyond this specific group.

Second, the study adopted a psychophysiological index (rMSSD) assessed at rest, which is a trait measure. It was implemented before the interview to estimate the participants' baseline disposition to a given level of affective intensity and capacity for affective regulation, respectively. Thus, it is not a direct measure of the level of affective intensity—and its variation—during the interview. We did this because the adoption of a state measure, mapping the ongoing physiological state of the participants alongside the interview, would have involved a level of computational complexity (e.g., the necessity to match the ASI values that took the whole interview as a unit of analysis and the instant-by-instant physiological values) which would have been outside the scope of the current study, which is aimed at the first stage of the validation of the ASI. Furthermore, while it must be recognized that the HRV index only partially captures the affective state occurring during the interview, it must also be highlighted that the use of this index is a conservative choice, which underestimates the relationship between the ASI and the participants' affective intensity manifested during the interview. This is because it is plausible to think that, in response to the interview, individuals would differ in how much their affective state varies from the baseline, as a result of individual differences in personality and other psychological characteristics. Thus, insofar as the ASI is a measure of the current affective intensity on the text, the use of a trait measure weakens the chance to detect the capacity of the ASI to estimate the physiological state underpinning text production.

The third hypothesis was based on an indirect proxy of the impact of affective intensity on meaning-making—the text's lexical-syntactic complexity. Thus, hypothesis 3 of the study—i.e., the fact that the ASI is able to detect the impact of affective intensity on meaning-making—must be considered only indirectly tested, insofar as one accepts the assumption that lexical-syntactic complexity is a valid marker of the efficiency of meaning-making. This assumption is fostered by the findings of the study (i.e., the high correlation between the FVI and the rMSSD), but is not systematically supported by the literature, which lacks specific studies on this issue.

These issues need to be addressed by the next steps of the ASI's validation. Further studies will be implemented to test the ASI's validity on other groups (e.g., lower educated people and clinical populations) and other kinds of texts (e.g., highly affect-laden communications and texts characterized by positive vs. negative affective valence). Finally, deeper analyses of the specific mechanisms underpinning the relationship between the textual and physiological components of affect, as well as the role played by the regulative cognitive processes in that relationship, will be brought into focus.

Author Contributions: Conceptualization, S.S., A.G. and V.C.; methodology, A.G., S.S., V.C. and C.O.; formal analysis, A.G., C.O. and S.S.; resources, V.C. and C.O.; data curation, C.P.; writing—original draft preparation, S.S. and A.P; writing—review and editing, A.P., A.G., V.C. and C.O.; supervision, S.S. All authors have read and agreed to the published version of the manuscript.

Funding: This research received no external funding.

Institutional Review Board Statement: The study was conducted according to the guidelines of the Declaration of Helsinki, and approved by the Ethics Committee of the Department of Dynamic and Clinical Psychology and Health studies, Sapienza, University of Rome; Prot. n. 0000453.

Informed Consent Statement: Informed consent was obtained from all subjects involved in the study.

Data Availability Statement: Not applicable.

Conflicts of Interest: The authors declare no conflict of interest.

Appendix A

Table A1. An example of a digital matrix obtained by the text comprising the title and first sentence of the current paper (note that, for simplicity, we selected all but instrumental lemmas, regardless of their frequency).

	Saturation	Index	Lexical	Measure	Measurement	Level	Affective	Intensity	Relevant	Issue	Theoretical	Practical	Reasons	Affect	To Play	Role	Individual	To Make Sense	Experience
Affective Saturation Index: A lexical measure of affect	1	1	1	1	0	0	1	0	0	0	0	0	0	1	0	0	0	0	0
The measurement of the level of affective intensity is a relevant issue for both theoretical and practical reasons	0	0	0	0	1	1	1	1	1	1	1	1	1	1	1	1	1	1	1

References

1. Barrett, L.F. The theory of constructed emotion: An active inference account of interoception and categorization. *Soc. Cogn. Affect. Neurosci.* **2016**, *12*, 1–23. [CrossRef]
2. Barrett, L.F.; Bliss-Moreau, E.; Duncan, S.L.; Rauch, S.L.; Wright, C.I. The amygdala and the experience of affect. *Soc. Cogn. Affect. Neurosci.* **2007**, *2*, 73–83. [CrossRef]
3. Valsiner, J. *General Human Psychology*; Springer: New York, NY, USA, 2021.
4. Schweizer, S.; Satpute, A.B.; Atzil, S.; Field, A.P.; Hitchcock, C.; Black, M.; Barrett, L.F.; Dalgleish, T. The impact of affective information on working memory: A pair of meta-analytic reviews of behavioral and neuroimaging evidence. *Psychol. Bull.* **2019**, *145*, 566–609. [CrossRef]
5. Salvatore, S.; Avdi, E.; Battaglia, F.; Bernal-Marcos, M.J.; Buhagiar, L.J.; Ciavolino, E.; Fini, V.; Kadianaki, I.; Kullasepp, K.; Mannarini, T.; et al. Distribution and Characteristics of Symbolic Universes over the European Societies. In *Symbolic Universes in Time of (Post)Crisis*; Springer: Cham, Switzerland, 2019; pp. 135–170.
6. Kleinbub, J.R.; Testolin, A.; Palmieri, A.; Salvatore, S. The phase space of meaning model of psychopathology: A computer simulation modelling study. *PLoS ONE* **2021**, *16*, e0249320. [CrossRef]
7. Menin, D.S.; Farach, F.J. Emotions and evolving treatments for adult psychopathology. *Clin. Psychol. Sci. Pract.* **2007**, *14*, 329–353. [CrossRef]
8. Smith, K.E.; Mason, T.B.; Anderson, N.L.; Lavender, J.M. Unpacking cognitive emotion regulation in eating disorder psychopathology: The differential relationships between rumination, thought suppression, and eating disorder symptoms among men and women. *Eat. Behav.* **2019**, *32*, 95–100. [CrossRef]
9. Rogier, G.; Velotti, P. Conceptualizing gambling disorder with the process model of emotion regulation. *J. Behav. Addict.* **2018**, *7*, 239–251. [CrossRef] [PubMed]
10. Velotti, P.; Rogier, G.; Zobel, S.B.; Billieux, J. Association between gambling disorder and emotion (dys)regulation: A systematic review and meta-analysis. *Clin. Psychol. Rev.* **2021**, *87*, 102037. [CrossRef]
11. Venuleo, C.; Rollo, S.; Marinaci, T.; Calogiuri, S. Towards a cultural understanding of addictive behaviours. The image of the social environment among problem gamblers, drinkers, internet users and smokers. *Addict. Res. Theory* **2016**, *24*, 274–287. [CrossRef]
12. Venuleo, C.; Mossi, P.; Calogiuri, S. Combining Cultural and Individual Dimensions in the Analysis of Hazardous Behaviours: An Explorative Study on the Interplay Between Cultural Models, Impulsivity, and Depression in Hazardous Drinking and Gambling. *J. Gambl. Issues* **2018**, *40*, 70–115. [CrossRef]
13. Venuleo, C.; Mossi, P.; Salvatore, S. Educational subculture and dropping out in higher education: A longitudinal case study. *Stud. High. Educ.* **2016**, *41*, 321–342. [CrossRef]
14. Bucci, W. *Psychoanalysis and Cognitive Science*; Guilford Press: New York, NY, USA, 1997.
15. Bucci, W. The interplay of subsymbolic and symbolic processes in psychoanalytic treatment: It takes two to tango—But who knows the steps, who's the leader? The choreography of the psychoanalytic interchange. *Psychoanal. Dialogues* **2011**, *21*, 45–54. [CrossRef]
16. Mergenthaler, E. Emotion-abstraction patterns in verbatim protocols: A new way of describing psychotherapeutic processes. *J. Consult. Clin. Psychol.* **1996**, *64*, 1306–1315. [CrossRef]
17. Mergenthaler, E.; Bucci, W. Linking verbal and non-verbal representations: Computer analysis of referential activity. *Br. J. Med Psychol.* **1999**, *72*, 339–354. [CrossRef] [PubMed]
18. Rocco, D.; Gennaro, A.; Salvatore, S.; Stoycheva, V.; Bucci, W. Clinical Mutual Attunement and the Development of Therapeutic Process: A Preliminary Study. *J. Constr. Psychol.* **2016**, *30*, 371–387. [CrossRef]
19. Mazzara, B.M.; Avdi, E.; Kadianaki, I.; Koutri, I.; Lancia, F.; Mannarini, T.; Mylona, A.; Pop, A.; Rochira, A.; Redd, R.E.; et al. The Representation of Immigration. A Retrospective Newspaper Analysis. *J. Immigr. Refug. Stud.* **2020**, 1–20. [CrossRef]
20. Salvatore, S.; Avdi, E.; Battaglia, F.; Bernal-Marcos, M.J.; Buhagiar, L.J.; Ciavolino, E.; Fini, V.; Kadianaki, I.; Kullasepp, K.; Mannarini, T.; et al. The Cultural Milieu and the Symbolic Universes of European Societies. In *Symbolic Universes in Time of (Post)Crisis*; Springer: Cham, Switzerland, 2019; pp. 53–133.
21. Veltri, G.A.; Redd, R.; Mannarini, T.; Salvatore, S. The identity of Brexit: A cultural psychology analysis. *J. Community Appl. Soc. Psychol.* **2019**, *29*, 18–31. [CrossRef]
22. Greenberg, J.; Solomon, S.; Arndt, J. A basic but uniquely human motivation: Terror management. In *Handbook of Motivation Science*; Shah, J., Gardner, W., Eds.; Guilford Press: New York, NY, USA, 2008; pp. 114–134.
23. Landau, M.J.; Johns, M.; Greenberg, J.; Pyszczynski, T.; Martens, A.; Goldenberg, J.L.; Solomon, S. A Function of Form: Terror Management and Structuring the Social World. *J. Pers. Soc. Psychol.* **2004**, *87*, 190–210. [CrossRef]
24. Proulx, T.; Heine, S.J. Death and Black Diamonds: Meaning, Mortality, and the Meaning Maintenance Model. *Psychol. Inq.* **2006**, *17*, 309–318. [CrossRef]
25. Proulx, T.; Inzlicht, M. The Five "A"s of Meaning Maintenance: Finding Meaning in the Theories of Sense-Making. *Psychol. Inq.* **2012**, *23*, 317–335. [CrossRef]
26. Solomon, S.; Greenberg, J.; Pyszczynski, T. A Terror Management Theory of Social Behavior: The Psychological Functions of Self-Esteem and Cultural Worldviews. In *Advances in Experimental Social Psychology*; Elsevier: Amsterdam, The Netherlands, 1991; Volume 24, pp. 93–159.
27. Cremaschi, M.; Fioretti, C.; Mannarini, T.; Salvatore, S. *Semiotic Capital*; Springer: Cham, Switzerland, 2021; pp. 173–190.

28. Egger, M.; Ley, M.; Hanke, S. Emotion Recognition from Physiological Signal Analysis: A Review. *Electron. Notes Theor. Comput. Sci.* **2019**, *343*, 35–55. [CrossRef]
29. Barrett, L.F. Are Emotions Natural Kinds? *Perspect. Psychol. Sci.* **2006**, *1*, 28–58. [CrossRef] [PubMed]
30. Nisett, R.E.; De Camp Wilson, T. The halo effect: Evidence for unconscious alteration judgements. *J. Pers. Soc. Psych.* **1997**, *35*, 250–256. [CrossRef]
31. Mauss, I.B.; Robinson, M.D. Measures of emotion: A review. *Cogn. Emot.* **2009**, *23*, 209–237. [CrossRef]
32. Dang, J.; Ekim, Z.E.; Ohlsson, S.; Schiöth, H.B. Is there prejudice from thin air? Replicating the effect of emotion on automatic intergroup attitudes. *BMC Psychol.* **2020**, *8*, 47. [CrossRef]
33. Protopapas, A.; Lieberman, P. Fundamental frequency of phonation and perceived emotional stress. *J. Acoust. Soc. Am.* **1997**, *101*, 2267–2277. [CrossRef]
34. Ververidis, D.; Kotropoulos, C. Emotional speech recognition: Resources, features, and methods. *Speech Commun.* **2006**, *48*, 1162–1181. [CrossRef]
35. Mauss, I.B.; Levenson, R.W.; McCarter, L.; Wilhelm, F.H.; Gross, J.J. The Tie That Binds? Coherence among Emotion Experience, Behavior, and Physiology. *Emotion* **2005**, *5*, 175–190. [CrossRef]
36. Coulson, M. Attributing Emotion to Static Body Postures: Recognition Accuracy, Confusions, and Viewpoint Dependence. *J. Nonverbal Behav.* **2004**, *28*, 117–139. [CrossRef]
37. Aviezer, H.; Hassin, R.R.; Ryan, J.; Grady, C.; Susskind, J.; Anderson, A.; Moscovitch, M.; Bentin, S. Angry, Disgusted, or Afraid? *Psychol. Sci.* **2008**, *19*, 724–732. [CrossRef] [PubMed]
38. Calbi, M.; Siri, F.; Heimann, K.; Barratt, D.; Gallese, V.; Kolesnikov, A.; Umiltà, M.A. How context influences the interpretation of facial expressions: A source localization high-density EEG study on the "Kuleshov effect". *Sci. Rep.* **2019**, *9*, 1–16. [CrossRef]
39. Barrett, L.F.; Mesquita, B.; Gendron, M. Context in Emotion Perception. *Curr. Dir. Psychol. Sci.* **2011**, *20*, 286–290. [CrossRef]
40. Lee, H.; Heller, A.; van Reekum, C.M.; Nelson, B.; Davidson, R.J. Amygdala–prefrontal coupling underlies individual differences in emotion regulation. *NeuroImage* **2012**, *62*, 1575–1581. [CrossRef] [PubMed]
41. Stock, J.V.D.; Vandenbulcke, M.; Sinke, C.B.A.; Goebel, R.; De Gelder, B. How affective information from faces and scenes interacts in the brain. *Soc. Cogn. Affect. Neurosci.* **2013**, *9*, 1481–1488. [CrossRef]
42. Wieser, M.J.; Brosch, T. Faces in Context: A Review and Systematization of Contextual Influences on Affective Face Processing. *Front. Psychol.* **2012**, *3*, 471. [CrossRef] [PubMed]
43. Xu, Q.; Yang, Y.; Tan, Q.; Zhang, L. Facial Expressions in Context: Electrophysiological Correlates of the Emotional Congruency of Facial Expressions and Background Scenes. *Front. Psychol.* **2017**, *8*. [CrossRef]
44. Salvatore, S.; Freda, M.F. Affect, unconscious and sensemaking. A psychodynamic, semiotic and dialogic model. *New Ideas Psychol.* **2011**, *29*, 119–135. [CrossRef]
45. Salvatore, S.; Picione, R.D.L.; Cozzolino, M.; Bochicchio, V.; Palmieri, A. The Role of Affective Sensemaking in the Constitution of Experience. The Affective Pertinentization Model (APER). *Integr. Physiol. Behav. Sci.* **2021**, 1–19. [CrossRef]
46. Fornari, F. *I Fondamenti di una Teoria Psicoanalitica del Linguaggio*; [Foundations for a psychoanalytic theory of the language]; Bollati Boringhieri: Torino, Italy, 1979.
47. Peirce, C.S. On Sign. In *Collected Papers of Charles Sanders Peirce (Volume II)*; Hartshorne, C., Weiss, P., Eds.; Harvard University Press: Cambridge, MA, USA, 1932; [Original version: 1987].
48. Hartshorne, C.; Weiss, P.; Burks, A.W. *Collected Papers of Charles Sanders Peirce*; Harvard University Press: Cambridge, MA, USA, 1931.
49. Barsalou, L.W. Perceptual symbol systems. *Behav. Brain Sci.* **1999**, *22*, 577–660. [CrossRef] [PubMed]
50. Cuccio, V.; Gallese, V. A Peircean account of concepts: Grounding abstraction in phylogeny through a comparative neuroscientific perspective. *Philos. Trans. R. Soc. B Biol. Sci.* **2018**, *373*, 20170128. [CrossRef] [PubMed]
51. Gallagher, S. *How the Body Shapes the Mind*; University Press: Oxford, UK, 2005.
52. Lindquist, K.A.; Satpute, A.; Wager, T.D.; Weber, J.; Barrett, L.F. The Brain Basis of Positive and Negative Affect: Evidence from a Meta-Analysis of the Human Neuroimaging Literature. *Cereb. Cortex* **2016**, *26*, 1910–1922. [CrossRef] [PubMed]
53. Lindblom, J. *Embodied Social Cognition*; Springer: Berlin, Germany, 2015.
54. Klein, M. *Contribution to Psychoanalysis*; Mac Graw-Hill: New York, NY, USA, 1967.
55. Salvatore, S.; Zuitton, T. *Cultural Psychology and Psychoanalysis*; Information Age Publishing: Charlotte, NC, USA, 2011.
56. Valsiner, J. *Culture in Minds and Societies: Foundations of Cultural Psychology Culture in Minds and Societies: Foundations of Cultural Psychology*; SAGE Publications: New Dheli, India, 2007.
57. Stein, J. The influence of theory on the psychoanalyst's countertransference. *Int. J. Psychoanal.* **1991**, *72*, 325–334.
58. Zuitton, T. *Transitions. Development through Symbolic Resources*; Information Age Publishing: Charlotte, NC, USA, 2011.
59. Matte Blanco, I. *The Unconscious as Infinite Sets. An Essay in Bi-Logic*; Gerald Duckworth & Comp. Ltd.: London, UK, 1975.
60. Ciavolino, E.; Redd, R.; Evrinomy, A.; Falcone, M.; Fini, V.; Kadianaki, I.; Kullasepp, K.; Mannarini, T.; Matsopoulos, A.; Mossi, P.; et al. Views of context. An instrument for the analysis of the cultural milieu. A first validation study. *Electron. J. App. Stat. Anal.* **2017**, *10*, 599–628.
61. Barrett, L.F.; Russell, J.A. Independence and bipolarity in the structure of current affect. *J. Pers. Soc. Psychol.* **1998**, *74*, 967–984. [CrossRef]
62. Osgood, C.E.; Suci, G.J.; Tannenbaum, P.H. *The Measurement of Meaning*; Univerity Press: Champaign, IL, USA, 1957.

63. Tonti, M.; Salvatore, S. Homogenization of Classification Functions Measurement (HOCFUN): A Method for Measuring the Salience of Emotional Arousal in Thinking. *Am. J. Psychol.* **2015**, *128*, 469. [CrossRef]
64. Mannarini, T.; Rochira, A.; Talò, C. How Identification Processes and Inter-Community Relationships Affect Sense of Community. *J. Community Psychol.* **2012**, *40*, 951–967. [CrossRef]
65. Murphy, S.T.; Zajonc, R.B. Affect, cognition, and awareness: Affective priming with optimal and suboptimal stimulus exposures. *J. Pers. Soc. Psychol.* **1993**, *64*, 723–739. [CrossRef]
66. Turvey, M.; Fertig, J. Polarity on the semantic differential and release from proactive interference in short-term memory. *J. Verbal Learn. Verbal Behav.* **1970**, *9*, 439–443. [CrossRef]
67. Niedenthal, P.M.; Halberstadt, J.B.; Innes-Ker, Å.H. Emotional response categorization. *Psychol. Rev.* **1999**, *106*, 337–361. [CrossRef]
68. Freud, S. *The Interpretation of Dreams. Standard Edition IV and V*; Hogarth: London, UK, 1900.
69. Carroll, J.B.; Osgood, C.E.; May, W.H.; Miron, M.S. Cross-Cultural Universals of Affective Meaning. *Am. J. Psychol.* **1976**, *89*, 172. [CrossRef]
70. Russell, J.A. Culture and the categorization of emotions. *Psychol. Bull.* **1991**, *110*, 426–450. [CrossRef] [PubMed]
71. Salvatore, S.; Fini, V.; Mannarini, T.; Suerdem, A.; Veltri, G.A. The Salience of Otherness. In *Media and Social Representations of Otherness. Psycho-Social-Cultural Implications*; Mannarini, T., Veltri, G.A., Salvatore, S., Eds.; Springer: Cham, Switzerland, 2020; pp. 103–131.
72. Wierzbicka, A. *Emotions across Languages and Cultures: Diversity and Universals*; Cambridge University Press: Cambridge, MA, USA, 1999. [CrossRef]
73. Salvatore, S.; Gelo, O.; Gennaro, A.; Manzo, S. Looking at the psychotherapy process as an intersubjective dynamic of meaning-making. A case study with Discourse Flow Analysis. *J. Construct. Psychol.* **2010**, *23*, 195–230. [CrossRef]
74. Venuleo, C.; Salvatore, G.; Ruggieri, R.A.; Marinaci, T.; Cozzolino, M.; Salvatore, S. Steps towards a unified theory of psychopathology: The Phase Space of Meaning model. *Clin. Neuropsych.* **2020**, *17*, 236–252.
75. Bucci, W.; Maskit, B. A weighted dictionary for Referential Activity. In *Computing Attitude and Affect in Text*; Shanahan, J.G., Qu, Y., Wiebe, J., Eds.; Springer: Dordrecht, The Netherlands, 2006; pp. 49–60.
76. Pennebaker, J.W.; Francis, M.E.; Booth, R.J. *Linguistic Inquiry and Word Count (LIWC): LIWC2001*; Lawrence Erlbaum Associates: Mahwah, NJ, USA, 2001.
77. Gennaro, A.; Kipp, S.; Viol, K.; De Felice, G.; Andreassi, S.; Aichhorn, W.; Salvatore, S.; Schiepek, G. A Phase Transition of the Unconscious: Automated Text Analysis of Dreams in Psychoanalytic Psychotherapy. *Front. Psychol.* **2020**, *11*, 1667. [CrossRef]
78. Gennaro, A.; Goncalves, M.; Mendes, I.; Ribeiro, A.; Salvatore, S. Dynamics of sense-making and development of the narrative in the clinical exchange. *Res. Psychother. Psychopathol. Process. Outcome* **2011**, *14*, 90–120. [CrossRef]
79. Keller, T.A.; Carpenter, P.A.; Just, M. The Neural Bases of Sentence Comprehension: A fMRI Examination of Syntactic and Lexical Processing. *Cereb. Cortex* **2001**, *11*, 223–237. [CrossRef] [PubMed]
80. Prat, C.S.; Keller, T.A.; Just, M.A. Individual Differences in Sentence Comprehension: A Functional Magnetic Resonance Imaging Investigation of Syntactic and Lexical Processing Demands. *J. Cogn. Neurosci.* **2007**, *19*, 1950–1963. [CrossRef]
81. Allen, M.; Frank, D.; Schwarzkopf, D.S.; Fardo, F.; Winston, J.S.; Hauser, T.U.; Rees, G. Unexpected arousal modulates the influence of sensory noise on confidence. *eLife* **2016**, *5*, 18103. [CrossRef]
82. Jung, N.; Wranke, C.; Hamburger, K.; Knauff, M. How emotions affect logical reasoning: Evidence from experiments with mood-manipulated participants, spider phobics, and people with exam anxiety. *Front. Psychol.* **2014**, *5*, 570. [CrossRef]
83. Buck, B.; Penn, D.L. Lexical Characteristics of Emotional Narratives in Schizophrenia. *J. Nerv. Ment. Dis.* **2015**, *203*, 702–708. [CrossRef]
84. DePaulo, B.M.; Lindsay, J.J.; Malone, B.E.; Muhlenbruck, L.; Charlton, K.; Cooper, H. Cues to deception. *Psychol. Bull.* **2003**, *129*, 74–118. [CrossRef] [PubMed]
85. Zuckerman, M.; De Paulo, B.M.; Rosenthal, R. Verbal and non verbal communication od deception. *Adv. Exp. Soc. Psychol.* **1981**, *14*, 1–59. [CrossRef]
86. Greenberg, J.; Pyszczynski, T.; Solomon, S. The Causes and Consequences of a Need for Self-Esteem: A Terror Management Theory. In *Public Self and Private Self*; Springer: New York, NY, USA, 1986; pp. 189–212.
87. Derogatis, L.R. *Symptom Checklist-90-R: Administration, Scoring & Procedure Manual for the Revised Version of the SCL-90*; National Computer Systems: Minneapolis, MN, USA, 1994.
88. Gratz, K.L.; Roemer, L. Multidimensional assessment of emotion regulation and dysregulation: Development, factor structure, and initial validation of the difficulties in emotion regulation scale. *J. Psychopath Behav. Assess.* **2004**, *26*, 41–54. [CrossRef]
89. Vacca, R.; Franchina, V. Taratura dell'indice di Flesch su testo bilingue italiano-inglese di unico autore. *Linguaggi* **1986**, *3*, 47–49.
90. Williams, D.P.; Cash, C.; Rankin, C.; Bernardi, A.; Koenig, J.; Thayer, J.F. Resting heart rate variability predicts self-reported difficulties in emotion regulation: A focus on different facets of emotion regulation. *Front. Psychol.* **2015**, *6*, 261. [CrossRef] [PubMed]
91. Preti, A.; Carta, M.G.; Petretto, D.R. Factor structure models of the SCL-90-R: Replicability across community samples of adolescents. *Psychiatry Res.* **2019**, *272*, 491–498. [CrossRef]
92. Lundqvist, L.-O.; Schröder, A. Evaluation of the SCL-9S, a short version of the symptom checklist-90-R, on psychiatric patients in Sweden by using Rasch analysis. *Nord. J. Psychiatry* **2021**, *75*, 538–546. [CrossRef] [PubMed]

93. Achenbach, T.M.; Ivanova, M.Y.; Rescorla, L.A.; Turner, L.V.; Althoff, R.R. Internalizing/Externalizing Problems: Review and Recommendations for Clinical and Research Applications. *J. Am. Acad. Child Adolesc. Psychiatry* **2016**, *55*, 647–656. [CrossRef]
94. Prunas, A.; Sarno, I.; Preti, E.; Madeddu, F.; Perugini, M. Psychometric properties of the Italian version of the SCL-90-R: A study on a large community sample. *Eur. Psychiatry* **2012**, *27*, 591–597. [CrossRef] [PubMed]
95. Schauenburg, H.; Strack, M. Measuring psychotherapeutic change with the symptom checklist SCL 90 R. *Psychother. Psychosom.* **1999**, *68*, 199–206. [CrossRef] [PubMed]
96. Schmitz, N. Assessing Clinically Significant Change: Application to the SCL-90-R. *Psychol. Rep.* **2000**, *86*, 263–274. [CrossRef]
97. Bardeen, J.R.; Fergus, T.A.; Orcutt, H.K. An Examination of the Latent Structure of the Difficulties in Emotion Regulation Scale. *J. Psychopathol. Behav. Assess.* **2012**, *34*, 382–392. [CrossRef]
98. Kökönyei, G.; Urbán, R.; Reinhardt, M.; Józan, A.; Demetrovics, Z. The Difficulties in Emotion Regulation Scale: Factor Structure in Chronic Pain Patients. *J. Clin. Psychol.* **2013**, *70*, 589–600. [CrossRef] [PubMed]
99. Fowler, J.C.; Charak, R.; Elhai, J.D.; Allen, J.G.; Frueh, B.C.; Oldham, J.M. Construct validity and factor structure of the difficulties in Emotion Regulation Scale among adults with severe mental illness. *J. Psychiatr. Res.* **2014**, *58*, 175–180. [CrossRef]
100. Osborne, T.L.; Michonski, J.; Sayrs, J.; Welch, S.S.; Anderson, L.K. Factor Structure of the Difficulties in Emotion Regulation Scale (DERS) in Adult Outpatients Receiving Dialectical Behavior Therapy (DBT). *J. Psychopathol. Behav. Assess.* **2017**, *39*, 355–371. [CrossRef]
101. Becerra, R.; Cruise, K.; Murray, G.; Bassett, D.; Harms, C.; Allan, A.; Hood, S. Emotion regulation in bipolar disorder: Are emotion regulation abilities less compromised in euthymic bipolar disorder than unipolar depressive or anxiety disorders? *Open J. Psychiatry* **2013**, *3*, 1–7. [CrossRef]
102. Ehring, T.; Quack, D. Emotion Regulation Difficulties in Trauma Survivors: The Role of Trauma Type and PTSD Symptom Severity. *Behav. Ther.* **2010**, *41*, 587–598. [CrossRef]
103. Van Rheenen, T.E.; Murray, G.; Rossell, S. Emotion regulation in bipolar disorder: Profile and utility in predicting trait mania and depression propensity. *Psychiatry Res.* **2015**, *225*, 425–432. [CrossRef]
104. Hallion, L.S.; Steinman, S.; Tolin, D.F.; Diefenbach, G.J. Psychometric Properties of the Difficulties in Emotion Regulation Scale (DERS) and Its Short Forms in Adults With Emotional Disorders. *Front. Psychol.* **2018**, *9*, 539. [CrossRef]
105. Gratz, K.L.; Gunderson, J.G. Preliminary Data on an Acceptance-Based Emotion Regulation Group Intervention for Deliberate Self-Harm Among Women with Borderline Personality Disorder. *Behav. Ther.* **2006**, *37*, 25–35. [CrossRef] [PubMed]
106. Gratz, K.L.; Lacroce, D.M.; Gunderson, J.G. Measuring Changes in Symptoms Relevant to Borderline Personality Disorder Following Short-Term Treatment across Partial Hospital and Intensive Outpatient Levels of Care. *J. Psychiatr. Pract.* **2006**, *12*, 153–159. [CrossRef] [PubMed]
107. Gratz, K.L.; Rosenthal, M.Z.; Tull, M.T.; Lejuez, C.W.; Gunderson, J.G. An experimental investigation of emotion dysregulation in borderline personality disorder. *J. Abnorm. Psychol.* **2006**, *115*, 850–855. [CrossRef] [PubMed]
108. Salvatore, S.; Gennaro, A.; Auletta, A.F.; Tonti, M.; Nitti, M. Automated method of content analysis: A device for psychotherapy process research. *Psychother. Res.* **2012**, *22*, 256–273. [CrossRef] [PubMed]
109. Salvatore, S.; Gelo, O.C.G.; Gennaro, A.; Metrangolo, R.; Terrone, G.; Pace, V.; Venuleo, C.; Venezia, A.; Ciavolino, E. An automated method of content analysis for psychotherapy research: A further validation. *Psychother. Res.* **2015**, *27*, 38–50. [CrossRef] [PubMed]
110. Lancia, F. Word Co-Occurrence and Theory of Meaning. Available online: http://www.mytlab.com/wctheory.pdf.2005 (accessed on 16 December 2012).
111. Lancia, F. User's Manual: Tools for Text Analysis. T-Lab Version Plus 2020. Available online: https://www.tlab.it/?lang=it (accessed on 20 June 2020).
112. Shannon, C.E. A mathematical theory of communication. *Bell Syst. Tech. J.* **1948**, *27*, 379–423. [CrossRef]
113. Melamed, I.D. Measuring semantic entropy. In Proceedings of the ACL-SIGLEX Workshop on Tagging Text with Lexical Semantics: Why, What and How, Washington, DC, USA, 4–5 April 1997; pp. 41–46.
114. Venhuizen, N.J.; Crocker, M.W.; Brouwer, H. Expectation-based Comprehension: Modeling the Interaction of World Knowledge and Linguistic Experience. *Discourse Process.* **2019**, *56*, 229–255. [CrossRef]
115. De Deyne, S.; Navarro, D.J.; Storms, G. Associative strength and semantic activation in the mental lexicon: Evidence from continued word associations. In Proceedings of the 35th Annual Conference of the Cognitive Science Society, Pasadena, CA, USA, 23–25 July 2015.
116. Flesch, H.R. A new readability yardstick. *J. Appl. Psychol.* **1948**, *32*, 221–233. [CrossRef]
117. Flesch, H.R. *The Anofplain Talk*; Harper and Row: New York, NY, USA, 1962.
118. Thayer, J.F.; Hansen, A.L.; Saus-Rose, E.; Johnsen, B.H. Heart Rate Variability, Prefrontal Neural Function, and Cognitive Performance: The Neurovisceral Integration Perspective on Self-regulation, Adaptation, and Health. *Ann. Behav. Med.* **2009**, *37*, 141–153. [CrossRef]
119. Thayer, J.F.; Lane, R.D. Claude Bernard and the heart–brain connection: Further elaboration of a model of neurovisceral integration. *Neurosci. Biobehav. Rev.* **2009**, *33*, 81–88. [CrossRef]
120. Jennings, J.R.; Kamarck, T.; Stewart, C.; Eddy, M.; Johnson, P. Alternate Cardiovascular Baseline Assessment Techniques: Vanilla or Resting Baseline. *Psychophysiology* **1992**, *29*, 742–750. [CrossRef]

121. Task Force of the European Society of Cardiology the North American Society of Pacing Electrophysiology. Heart Rate Variability. *Circulation* **1996**, *93*, 1043–1065. [CrossRef]
122. Weippert, M.; Kumar, M.; Kreuzfeld, S.; Arndt, D.; Rieger, A.; Stoll, R. Comparison of three mobile devices for measuring R–R intervals and heart rate variability: Polar S810i, Suunto t6 and an ambulatory ECG system. *Graefe's Arch. Clin. Exp. Ophthalmol.* **2010**, *109*, 779–786. [CrossRef] [PubMed]
123. Tarvainen, M.P.; Niskanen, J.-P.; Lipponen, J.; Ranta-Aho, P.O.; Karjalainen, P. Kubios HRV—Heart rate variability analysis software. *Comput. Methods Programs Biomed.* **2014**, *113*, 210–220. [CrossRef] [PubMed]
124. Park, G.; Vasey, M.; Van Bavel, J.J.; Thayer, J.F. When tonic cardiac vagal tone predicts changes in phasic vagal tone: The role of fear and perceptual load. *Psychophysiology* **2014**, *51*, 419–426. [CrossRef] [PubMed]
125. Tarvainen, M.P.; Ranta-Aho, P.O.; Karjalainen, P. An advanced detrending method with application to HRV analysis. *IEEE Trans. Biomed. Eng.* **2002**, *49*, 172–175. [CrossRef] [PubMed]
126. Appelhans, B.M.; Luecken, L.J. Heart Rate Variability as an Index of Regulated Emotional Responding. *Rev. Gen. Psychol.* **2006**, *10*, 229–240. [CrossRef]
127. Butler, E.A.; Wilhelm, F.H.; Gross, J.J. Respiratory sinus arrhythmia, emotion, and emotion regulation during social interaction. *Psychophysiology* **2006**, *43*, 612–622. [CrossRef] [PubMed]
128. Fujimura, T.; Okanoya, K. Heart Rate Variability Predicts Emotional Flexibility in Response to Positive Stimuli. *Psychophysiology* **2012**, *3*, 578–582. [CrossRef]
129. Hildebrandt, L.K.; McCall, C.; Engen, H.G.; Singer, T. Cognitive flexibility, heart rate variability, and resilience predict fine-grained regulation of arousal during prolonged threat. *Psychophysiology* **2016**, *53*, 880–890. [CrossRef] [PubMed]
130. Radmanesh, M.; Jalili, M.; Kozlowska, K. Activation of Functional Brain Networks in Children With Psychogenic Non-epileptic Seizures. *Front. Hum. Neurosci.* **2020**, *14*, 339. [CrossRef]
131. Lohani, M.; Cooper, J.M.; Erickson, G.G.; Simmons, T.G.; McDonnell, A.S.; Carriero, A.E.; Crabtree, K.W.; Strayer, D.L. Driver Arousal and Workload Under Partial Vehicle Automation: A Pilot Study. In *Proceedings of the Human Factors and Ergonomics Society Annual Meeting*; SAGE Journals: Thousand Oaks, CA, USA, 2020; Volume 64, pp. 1955–1959.

Article

Beyond Dyadic Coupling: The Method of Multivariate Surrogate Synchrony (mv-SUSY)

Deborah Meier * and Wolfgang Tschacher

University Hospital of Psychiatry and Psychotherapy, University of Bern, 3000 Bern 60, Switzerland; wolfgang.tschacher@upd.unibe.ch
* Correspondence: deborah.meier@upd.unibe.ch

Abstract: Measuring interpersonal synchrony is a promising approach to assess the complexity of social interaction, which however has been mostly limited to dyads. In this study, we introduce multivariate Surrogate Synchrony (mv-SUSY) to extend the current set of computational methods. Methods: mv-SUSY was applied to eight datasets consisting of 10 time series each, all with n = 9600 observations. Datasets 1 to 5 consist of simulated time series with the following characteristics: white noise (dataset 1), non-stationarity with linear time trends (dataset 2), autocorrelation (dataset 3), oscillation (dataset 4), and multivariate correlation (dataset 5). Datasets 6 to 8 comprise empirical multivariate movement data of two individuals (datasets 6 and 7) and between members of a group discussion (dataset 8.) Results: As hypothesized, findings of mv-SUSY revealed absence of synchrony in datasets 1 to 4 and presence of synchrony in dataset 5. In the empirical datasets, mv-SUSY indicated significant movement synchrony. These results were predominantly replicated by two well-established dyadic synchrony approaches, Surrogate Synchrony (SUSY) and Surrogate Concordance (SUCO). Conclusions: The study applied and evaluated a novel synchrony approach, mv-SUSY. We demonstrated the feasibility and validity of estimating multivariate nonverbal synchrony within and between individuals by mv-SUSY.

Keywords: surrogate synchrony; multivariate analysis; simulation; movement synchrony

Citation: Meier, D.; Tschacher, W. Beyond Dyadic Coupling: The Method of Multivariate Surrogate Synchrony (mv-SUSY). *Entropy* **2021**, *23*, 1385. https://doi.org/10.3390/e23111385

Academic Editor: José F. F. Mendes

Received: 17 August 2021
Accepted: 19 October 2021
Published: 22 October 2021

1. Introduction

Interpersonal synchrony has become a growing field of empirical research in social psychology and psychotherapy. Synchrony, composed of the Greek words syn (together) and chronos (time), denotes the coordination of variables that each represent the temporal evolution of the momentary state of a system. The bulk of applications has focused on the synchrony of two individuals A and B in social interaction. If these individuals act "together in time", it is expected that their behavior is coupled and mutually entrained to a degree exceeding random correlations: A and B are then said to be in synchrony. The basic level of synchrony occurs at a phasic time scale of a few seconds; this is the time scale of the "Now", the experienced present moment, during which present-time consciousness of the "here-and-now" arises [1]. This "Now" has a temporal duration, which provides a reason for including lagged correlations. At this time scale, synchrony can be detected in A and B's body movement [2,3], neural activity [4], or physiological activations [5,6]. Accordingly, to quantify synchrony the variable of interest is observed continuously over time and the corresponding time series must be sampled at high temporal resolution. Temporal delays (lags) of the time series may be considered to account for the durational aspects of the "Now" as well as the psychological reality of response times in social interaction.

Various fields of psychological research have used synchrony measures. Since each field has its own terminology, several synonyms for the phenomenon of synchrony are used, such as attunement, interpersonal physiology, concordance, coupling or entrainment. The largest number of applications is in clinical psychology, where the coordination of therapist and patient in a therapy session is a phenomenon of major interest. Many studies were conducted addressing nonverbal measures, as participants' body movements can be economically sampled by Motion Energy Analysis [7,8], a software package for automated motion capture from video recordings. Generally, studies have found that levels of movement synchrony significantly exceeded random control conditions [8–10]. It was often reported that synchrony had pro-social effects [11], predicted better outcomes, such as reduced interpersonal problems of patients, and was associated with the quality of the therapeutic alliance. Prosodic aspects of speech, especially the attunement of pitch and loudness of verbal utterances, can also be monitored in a non-invasive way in psychotherapy sessions [12]. Increasingly, synchrony is studied also on the basis of physiological variables of two participants. For an extensive literature review see Palumbo et al. [13]. This field has seen early forerunners in DiMascio's group, who focused on interpersonal synchrony of heart rate in psychiatric interviews [14]. Marci & Orr [15] found concordance of electrodermal activity in clinical interviews, and this approach was also chosen in a setting of naturalistic couple therapies with two therapists [16]. Tschacher and Meier [6] investigated the synchrony of heart rate, heart-rate variability and respiration in psychotherapy sessions.

Fields of research outside clinical applications concerned social psychology [17], such as instructing dyads to have topical conversations [3], or cooperation tasks [18]. Synchrony of team performance has been studied, for instance, by Guastello, Mirabiti, and Peressini [19]. Findings have been that synchrony, also labeled mimicry, arises unintentionally between interacting participants, and increases the smoothness of social interactions, mutual liking, and positive affect. Furthermore, researchers in the field of synchrony have focused on close relationships such as romantic couples [20,21], co-parenting spouses [22], and mother-infant dyads [23]. Some applications have addressed music psychology and analyzed movement synchrony of musicians in concerts [24], physiological synchrony in concert audiences [25] and audiences of dance performance [26].

Synchrony was analyzed using a variety of computational methods, which generally consider the synchrony between two univariate time series. There are three clusters of such methods: Correlational methods (time-based), Fourier analysis (frequency-based) and cross-recurrence quantification (nonparametric). The time-based and frequency-based approaches are mathematically related because Fourier spectra in the limit can be transformed to correlational terms. Frequency-based methods have made use of wavelet transforms of the dual time series, and cross-wavelet coherence was then used as a synchrony measure, e.g., by Fujiwara and Daibo [2]. Cross-recurrence plots illustrate the dyadic time series A, B in the two-dimensional coordinate system, where the axes are given by 'A plus/minus a time lag' and 'B plus/minus a time lag'. Regular behavior of A and B such as corresponding oscillations then generates regular patterns in the cross-recurrence plot that can be quantified to provide a measure of synchrony [27]. Published research on the quantification of multivariate synchrony is very limited. Multivariate methods include phase synchronization based on Kuramoto order [28,29], multidimensional recurrence quantification analysis [30], and matrix calculations to determine the "driver" and "empath" of a group of interacting persons, and based on this, to estimate a synchronization coefficient [31].

Correlational methods underlie a majority of applications in the synchrony literature; they are either based on the cross-correlation function (CCF) of A and B, which denotes the correlation between A and B depending on their lags, or on the correlation of A and B's local slopes, which defines the concordance measure [15]. An elaborated variant of the cross-correlational approach is computing the CCF segment-wise to account for non-stationarity due to overarching trends or seasonality of the time series, which may lead to inflated correlations. This principle forms the core of windowed cross-correlation (WCC) analysis, where "segment-wise" is referred to as "windowed". WCC analysis requires the specification of multiple parameters: segment size, segment increment, maximum lag, and lag increment. For statistical and theoretical considerations in WCC analysis see for example Behrens et al. [32]. Accordingly, differences in results between computational approaches often arise from different choices of one or more of these parameters. In this respect, surrogate synchrony (SUSY) has fixed segment increment and lag increment, which keeps the dependency on parameter settings low. In SUSY the time series is cut into segments (e.g., of 30 s duration) and all cross-correlations of this segment's CCF within a certain range of lags (e.g., lags of up to ± 3 s) are computed. Aggregation is done by transforming cross-correlations to (absolute) values of Fisher's Z and computing the mean Z for each lag across segments. Aggregation is repeated across all lags, thus delivering a signature of synchrony by the mean Z ($\overline{Z}_{\mathrm{noabs}}$) or by the mean of all absolute values of Z ($\overline{Z}_{\mathrm{abs}}$). SUSY does not use overlapping segments (hence, segment increment = segment size) or lag increment (hence, lag increment is determined by the sampling rate of the time series).

Well-established approaches share the common characteristic of keeping parameter settings constant throughout the analysis. Thus, these methods do not interfere with the equidistance between time points. A very recent approach, dynamic time warping, is based on varying lags which allows stretching the time series to accordingly estimate synchrony [33]. Since parameter settings have the potential to influence the results to a considerable degree, researchers in the field of synchrony aimed to define universal guidelines [34].

Surrogate analysis is an appropriate tool to establish the significance of aggregated cross-correlations, which indicate the measured 'real' synchrony, by comparing this synchrony to a control condition, called surrogate synchrony. Measured synchrony may be due to random fluctuations, and surrogate controls help to determine the level of randomness. Measured synchrony may even be inflated because of trends and/or autocorrelation in the time series. For example, we would expect to estimate a certain level of synchrony in the respiration of individuals, even though they did not even interact. This would be simply due to the periodic rhythm of inhaling and exhaling. Furthermore, elevated synchrony levels may derive from coordinated dynamics in human interaction (e.g., linguistic turn-taking) and therefore merely reflect patterns of shared task context, whereas we are interested in the unique levels of synchrony above what can be expected due to that particular task. Segment shuffling tests the null hypothesis that there is no synchrony difference between the measured time series and the sample of randomly shuffled time series, where the temporal intercorrelations are lost. Assuming the null hypothesis is rejected, temporal intercorrelations between the measured time series exist above chance level. Surrogate analysis must be adapted to the respective null hypothesis [35]. Accordingly, surrogate data may be generated by various methods such as data shuffling, segment shuffling, data sliding, or participant shuffling [36].

With the present study, we aim to extend the synchrony methodology to multivariate time series and thus go beyond the predominant analysis of dyadic synchronies toward more complex datasets. To do this, we will propose different methods of multivariate surrogate synchrony (mv-SUSY), and all methods use segment shuffling for surrogate analysis. The extension to multivariate time series was motivated by the need for synchrony measures when more than just two variables are considered coupled. In interpersonal contexts, we may be interested in analyzing the synchrony in groups of people rather than dyads only; in the single person, we may be interested in the joint coupling of several physiological variables at the same time, or in the coordinated movement of several body parts of the same person. To demonstrate its validity, mv-SUSY must accurately indicate the existence or non-existence of the synchrony phenomenon it is supposed to capture.

The goals of the validity tests conducted in the present study were threefold: First, we aimed to establish the validity of mv-SUSY in simulated data. We hypothesized (H1a) to demonstrate the absence of multivariate synchrony in stationary random time series (dataset 1), non-stationary random time series with linear trends (dataset 2), autocorrelated time series (dataset 3), and time series of regular oscillations (dataset 4). We further hypothesized (H1b) that the presence of synchrony would be detected and confirmed in mutually correlated time series (dataset 5). Second, we extended the application of mv-SUSY to examples of empirical movement data. We hypothesized (H2) to demonstrate multivariate synchrony in movement time series captured using Kinect (datasets 6 and 7) and motion energy analysis (dataset 8). Third, well-established methods for dyadic synchrony (Surrogate Synchrony; SUSY, Surrogate Concordance; SUCO) were applied as a control methodology. We hypothesized (H3) that the repeated application of dyadic SUSY and SUCO would validate the mv-SUSY findings of the first and second goals.

2. Materials and Methods

2.1. Simulated Multivariate Datasets

To illustrate the estimation of multivariate synchrony, we generated five datasets that simulate different types of temporal behavior. Each dataset comprises 10 time series, y_m ($1 \leq m \leq 10$), each consisting of 9600 observations y_{mt} ($1 \leq m \leq 10; 1 \leq t \leq 9600$).

Dataset 1 was constructed to represent stationary random time series, i.e., white noise. The time series y_m consist of normally distributed random values, and time series are independent of each other. The distributions of values have mean = 0 and standard deviation = 1.

$$\text{Dataset 1: } y_{mt} \sim N(0,1) \quad (1)$$

The next three datasets simulate characteristics and non-stationarities frequently found in empirical time series, such as stable time trends, autocorrelated evolutions, and oscillatory behavior. These characteristics can lead to increased levels of detected synchrony, which should be identified as spurious by the respective surrogate analyses leading to support of null hypotheses.

Dataset 2 was generated to express white noise data with varying but stable trends (positive and negative), thus non-stationary time series. We simulated 9600 observations as a linear function of time. For each time series in dataset 2 the respective intercepts, β_0, and slope coefficients, β_1, were sampled individually, where $\beta_0 \sim N(0,25)$ and $\beta_1 \sim N(0,0.000001)$. We added some noise, ε_t, which was sampled from a distribution with mean = 0 and standard deviation = 1.

$$\text{Dataset 2: } y_{mt} = \beta_0 + \beta_{1t} + \varepsilon_t \quad (2)$$

Dataset 3 comprises autocorrelated time series with high lag 1 correlations between observations (so-called AR(1) processes). Again, we added an error term, where $\varepsilon_t \sim N(0,1)$.

$$\text{Dataset 3: } y_{mt} = 0.9 \times y_{t-1} + \varepsilon_t \quad (3)$$

Dataset 4 consists of oscillatory time series produced by sine functions of varying frequencies and vertical shifts. For each time series in dataset 4, periods given by $2\pi/b$ were individually sampled with $b \sim N(0.4,0.0064)$ and vertically shifted with $d \sim N(0,1)$.

$$\text{Dataset 4: } y_{mt} = \sin(b \times t) + d \quad (4)$$

Dataset 5: Multivariate correlations among time series in this dataset were realized by creating a white noise baseline time series, y_1, for which we assumed $y_1 \sim N(0,1)$. Based on y_1, nine dependent time series, y_2 to y_{10}, were generated by repeatedly adding a certain degree of noise, where $\varepsilon_t \sim N(0,0.25)$. The intercorrelated times series of this dataset was expected to entail rejection of null hypotheses.

$$\text{Dataset 5: } y_t = y_1 + \varepsilon_t \quad (5)$$

Visualizations of these datasets are given in Figure 1. The left panels correspond to the time series plots. The first five rows are time series plots of short sections (100 observations) of the simulated time series. The time series plot of dataset 2 shows 7000 observations to illustrate the long-range trends. The right panels of Figure 1 provide the ten autocorrelation functions (ACF) of each dataset up to lag 10. Autocorrelations indicate the different patterns of temporal dependence between single observations of the time series. Across datasets, time series were not autocorrelated (datasets 1 and 5), had stable autocorrelations (dataset 2). Dataset 3 showed the slowly decaying autocorrelation functions of autoregressive processes, dataset 4 the regular cyclical autocorrelation patterns of periodic data. Furthermore, the mean intercorrelations r_m between the respective ten time series of each dataset demonstrated varying degrees of relationship within these samples: r_m= 0.002 (dataset 1), r_m = 0.12 (dataset 2), r_m = 0.002 (dataset 3), r_m = −0.00009 (dataset 4), and r_m = 0.82 (dataset 5). Higher values for mean absolute intercorrelations indicated positive and negative pairwise correlations (dataset 1: r_m = 0.01, dataset 2: r_m = 0.65, dataset 3: r_m = 0.03, dataset 4: r_m = 0.002, dataset 5: r_m = 0.82).

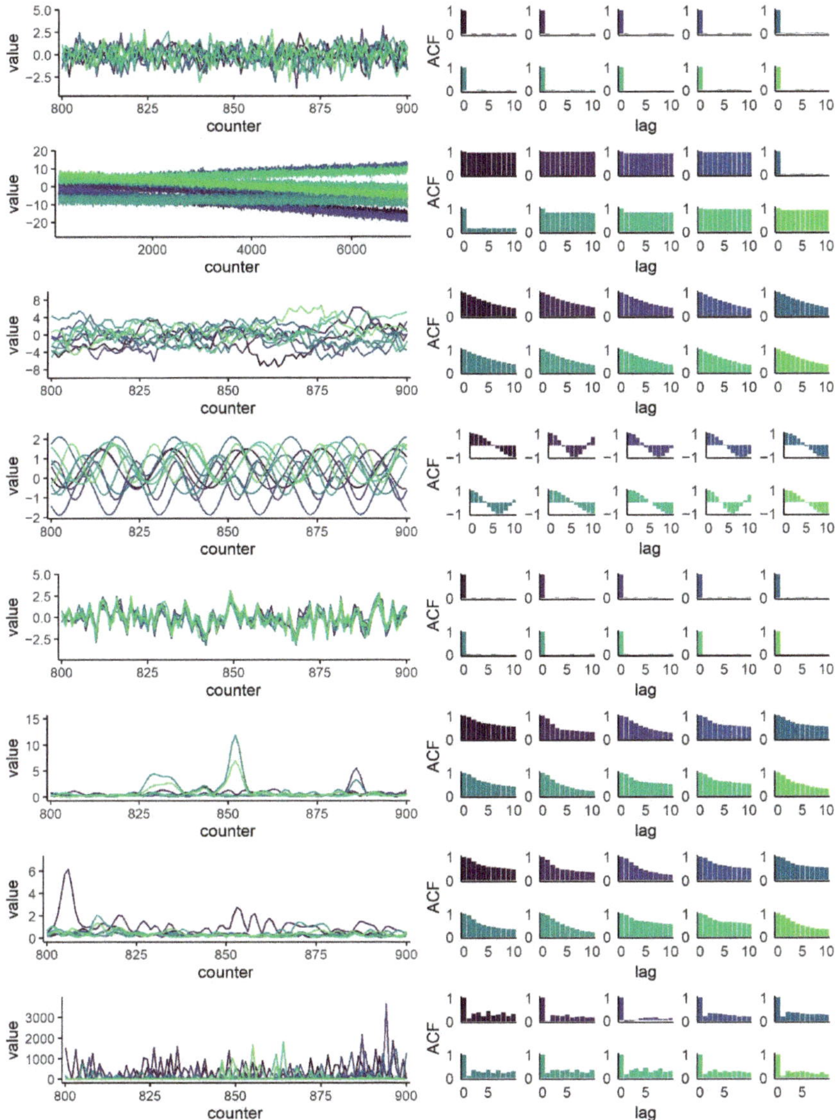

Figure 1. Multivariate datasets, each consisting of m = 10 time series (left panels, characteristic cutout plots) and autocorrelation functions (ACF, right panels). From top to bottom: simulated data (datasets 1 to 5), empirical data (datasets 6 to 8).

2.2. Empirical Multivariate Datasets

Datasets 6 to 8 represent empirical movement data. The Kinect datasets 6 and 7 originated from an experimental study, in which two participants were instructed to have a conversation exclusively with nonverbal expressions and dance movements in order "to try to get to know each other without words" [37]. Two previously unacquainted adults were allocated two non-overlapping areas marked on the floor, in which they could move freely without touching. All movements during the 5:20 minutes of this "body conversation task" were motion-captured using Kinect cameras, and time series consisted of 10 limb positions per person (head, chest, both upper arms, both lower arms, both upper legs, and both lower legs) at a sampling rate of 30 Hz.

The Zoom video data (dataset 8) consists of movement data recorded during an online group discussion among 10 individuals, whose body movements were captured by the automatized method MEA. MEA was based on pixel changes in assigned regions of interest of the Zoom video recording. The regions chosen were the panels that showed the respective head and upper body of the participants of the group discussion. Thus, the time series (sampling rate: 25 Hz) represent nonverbal movements of head, face, and upper body. The scholarly discussion was on the topic of "Embodiment, Physical Distancing and Treatment" (Virtual Structured Discussion, organized by the special interest group "Complexity in Psychotherapy" of the Society for Psychotherapy Research (SPR), 26 June 2020). All subjects agreed to the video recording. We selected the especially engaging final section of the Zoom discussion for further analysis. The section had a duration of 6 min 24 s (i.e., 9600 data points in the time series).

Visualizations of the empirical datasets are provided in Figure 1 (rows 6 to 8). On the left panels, there are time series plots of short sections (100 observations) of the empirical datasets. The right panels of Figure 1 provide the autocorrelation functions up to lag 10. Autocorrelations functions of the Kinect data (rows 6 and 7) resemble those of AR(1) autoregressive processes, whereas the Zoom dataset (row 8) had a pattern of smaller autocorrelations. The mean intercorrelations between the time series of each dataset were $r_m = 0.47$ (dataset 6), $r_m = 0.48$ (dataset 7), and $r_m = 0.01$ (dataset 8). Negative correlations were only found in dataset 8, where the mean absolute intercorrelations $r_m = 0.02$.

2.3. Dyadic Synchrony Computation

Surrogate Synchrony (SUSY, cf. www.embodiment.ch, accessed on 11 October 2021) estimates dyadic synchrony defined as cross-correlations between two time series A and B. The core procedure lies in the control of real synchrony by surrogate synchrony. Therefore, time series A and B are cut into segments according to 'segment size'. First, SUSY computes cross-correlations within each segment across a certain range of lags. For example, for a parameter setting 'maximum lag' = ± 3 s, all cross-correlations within a six-second window are considered. Twofold aggregation of these cross-correlations (across all segments and lags) yields a measure for real synchrony. Beforehand, cross-correlations are Fisher's Z-transformed to allow for aggregation. SUSY provides two indices based on absolute and non-absolute Z values: $\overline{Z}_{noabs}$ (SUSY) and $\overline{Z}_{abs}$ (SUSY). Whereas $\overline{Z}_{abs}$ (SUSY) indicates overall synchrony, $\overline{Z}_{noabs}$ (SUSY) distinguishes between in-phase and anti-phase synchrony. The complete procedure of dyadic SUSY generates a surrogate control condition for $\overline{Z}_{noabs}$ and $\overline{Z}_{abs}$ by shuffling the sequence of segments of the original time series, so that segments of A are 'falsely' aligned with segments of B. Shuffling can be repeated and produces many different surrogates. Then $\overline{Z}_{noabs-surr}$ (SUSY) and $\overline{Z}_{abs-surr}$ (SUSY) as markers of surrogate synchrony are computed. Mathematical details of SUSY methodology were described by Tschacher and Haken [38] and Tschacher and Meier [6].

Surrogate Concordance (SUCO, cf. www.embodiment.ch, accessed on 11 October 2021) has its origins in the concordance approach by Marci and Orr [15] and was first implemented in Tschacher and Meier [6]. SUCO requires four basic parameters: 'segment size', 'window size', 'lag', and 'increment'. A local slope is computed inside a window by least squares regression in the corresponding segments of A and B. Each window is shifted according to 'increment' until all windows per segment are taken into account. Synchrony is defined as correlation of A's and B's local slopes. Then, after Fisher's Z transformation of correlations, these are aggregated across all segments of the time series. All procedures can be performed with lagged windows. SUCO yields two dyadic synchrony indices, $\overline{Z}_{abs}$ (SUCO) and $\overline{Z}_{noabs}$ (SUCO) representing real synchrony. Again, surrogate analysis is performed by random shuffling of segments, defining the indices for surrogate synchronies.

2.4. Multivariate Surrogate Synchrony Computation

Multivariate Surrogate Synchrony (mv-SUSY) estimates the synchrony within datasets that contain more than two time series. The number of time series is denoted by m. mv-SUSY was adapted and extended from SUCO and SUSY with respect to implementing the surrogate controls [3,8]. As in dyadic synchrony, two computation steps are conducted to compare synchrony of the time series to surrogate synchrony of segment-shuffled surrogate time series. According to 'segment size', all m time series of the m-variate dataset are cut into equal-sized segments. The number of segments follows from the duration (number of observations divided by sampling rate) for each time series divided by segment size. For example, a dataset with m = 10 time series of 5 min (9000 observations at 30 Hz) and 'segment size' set to five seconds contains 60 segments. Each segment renders one synchrony value.

We developed two methods to assess mv-SUSY: omega and lambda$_{max}$. omega is a measure of multivariate synchrony that makes use of the actually measured degree of entropy H_{act} (actual entropy). Entropy is a measure of disorder of a dataset in thermodynamics, with its equivalent Shannon information in information theory [39]. Landsberg [40] suggested to normalize entropy by the maximum entropy possible in a system H_{pot} (potential entropy), thus H_{act}/H_{pot}. As this ratio assumes values between 0 and 1, 'Landsberg order' consequently becomes omega = $1 - H_{act}/H_{pot}$. Banerjee et al. [41] proposed to estimate these entropies based on the variance-covariance matrix, which avoids specific problems of assessing entropies in psychological and biological systems [42,43]. Thus, entropies are derived from the covariance matrix of a dataset. We used the absolute values of all covariances to indicate that also negative covariances contribute positively to the overall covariance of the dataset. The variance-covariance matrix is a $m \times m$ symmetrical matrix whose cells contain the covariance coefficients, measures of linear relationship, of each pair of variables, and whose main diagonal represents the univariate variances of each variable. H_{act} is estimated from the determinant of this matrix as a measure of generalized variance. The normalized entropy is therefore the actual entropy in relation to the maximum potential entropy of a system. Maximum entropy corresponds to a state of total independence between the elements of a system, where the covariance coefficients would be zero and the determinant becomes the product of the diagonal. Thus, maximum entropy H_{pot} is the product of all univariate variances. In mv-SUSY, omega is computed as $1 - (H_{act}/H_{pot})$ in each segment of the dataset, and these values are aggregated across all segments.

The second mv-SUSY method, lambda$_{max}$, is computed by eigendecomposition of the correlation matrix for each segment. The correlation matrix is a $m \times m$ matrix where m corresponds to the number of dimensions of a dataset, in the present case the number of time series. The correlation matrix contains the correlation coefficients between two variables ($-1 \leq r \leq 1$) in the upper and lower triangle. All diagonal elements are equal to 1, the correlation of each variable with itself. The $m \times m$ correlation matrix can be decomposed into m eigenvalues λ, and m corresponding eigenvectors v. For details on the calculation of eigenvalues and eigenvectors see for example Fischer [44]. Eigenvalues are associated with the variances of the variables on which the correlation matrix is based. Geometrically, eigenvectors are orthogonal vectors scaled by their corresponding eigenvalues, indicating the multidimensional dispersion of the data. In detail, the proportion of variance associated with a particular dimension is equal to the corresponding eigenvalue divided by the sum of all eigenvalues. The sum of the eigenvalues refers to the sum of the diagonal elements, which in the case of the correlation matrix is always m [45]. We propose that multivariate synchrony can be defined by one or a few large eigenvalues, when the remaining eigenvalues are small. In this case, there are only a few dominant dimensions that account for most of the variance in the data. Thus, lambda_{max} in mv-SUSY is computed by the proportion of the largest eigenvalue to the sum of all eigenvalues for each segment. Eigenvalues and eigenvectors are also the fundament of principle component analysis (PCA). PCA prioritizes key dimensions, so-called principal components, for the use of dimensionality reduction. In the field of self-organization, Tschacher and Grawe [46] previously used the variance explained by the first component as a measure of order in continuously rated therapy session reports.

Finally, the estimated multivariate synchronies of both omega and lambda_{max} are controlled for random or spuriously inflated synchrony using surrogate analysis. Time series segmentation formed the base of surrogate analysis (step two) in order to reduce potential effects of non-stationarity of time series data. In the surrogate step, mv-SUSY randomly shuffles the segments of all m time series independently of each other. Only those surrogates are allowed that do not include segments that matched in the original dataset. From m time series with s segments, $s!/(s-m)!$ surrogates can be generated. The parameter 'number of surrogates' determines the number of surrogates randomly drawn from the pool of surrogates to limit processing time. Randomized shuffling in step two provides surrogate time series that share important characteristics with the measured time series segments such as the length of segments, their means and standard deviations.

Synchrony of surrogates is computed in the same way as real synchrony, using the methods omega or lambda_{max}. Finally, to obtain global measures for synchrony, we compute effect sizes, ES (mv), of omega and lambda_{max} as the difference between the respective synchrony and mean surrogate synchrony standardized by the standard deviation of surrogate synchronies.

2.5. Statistical Analysis

To investigate the first goal, mv-SUSY was applied to simulated datasets 1 to 5. The segment size parameter was set to five seconds. Assuming a sampling rate of 30 Hz, each segment comprised 1500 observations. The number of surrogates was limited to 1000. Addressing the second goal, we estimated synchrony for the empirical datasets 6 to 8, using mv-SUSY. The same parameter settings as for the simulated data were adopted for datasets 6 and 7. To account for the lower sampling rate in dataset 8, segment size was set to six seconds, thus the number of observations per segment was kept stable across datasets. Again, we randomly chose 1000 surrogate segments as a control for real synchrony. For goal one and two, we tested the null hypotheses that there were no differences between synchrony of the original time series and synchrony of surrogates using Wilcoxon rank tests.

To explore the third goal of the present article, dyadic SUCO and dyadic SUSY were applied to all datasets. To account for the multivariate nature of the datasets, we computed synchronies of all dyadic combinations of the m time series of each dataset (for $m = 10$, this yields $(10 \times 9)/2 = 45$ dyads). With regard to parameter settings, a segment size of 20 s (datasets 1 to 7) and 24 s (dataset 8) was chosen for both dyadic approaches. We set the number of surrogates to the maximum (240 surrogates). The lag parameter in dyadic SUSY was fixed at ± 3 s across datasets. For SUCO, linear slopes were computed within a three-second window, and the increment was one second. To obtain global synchrony measures, absolute and non-absolute effect sizes were aggregated across all 45 dyads for both approaches yielding ES_{abs} (SUCO), ES_{noabs} (SUCO), ES_{abs} (SUSY), and ES_{noabs} (SUSY). The general formula for effect sizes is $ES = (\text{mean}(\overline{Z}) - \text{mean}(\overline{Z}\text{surr}))/\text{SD}(\overline{Z}\text{surr})$. For each dataset, we performed one-sample t-tests against the null hypothesis that the respective aggregated effect sizes, ES_{noabs}, were not different from zero. Paired t-tests of $\overline{Z}_{abs}$ and $\overline{Z}_{abs-surr}$ were conducted to test whether synchrony was present based on absolute values. In the case of negative absolute effect sizes, additional paired t-tests were considered redundant because surrogate synchrony exceeded real synchrony if $ES_{abs} < 0$. Statistical analyses and plots were performed using the software environment R [47].

3. Results

3.1. Simulated Data

The simulated data in datasets 1 to 5 were analyzed using the synchrony algorithms mv-SUSY, SUSY, and SUCO. Results are presented in Table 1.

Table 1. Synchrony results for simulated data (datasets 1 to 5).

Dataset 1: stationary random time series				
	mv-SUSY		**SUCO**	**SUSY**
method	$lambda_{max}$	*omega*	window-wise slopes	cross-correlation
Hertz	30	30	30	30
number of surrogates	1000	1000	240	240
segment size [s]	5	5	20	20
window size [s]			3	
lag [s]				$-3 \leq lag \leq 3$
real synchrony	14.22	0.23	$\overline{Z}_{abs} = 0.23$, $\overline{Z}_{noabs} = 0.01$	$\overline{Z}_{abs} = 0.03$, $\overline{Z}_{noabs} = 0.00$
surrogate synchrony	14.26	0.23	$\overline{Z}_{abs-surr} = 0.23$, $\overline{Z}_{noabs-surr} = 0.00$	$\overline{Z}_{abs-surr} = 0.03$, $\overline{Z}_{noabs-surr} = 0.00$
ES (mv)	−0.05, $W = 33315.00$ (ns)	−0.03, $W = 33379.00$ (ns)		
ES_{noabs} (SUCO)			0.02, $t = 0.11$ (ns)	
ES_{abs} (SUCO)			0.09	
ES_{noabs} (SUSY)				0.01, $t = 0.19$ (ns)
ES_{abs} (SUSY)				−0.03
Dataset 2: non-stationary time series with linear trends				
	mv-SUSY		**SUCO**	**SUSY**
method	$lambda_{max}$	*omega*	window-wise slopes	cross-correlation
Hertz	30	30	30	30
number of surrogates	1000	1000	240	240
segment size [s]	5	5	20	20
window size [s]			3	
lag [s]				$-3 \leq lag \leq 3$
real synchrony	14.31	0.23	$\overline{Z}_{abs} = 0.24$, $\overline{Z}_{noabs} = 0.01$	$\overline{Z}_{abs} = 0.04$, $\overline{Z}_{noabs} = 0.01$
surrogate synchrony	14.26	0.23	$\overline{Z}_{abs-surr} = 0.23$, $\overline{Z}_{noabs-surr} = 0.00$	$\overline{Z}_{abs-surr} = 0.04$, $\overline{Z}_{noabs-surr} = 0.01$
ES (mv)	0.07, $W = 31310.00$ (ns)	0.03, $W = 30965.00$ (ns)		
ES_{noabs} (SUCO)			0.10, $t = 0.55$ (ns)	
ES_{abs} (SUCO)			0.17	
ES_{noabs} (SUSY)				0.05, $t = 0.42$ (ns)
ES_{abs} (SUSY)				−0.01
Dataset 3: lag 1 autocorrelated time series				
	mv-SUSY		**SUCO**	**SUSY**
method	$lambda_{max}$	*omega*	window-wise slopes	cross-correlation
Hertz	30	30	30	30
number of surrogates	1000	1000	240	240
segment size [s]	5	5	20	20
window size [s]			3	
lag [s]				$-3 \leq lag \leq 3$
real synchrony	24.77	0.87	$\overline{Z}_{abs} = 0.27$, $\overline{Z}_{noabs} = 0.00$	$\overline{Z}_{abs} = 0.09$, $\overline{Z}_{noabs} = 0.00$
surrogate synchrony	24.84	0.88	$\overline{Z}_{abs-surr} = 0.25$, $\overline{Z}_{noabs-surr} = 0.01$	$\overline{Z}_{abs-surr} = 0.01$, $\overline{Z}_{noabs-surr} = 0.00$
ES (mv)	−0.02, $W = 32714.00$ (ns)	−0.14, $W = 33537.00$ (ns)		
ES_{noabs} (SUCO)			−0.06, $t = -0.34$ (ns)	
ES_{abs} (SUCO)			0.42	
ES_{noabs} (SUSY)				0.25, $t = 0.98$ (ns)
ES_{abs} (SUSY)				−0.10

Table 1. *Cont.*

	Dataset 4: sine oscillations			
	mv-SUSY		**SUCO**	**SUSY**
method	*$lambda_{max}$*	*omega*	window-wise slopes	cross-correlation
Hertz	30	30	30	30
number of surrogates	1000	1000	240	240
segment size [s]	5	5	20	20
window size [s]			3	
lag [s]				$-3 \leq \text{lag} \leq 3$
real synchrony	24.79	0.98	$\overline{Z}_{abs} = 0.22$, $\overline{Z}_{noabs} = 0.01$	$\overline{Z}_{abs} = 0.04$, $\overline{Z}_{noabs} = 0.00$
surrogate synchrony	24.83	0.98	$\overline{Z}_{abs-surr} = 0.22$, $\overline{Z}_{noabs-surr} = 0.00$	$\overline{Z}_{abs-surr} = 0.04$, $\overline{Z}_{noabs-surr} = 0.00$
ES (mv)	−0.01, $W = 32178.00$ (ns)	0.13, $W = 31729.00$ (ns)		
ES_{noabs} (SUCO)			−0.20, $t = -0.19$ (ns)	
ES_{abs} (SUCO)			−0.04	
ES_{noabs} (SUSY)				−0.02, $t = -0.25$ (ns)
ES_{abs} (SUSY)				0.00
	Dataset 5: correlated random time series			
	mv-SUSY		**SUCO**	**SUSY**
method	*$lambda_{max}$*	*omega*	window-wise slopes	cross-correlation
Hertz	30	30	30	30
number of surrogates	1000	1000	240	240
segment size [s]	5	5	20	20
window size [s]			3	
lag [s]				$-3 \leq \text{lag} \leq 3$
real synchrony	83.67	1.00	$\overline{Z}_{abs} = 1.19$, $\overline{Z}_{noabs} = 1.19$	$\overline{Z}_{abs} = 0.04$, $\overline{Z}_{noabs} = 0.01$
surrogate synchrony	14.33	0.24	$\overline{Z}_{abs-surr} = 0.23$, $\overline{Z}_{noabs-surr} = 0.04$	$\overline{Z}_{abs-surr} = 0.03$, $\overline{Z}_{noabs-surr} = 0.00$
ES (mv)	96.47, $W = 0.00$ ****	17.33, $W = 0.00$ ****		
ES_{noabs} (SUCO)			11.93, $t = 38.19$ ****	
ES_{abs} (SUCO)			23.78	
ES_{noabs} (SUSY)				2.06, $t = 46.04$ ****
ES_{abs} (SUSY)				4.12

Note. **** $p < 0.0001$; ns, non-significant.

Dataset 1 comprised stationary random time series. Multivariate synchrony estimated by the standardized differences between real synchrony and surrogate synchrony, ES (mv), was close to zero for $\text{lambda}_{\text{max}}$ and omega. This was further confirmed by Wilcoxon rank tests revealing non-significant differences between real synchrony and surrogate synchrony for both mv-SUSY methods. Analysis of aggregated dyadic synchrony with SUSY and SUCO required paired t-tests (absolute values) and one-sample t-tests (non-absolute values). Paired t-testing for $\overline{Z}_{abs}$ and $\overline{Z}_{abs-surr}$ revealed that real synchrony did not significantly exceed surrogate synchrony for SUCO ($t(44) = -1.04$, ns). Negative ES_{abs} (SUSY) showed that surrogate synchrony was even higher compared to real synchrony. Thus, comparison of means by paired t-testing was redundant. Both non-absolute effect sizes, ES_{noabs} (SUSY) and ES_{noabs} (SUCO), did not differ from zero. Accordingly, mv-SUSY, SUSY, and SUCO did not indicate synchrony in dataset 1.

The same procedure was used to investigate non-stationary time series with linear trends (dataset 2). With regard to mv-SUSY, Wilcoxon rank tests revealed no significant differences between real synchrony and surrogate synchrony ($\text{lambda}_{\text{max}}$ and omega). SUCO and SUSY confirmed these results for dyadic synchrony. Non-absolute effect sizes revealed absence of synchrony. This was found for synchrony based on window-wise slopes (SUCO) as well as for synchrony estimated by cross-correlations (SUSY). Results were the same for absolute values. The difference between real synchrony and surrogate synchrony was not significant in SUCO ($t(44) = 1.51$, ns) and SUSY ($t(44) = 0.68$, ns). Thus, the multivariate approaches as well as the dyadic approaches did not indicate presence of synchrony.

We evaluated autocorrelated time series in dataset 3. Results demonstrated no synchrony estimated by the mv-SUSY approach. In synchrony aggregated across dyads a less clear picture emerged. For SUCO, surrogate synchrony was significantly higher compared to real synchrony ($t(44) = 2.34$, $p = 0.02$). However, this was only the case for absolute values. ES_{noabs} (SUCO) did not indicate synchrony. SUSY showed absence of synchrony consistently. Absolute effect size, ES_{abs} (SUSY), was negative and non-absolute effect size, ES_{noabs} (SUSY), did not significantly differ from zero. Hence, we did not find synchrony for autocorrelated time series with one exception, namely SUCO (absolute values).

The SUCO, SUSY and mv-SUSY approaches were used to investigate the presence of synchrony in oscillatory time series (dataset 4). Both mv-SUSY methods, $lambda_{max}$ and omega, revealed high values for real synchrony. Effect sizes remained small and non-significant due to high surrogate synchrony. SUCO and SUSY showed similar results. Non-absolute effect sizes did not differ from zero for both approaches. Negative absolute effect sizes in SUCO indicated that real synchrony was lower compared to surrogate synchrony. Furthermore, real synchrony did not differ from surrogate synchrony estimated by SUSY ($t(44) = 0.56$, ns). For dataset 4, mv-SUSY, SUSY and SUCO demonstrated no synchrony.

Analysis of dataset 5 addressed multivariate correlation among time series. Difference of real synchrony and surrogate synchrony was significant for $lambda_{max}$ and omega indicated by high, positive effect sizes. The non-absolute effect size for SUCO did significantly differ from zero. In addition, the difference between real synchrony and surrogate synchrony estimated by SUCO was highly significant ($t(44) = 43.80$, $p < 0.0001$). Higher absolute effect sizes suggest that positive and negative correlations canceled each other out in the non-absolute effect sizes. Aggregated real synchrony exceeded surrogate synchrony in SUSY as well ($t(44) = 47.52$, $p < 0.0001$). The mean effect sizes across dyads differed significantly from zero (absolute values). Thus, all approaches demonstrated substantial synchrony for dataset 5.

In sum, hypothesis 1a ("no spurious synchrony detection with mv-SUSY in datasets 1 to 4") was supported. Hypothesis 1b ("synchrony detection with mv-SUSY in correlated dataset 5") was also supported. Hypothesis 3 ("validation of mv-SUSY by dyadic SUSY and SUCO") was supported in four of the five datasets. In dataset 3 the absolute effect size values and further t-test indicated significant differences between real synchrony and surrogate synchrony (SUCO).

3.2. Empirical Data

In Table 2, we present findings of mv-SUSY, SUSY, and SUCO for datasets 6 to 8. Results of synchrony analysis in dataset 6 (Kinect intraindividual movement data) showed significant differences between real synchrony and the surrogate synchrony using the mv-SUSY approach. This was found for both $lambda_{max}$ and omega. SUSY and SUCO suggested presence of high synchrony as well. The dyadic-based approaches revealed large and positive non-absolute effect sizes, which differed significantly from zero. Furthermore, paired t-tests revealed significant differences between real synchrony and surrogate synchrony for absolute values estimated by SUCO ($t(44) = 8.23$, $p < 0.0001$) as well as SUSY ($t(44) = 14.65$, $p < 0.0001$). Accordingly, presence of synchrony in intraindividual movement data (participant 1) was confirmed by mv-SUSY, SUSY and SUCO.

With regard to the second Kinect movement dataset, dataset 7, the same pattern emerged. mv-SUSY methods omega and $lambda_{max}$ showed lower synchrony in surrogates. Wilcoxon rank tests revealed that this difference was significant. As in dataset 6, multivariate synchrony reached a maximum with aggregated synchrony = 1 for omega. Non-absolute effect sizes pointed toward high synchrony based on window-wise slopes (SUCO) as well as the windowed cross-correlation approach (SUSY). Paired t-tests for absolute values indicated synchrony in SUCO ($t(44) = 8.80$, $p < 0.0001$). So did the results for SUSY comparing real synchrony to surrogate synchrony ($t(44) = 18.74$, $p < 0.0001$).

The findings for the intraindividual movement data (participant 2) consistently indicated multivariate synchrony across all approaches.

Table 2. Synchrony results for empirical movement data (datasets 6 to 8).

	Dataset 6: Kinect, coordinated intraindividual movement (participant 1)			
	mv-SUSY		**SUCO**	**SUSY**
method	*$lambda_{max}$*	*omega*	window-wise slopes	cross-correlation
Hertz	30	30	30	30
number of surrogates	1000	1000	240	240
segment size [s]	5	5	20	20
window size [s]			3	
lag [s]				$-3 \leq lag \leq 3$
real synchrony	47.38	1.00	$\overline{Z}_{abs} = 0.92$, $\overline{Z}_{noabs} = 0.90$	$\overline{Z}_{abs} = 0.12$, $\overline{Z}_{noabs} = 0.07$
surrogate synchrony	21.01	0.69	$\overline{Z}_{abs-surr} = 0.23$, $\overline{Z}_{noabs-surr} = -0.01$	$\overline{Z}_{abs-surr} = 0.08$, $\overline{Z}_{noabs-surr} = 0.00$
ES (mv)	9.77, $W = 51.00$ ****	2.97, $W = 0.00$ ****		
ES_{noabs} (SUCO)			14.80, $t = 10.63$ ****	
ES_{abs} (SUCO)			12.29	
ES_{noabs} (SUSY)				13.02, $t = 15.34$ ****
ES_{abs} (SUSY)				8.35
	Dataset 7: Kinect, coordinated intraindividual movement (participant 2)			
	mv-SUSY		**SUCO**	**SUSY**
method	*$lambda_{max}$*	*omega*	window-wise slopes	cross-correlation
Hertz	30	30	30	30
number of surrogates	1000	1000	240	240
segment size [s]	5	5	20	20
window size [s]			3	
lag [s]				$-3 \leq lag \leq 3$
real synchrony	47.46	1.00	$\overline{Z}_{abs} = 0.89$, $\overline{Z}_{noabs} = 0.88$	$\overline{Z}_{abs} = 0.13$, $\overline{Z}_{noabs} = 0.08$
surrogate synchrony	21.28	0.69	$\overline{Z}_{abs-surr} = 0.23$, $\overline{Z}_{noabs-surr} = 0.00$	$\overline{Z}_{abs-surr} = 0.09$, $\overline{Z}_{noabs-surr} = 0.00$
ES (mv)	9.49, $W = 0.00$ ****	2.91, $W = 0.00$ ****		
ES_{noabs} (SUCO)			12.56, $t = 10.86$ ****	
ES_{abs} (SUCO)			15.15	
ES_{noabs} (SUSY)				18.40, $t = 25.24$ ****
ES_{abs} (SUSY)				11.69
	Dataset 8: Zoom motion energy, group discussion			
	mv-SUSY		**SUCO**	**SUSY**
method	*$lambda_{max}$*	*omega*	window-wise slopes	cross-correlation
Hertz	25	25	25	25
number of surrogates	1000	1000	240	240
segment size [s]	6	6	24	24
window size [s]			3	
lag [s]				$-3 \leq lag \leq 3$
real synchrony	16.14	0.36	$\overline{Z}_{abs} = 0.22$, $\overline{Z}_{noabs} = 0.04$	$\overline{Z}_{abs} = 0.05$, $\overline{Z}_{noabs} = 0.01$
surrogate synchrony	15.59	0.31	$\overline{Z}_{abs-surr} = 0.21$, $\overline{Z}_{noabs-surr} = 0.01$	$\overline{Z}_{abs-surr} = 0.04$, $\overline{Z}_{noabs-surr} = 0.00$
ES (mv)	0.39, $W = 25525.00$ **	0.51, $W = 22495.00$ ****		
ES_{noabs} (SUCO)			0.52, $t = 2.82$ **	
ES_{abs} (SUCO)			0.25	
ES_{noabs} (SUSY)				1.86, $t = 3.69$ ***
ES_{abs} (SUSY)				0.58

Note. ** $p < 0.01$; *** $p < 0.001$; **** $p < 0.0001$.

Synchrony measures mv-SUSY, SUSY and SUCO were applied to dataset 8 (movement coordination of participants of a group discussion). Findings indicated higher effect sizes for omega compared to $lambda_{max}$. Nevertheless, both methods clearly showed presence of synchrony. Whereas the one-sample *t*-test revealed significant non-absolute effect sizes, the paired t-test of absolute values did not reach significance for SUCO ($t(44) = 1.38$, ns). SUSY results presented an unambiguous picture. Non-absolute effect sizes against zero as well as real synchrony compared to surrogate synchrony reached significance ($t(44) = 2.14$, $p < 0.05$).

Taken together, hypothesis 2 ("synchrony detection with mv-SUSY in empirical datasets 6 to 8") was accepted. In 2 of 3 datasets we further confirmed hypothesis 3 ("validation of mv-SUSY by dyadic SUSY and SUCO"). In dataset 8 the absolute effect size values and further t-test showed that real synchrony did not significantly exceed surrogate synchrony (SUCO).

4. Discussion

The methodology of synchrony research should move beyond dyadic measures alone. Dyadic synchronies are of course well-justified as signatures of the relationship between therapist and client, or spouses of a couple, or the participants of conversation between two people. Yet, how may we address the synchronization of more complex phenomena such as groups, multi-person therapies, or generally dynamical systems that comprise more than just two variables? The present article therefore introduced a multivariate approach to quantify synchrony in more complex multidimensional systems. We presented the novel mv-SUSY methodology to estimate multivariate synchrony, and examined its functionality in eight datasets that exemplify different types of longitudinal patterns.

Our first goal was to investigate the performance of mv-SUSY in five simulated time series. Analysis of simulated time series in the context of synchrony has been used before. This was done to introduce new computational methods [48] or in surrogate analysis [36]. The time series in the current study comprised autocorrelation, trends, and seasonality which are common properties of empirical time series, for example in physiology (e.g., electrodermal activity, respiration, heart beats) and movement data. Results confirmed the hypothesized detection and rejections of synchrony. Both mv-SUSY methods, lambda_{max} and omega, revealed similar patterns of real and surrogate synchrony across datasets. We found that real synchrony (cf Table 1) was rather low in stationary random time series (dataset 1) and non-stationary time series with linear trends (dataset 2). In dataset 1, small real synchrony values resulted from intercorrelations close to zero among white noise data (see intercorrelations in the methods section). Dataset 2 time series correlated substantially because of their stable linear trends. Segment-wise computation of real synchrony allowed to account for these overarching trends. Thus, we found comparable degrees of real synchrony in white noise and non-stationary linear trends of otherwise unrelated time series. mv-SUSY did correctly reject synchrony for stationary random data as well as non-stationary linear trend data despite its temporal dependency. Autocorrelated (dataset 3) and oscillation time series (dataset 4) showed higher levels of real synchrony compared to the first two datasets. The lack of significance here resulted from similarly high levels of synchrony in the surrogates. This finding can be explained by the covariance respectively correlation-based nature of mv-SUSY. Autocorrelation and oscillation represent high dependency between observations at different time points thus provoking substantial segment-wise correlations. Surrogate testing controlled successfully for these reoccurring data properties, thus avoiding false-positive synchrony findings for uncorrelated time series. Not only the absence but also the presence of synchrony was found as assumed. In concordance with our hypothesis, mv-SUSY indicated a considerable degree of synchrony in multivariate correlated data (dataset 5).

We examined mv-SUSY for movement data in goal two. Synchrony was present in a dyadic body conversation task. This experimental task was designed to limit interactive dynamics to one single modality, dance-like movements, while allowing focusing on the whole body [37]. Both participants (dataset 6 and dataset 7) demonstrated significant coordination between their limbs, respectively, which was also found in the original study. There, Galbusera and colleagues [37] investigated interpersonal and intrapersonal synchrony and its association with self-regulation of emotion in 66 adults undergoing the body conversation task. In the present study, mv-SUSY further revealed synchrony in body movement (upper body region) in a virtual group discussion of 10 individuals. Accordingly, we found above-random coordination between the members of this discussion.

The third goal was to compare mv-SUSY performance with well-established dyadic synchrony approaches. SUSY and SUCO replicated mv-SUSY findings to a large extent. There were few cases of divergent results that did not support the hypotheses. First, SUCO indicated synchrony for dataset 3 (autocorrelated time series). However, this was only the case for absolute values. Since we did not find similar results in oscillation or trend data, which also showed a reasonable degree of autocorrelation, there is little evidence that SUCO is prone to autocorrelation. Second, SUCO did not find synchrony in dataset 8. Again, this was only the case for absolute values. As surrogate synchrony was lower compared to real synchrony, the difference pointed toward synchrony but it did not reach significance. Overall, analysis based on mv-SUSY, SUSY, and SUCO resulted in similar estimations of synchrony. Rather than competing with each other, each algorithm covers unique aspects with regard to synchrony analysis. One main feature of SUSY is the computation of cross-correlations. Accordingly, delayed responses in a predefined window are taken into account. SUSY and SUCO allow differentiating between in-phase and anti-phase synchrony. This is of considerable relevance in social contexts where each member has a specific role, e.g., in patient-therapist interaction.

Findings of the present study underline the feasibility of estimating synchrony within groups of individuals by mv-SUSY. Multi-person synchrony, although based on the aggregation of dyadic synchrony, has been investigated across disciplines. For example, researchers found physiological synchrony in audience members of classical concerts [25], brain-to-brain synchrony in a classroom [49], and physiological synchrony in teams of three during a cooperative production task [50]. Other studies investigated the effects of experimentally manipulated synchrony compared to control conditions. In a meta-analysis of 42 studies, synchrony in groups was associated with prosocial behavior, perceived social bonding, social cognition, and positive affect with effect sizes ranging from 0.11 to 0.28 [51]. Recently, multivariate synchrony approaches have been proposed, yet these have limitations such as the absence of surrogate controls that should be addressed. mv-SUSY complements the available methods by providing two novel synchrony measures ($lambda_{max}$ and omega), which are applicable to oscillatory, non-oscillatory, and non-stationary time series, and are adequately controlled by surrogate analysis.

Analysis of synchrony corresponds to dynamical systems theory, which originated from mathematics and physics, but is increasingly used to focus on the change over time in biological, cognitive, or social systems [39,52]. At its core, the theory assumes that collective behavior may result from self-organization in interacting components of the respective system and thus exhibit processes of emergence. In line with these ideas, the study of synchrony accounts for the time-evolving pattern in individuals' time series. For an overview regarding complex dynamical systems in social and personality psychology, see for example Richardson et al. [53]. Furthermore, the meaning of nonverbal synchrony is associated with the embodiment concept. The embodiment paradigm is a vibrant field that can be seen as the preliminary endpoint of the historical development from behaviorism to cognitivism to embodied cognition. According to the 4E paradigm, cognition is embodied, enactive, embedded and extended [54]. Taken together, to study social interaction between individuals, one has to consider dynamic processes, i.e., change over time. These may lead to coordinated, self-organized patterns of behavior, where individuals form a coupled emerging system which is said to be coordinated.

4.1. Strengths and Limitations

Several strengths and limitations of the study are noteworthy. The methodological approach was appropriate to the research questions and allowed (a) the application of mv-SUSY to simulated and empirical data, and (b) the comparison of the results with established dyadic approaches. mv-SUSY is based on general mathematical principles, namely order analysis and eigendecomposition. These concepts are of widespread use in statistical analysis, for example in principle component analysis, which is commonly used for dimensionality reduction in large datasets.

Further strengths and weaknesses of the study are closely linked to the differentiation of dyadic and multivariate synchrony analysis. Analysis of multivariate time series with dyadic synchrony approaches generates increased number of comparisons and further leads to a hierarchical data structure. Yet mv-SUSY provides one global measure of synchrony for multiple variables thus minimizing the number of aggregation steps. In future research implementations, mv-SUSY however still allows for estimating the degree of individual contributions to group synchrony, for instance by stepwise exclusion of the respective group member.

In line with other dyadic approaches such as SUSY and SUCO, mv-SUSY provides an appropriate control condition to determine above-random synchrony. According to a recent systematic review, "To validate findings, a null hypothesis determining the potential for chance findings of PS [physiological synchrony] in contextually matched, randomized data is often necessary. Otherwise, it may be unclear whether results are valid, or due to chance" [13]. Since surrogate analysis based on segment shuffling uses inherent characteristics of the respective data set, it is a conservative form of a surrogate synchrony control condition. In contrast to participant shuffling, the procedure in mv-SUSY, SUSY, and SUCO does not require additional data.

A limitation of the presented approach lies in the dependency on parameter settings. mv-SUSY adds group size to the current set of parameters. The effect of group size needs to be investigated in future studies.

4.2. Implications and Future Research

The present study emphasizes the importance of considering multiple variables in synchrony research, whether these variables are assessed within or across individuals. Recent technological advances allow researchers to record behavior or physiology of more than two individuals to obtain high-frequency multivariate time series. We introduced mv-SUSY as a computational approach to estimate synchrony based on such multivariate datasets. This complements current auto- or cross-correlational methods, which were limited to two variables.

The application to multivariate time series makes mv-SUSY a versatile methodology. mv-SUSY can be used to investigate nonverbal synchrony across modalities. Current research mainly addresses traditional measures such as movement behavior or physiology (e.g., cardiac activity) in synchrony analysis. Other modalities have not received that much attention yet, for example, there is limited but promising evidence on hormonal synchrony [55] or synchrony in eye gaze [56]. These measures represent potential starting points to quantify multivariate synchrony by mv-SUSY. Future studies should not only focus on different measures but should strive for an integration of synchrony across multiple levels. Such studies would help to gain deeper insights in the embodiment of social interaction.

Furthermore, mv-SUSY can be applied to a variety of research fields to expand our understanding of interpersonal coordination. Future research might focus on the role of multivariate synchrony in clinical psychology and psychotherapy. Nonverbal synchrony has been associated with so-called common factors that represent the underlying change mechanisms in psychotherapy, such as the working alliance [57]. Advancing empirical work in this area includes shifting the perspective from single patients to multi-person settings (e.g., couple therapy, support groups). This will have important practical implications for group-based therapies leading to improved mental health care. In general, future studies should not only go beyond dyads but also beyond disciplines. At the intersection of psychology, musicology, and cultural studies particularly fruitful options arise. mv-SUSY may be applied to explore the presence of synchrony in ecological valid settings (e.g., school, sports, concerts). For example, mv-SUSY can help identify synchrony in concert audiences enriching the exploration of aesthetic experiencing [25]. Experimental research designs may also be used to investigate the association between synchrony measured by mv-SUSY and work-team cooperation under different cooperative task conditions.

In summary, with the introduction of mv-SUSY, we have provided a valid method based on the theoretical framework of dynamical systems theory to estimate intrapersonal and interpersonal synchrony in complex data.

Author Contributions: Conceptualization, D.M. and W.T.; Data curation, D.M. and W.T.; Formal analysis, D.M.; Methodology, D.M. and W.T.; Software, D.M.; Validation, D.M. and W.T.; Visualization, D.M.; Writing—original draft, D.M. and W.T.; Writing—review & editing, D.M. and W.T. All authors have read and agreed to the published version of the manuscript.

Funding: The authors received funding by the Volkswagen Foundation, grant 38963—Experimental Concert Research. There was no financial support specifically dedicated to the present article.

Institutional Review Board Statement: Ethical review and approval were waived for this study. Our research involved secondary use of data, which is provided without any identifier that would allow attribution of private information to an individual.

Informed Consent Statement: Research involved publicly accessible records, thus participant consent was waived for dataset 8. Datasets 6 and 7 was re-analyzed under basis of informed consent.

Data Availability Statement: The simulated datasets generated and analyzed during the current study are available from the corresponding author upon reasonable request. The empirical datasets 6 and 7 are not publicly available, for contact see [37].

Acknowledgments: The authors thank Laura Galbusera for providing us with empirical sample data assessed at the Technical University of Berlin (TUB).

Conflicts of Interest: The authors declare no conflict of interest.

References

1. Varela, F.J. Present-Time Consciousness. In *The View from within: First-Person Approaches to the Study of Consciosness*; Varela, F.J., Shear, J., Eds.; Imprint Academic: Thoverton, UK, 1999; pp. 111–140.
2. Fujiwara, K.; Daibo, I. Affect as an Antecedent of Synchrony: A Spectrum Analysis with Wavelet Transform. *Q. J. Exp. Psychol.* **2018**, *71*, 2520–2530. [CrossRef]
3. Tschacher, W.; Rees, G.M.; Ramseyer, F. Nonverbal Synchrony and Affect in Dyadic Interactions. *Front. Psychol.* **2014**, *5*, 1323. [CrossRef] [PubMed]
4. Lindenberger, U.; Li, S.-C.; Gruber, W.; Müller, V. Brains Swinging in Concert: Cortical Phase Synchronization While Playing Guitar. *BMC Neurosci.* **2009**, *10*, 22. [CrossRef] [PubMed]
5. Kleinbub, J.R. State of the Art of Interpersonal Physiology in Psychotherapy: A Systematic Review. *Front. Psychol.* **2017**, *8*, 2053. [CrossRef]
6. Tschacher, W.; Meier, D. Physiological Synchrony in Psychotherapy Sessions. *Psychother. Res.* **2020**, *30*, 558–573. [CrossRef] [PubMed]
7. Ramseyer, F. Motion Energy Analysis (MEA): A Primer on the Assessment of Motion from Video. *J. Couns. Psychol.* **2020**, *67*, 536–549. [CrossRef]
8. Ramseyer, F.; Tschacher, W. Nonverbal Synchrony in Psychotherapy: Coordinated Body Movement Reflects Relationship Quality and Outcome. *J. Consult. Clin. Psychol.* **2011**, *79*, 284–295. [CrossRef]
9. Altmann, U.; Schoenherr, D.; Paulick, J.; Deisenhofer, A.-K.; Schwartz, B.; Rubel, J.A.; Stangier, U.; Lutz, W.; Strauss, B. Associations between Movement Synchrony and Outcome in Patients with Social Anxiety Disorder: Evidence for Treatment Specific Effects. *Psychother. Res.* **2019**, *30*, 574–590. [CrossRef] [PubMed]
10. Paulick, J.; Deisenhofer, A.-K.; Ramseyer, F.; Tschacher, W.; Boyle, K.; Rubel, J.; Lutz, W. Nonverbal Synchrony: A New Approach to Better Understand Psychotherapeutic Processes and Drop-Out. *J. Psychother. Integr.* **2018**, *28*, 367–384. [CrossRef]
11. Chartrand, T.L.; Lakin, J.L. The Antecedents and Consequences of Human Behavioral Mimicry. *Annu. Rev. Psychol.* **2013**, *64*, 285–308. [CrossRef]
12. Imel, Z.E.; Barco, J.S.; Brown, H.J.; Baucom, B.R.; Baer, J.S.; Kircher, J.C.; Atkins, D.C. The Association of Therapist Empathy and Synchrony in Vocally Encoded Arousal. *J. Couns. Psychol.* **2014**, *61*, 146–153. [CrossRef]
13. Palumbo, R.V.; Marraccini, M.E.; Weyandt, L.L.; Wilder-Smith, O.; McGee, H.A.; Liu, S.; Goodwin, M.S. Interpersonal Autonomic Physiology: A Systematic Review of the Literature. *Personal. Soc. Psychol. Rev.* **2017**, *21*, 99–141. [CrossRef] [PubMed]
14. Di Mascio, A.; Boyd, R.W.; Greenblatt, M.; Solomon, H.C. The Psychiatric Interview: A Sociophysiologic Study. *Dis. Nerv. Syst.* **1955**, *16*, 4–9.

15. Marci, C.D.; Orr, S.P. The Effect of Emotional Distance on Psychophysiologic Concordance and Perceived Empathy Between Patient and Interviewer. *Appl. Psychophysiol. Biofeedback* **2006**, *31*, 115–128. [CrossRef]
16. Karvonen, A.; Kykyri, V.-L.; Kaartinen, J.; Penttonen, M.; Seikkula, J. Sympathetic Nervous System Synchrony in Couple Therapy. *J. Marital Fam. Ther.* **2016**, *42*, 383–395. [CrossRef] [PubMed]
17. Chartrand, T.L.; Bargh, J.A. The Chameleon Effect: The Perception–Behavior Link and Social Interaction. *J. Pers. Soc. Psychol.* **1999**, *76*, 893–910. [CrossRef]
18. Wiltermuth, S.S.; Heath, C. Synchrony and Cooperation. *Psychol. Sci.* **2009**, *20*, 1–5. [CrossRef]
19. Guastello, S.J.; Mirabito, L.; Peressini, A.F. Autonomic Synchronization under Three Task Conditions and Its Impact on Team Performance. *Nonlinear Dyn. Psychol. Life Sci.* **2020**, *24*, 79–104.
20. Coutinho, J.; Oliveira-Silva, P.; Fernandes, E.; Gonçalves, O.F.; Correia, D.; Perrone Mc-Govern, K.; Tschacher, W. Psychophysiological Synchrony During Verbal Interaction in Romantic Relationships. *Fam. Process* **2019**, *58*, 716–733. [CrossRef]
21. Coutinho, J.; Pereira, A.; Oliveira-Silva, P.; Meier, D.; Lourenço, V.; Tschacher, W. When Our Hearts Beat Together: Cardiac Synchrony as an Entry Point to Understand Dyadic Co-regulation in Couples. *Psychophysiology* **2020**, *58*, e13739. [CrossRef] [PubMed]
22. Azhari, A.; Lim, M.; Bizzego, A.; Gabrieli, G.; Bornstein, M.H.; Esposito, G. Physical Presence of Spouse Enhances Brain-to-Brain Synchrony in Co-Parenting Couples. *Sci. Rep.* **2020**, *10*, 7569. [CrossRef] [PubMed]
23. Feldman, R.; Magori-Cohen, R.; Galili, G.; Singer, M.; Louzoun, Y. Mother and Infant Coordinate Heart Rhythms through Episodes of Interaction Synchrony. *Infant Behav. Dev.* **2011**, *34*, 569–577. [CrossRef] [PubMed]
24. Seibert, C.; Greb, F.; Tschacher, W. Nonverbale Synchronie und Musik-Erleben im klassischen Konzert. *Jahrb. Musikpsychol.* **2019**, *28*, e18. [CrossRef]
25. Tschacher, W.; Greenwood, S.; Wald-Fuhrmann, M.; Czepiel, A.; Tröndle, M.; Meier, D. Physiological Synchrony in Audiences of Live Concerts. *Psychol. Aesthet. Creat. Arts* **2021**. Advance online publication. [CrossRef]
26. Bachrach, A.; Fontbonne, Y.; Joufflineau, C.; Ulloa, J.L. Audience Entrainment during Live Contemporary Dance Performance: Physiological and Cognitive Measures. *Front. Hum. Neurosci.* **2015**, *9*, 179. [CrossRef]
27. Coco, M.I.; Dale, R. Cross-Recurrence Quantification Analysis of Categorical and Continuous Time Series: An R Package. *Front. Psychol.* **2014**, *5*, 510. [CrossRef]
28. Frank, T.D.; Richardson, M.J. On a Test Statistic for the Kuramoto Order Parameter of Synchronization: An Illustration for Group Synchronization during Rocking Chairs. *Phys. Nonlinear Phenom.* **2010**, *239*, 2084–2092. [CrossRef]
29. Richardson, M.J.; Garcia, R.L.; Frank, T.D.; Gergor, M.; Marsh, K.L. Measuring Group Synchrony: A Cluster-Phase Method for Analyzing Multivariate Movement Time-Series. *Front. Physiol.* **2012**, *3*, 405. [CrossRef]
30. Wallot, S. Multidimensional Cross-Recurrence Quantification Analysis (MdCRQA)—A Method for Quantifying Correlation between Multivariate Time-Series. *Multivar. Behav. Res.* **2019**, *54*, 173–191. [CrossRef] [PubMed]
31. Guastello, S.J.; Peressini, A.F. Development of a Synchronization Coefficient for Biosocial Interactions in Groups and Teams. *Small Group Res.* **2017**, *48*, 3–33. [CrossRef]
32. Behrens, F.; Moulder, R.G.; Boker, S.M.; Kret, M.E. Quantifying Physiological Synchrony through Windowed Cross-Correlation Analysis: Statistical and Theoretical Considerations. *BioRxiv Prepr.* **2020**. [CrossRef]
33. Kleinbub, J.R.; Talia, A.; Palmieri, A. Physiological Synchronization in the Clinical Process: A Research Primer. *J. Couns. Psychol.* **2020**, *67*, 420–437. [CrossRef] [PubMed]
34. Schoenherr, D.; Paulick, J.; Strauss, B.M.; Deisenhofer, A.-K.; Schwartz, B.; Rubel, J.A.; Lutz, W.; Stangier, U.; Altmann, U. Identification of Movement Synchrony: Validation of Windowed Cross-Lagged Correlation and -Regression with Peak-Picking Algorithm. *PLoS ONE* **2019**, *14*, e0211494. [CrossRef] [PubMed]
35. Strang, A.J.; Funke, G.J.; Russell, S.M.; Dukes, A.W.; Middendorf, M.S. Physio-Behavioral Coupling in a Cooperative Team Task: Contributors and Relations. *J. Exp. Psychol. Hum. Percept. Perform.* **2014**, *40*, 145–158. [CrossRef]
36. Moulder, R.G.; Boker, S.M.; Ramseyer, F.; Tschacher, W. Determining Synchrony between Behavioral Time Series: An Application of Surrogate Data Generation for Establishing Falsifiable Null-Hypotheses. *Psychol. Methods* **2018**, *23*, 757–773. [CrossRef] [PubMed]
37. Galbusera, L.; Finn, M.T.M.; Tschacher, W.; Kyselo, M. Interpersonal Synchrony Feels Good but Impedes Self-Regulation of Affect. *Sci. Rep.* **2019**, *9*, 14691. [CrossRef]
38. Tschacher, W.; Haken, H. *The Process of Psychotherapy: Causation and Chance*, 1st ed.; Springer Nature: Cham, Switzerland, 2019; ISBN 978-3-030-12747-3.
39. Haken, H.; Portugali, J. Information and Selforganization: A Unifying Approach and Applications. *Entropy* **2016**, *18*, 197. [CrossRef]
40. Landsberg, P.T. Can Entropy and "Order" Increase Together? *Phys. Lett. A* **1984**, *102*, 171–173. [CrossRef]
41. Banerjee, S.; Sibbald, P.R.; Maze, J. Quantifying the Dynamics of Order and Organization in Biological Systems. *J. Theor. Biol.* **1990**, *143*, 91–112. [CrossRef]
42. Shiner, J.S.; Davison, M.; Landsberg, P.T. On measures for order and its relation to complexity. In *Dynamics, Synergetics, Autonomous Agents*; Tschacher, W., Dauwalder, J.-P., Eds.; World Scientific: Singapore, 1999; Volume 8, pp. 49–63, ISBN 978-981-02-3837-7.
43. Tschacher, W.; Scheier, C.; Grawe, K. Order and Pattern Formation in Psychotherapy. *Nonlinear Dyn. Psychol. Life Sci.* **1998**, *2*, 195–215. [CrossRef]

44. Fischer, G. Lineare Algebra. In *Vieweg Studium*, 14th ed.; Vieweg+Teubner Verlag: Wiesbaden, Germany, 2003; ISBN 978-3-528-03217-3.
45. Hadd, A.R.; Rodgers, J.L. *Understanding Correlation Matrices*, 1st ed.; SAGE Publications, Inc.: Thousand Oaks, CA, USA, 2020; ISBN 978-1-5443-4109-5.
46. Tschacher, W.; Grawe, K. Selbstorganisation in Therapieprozessen. *Z. Für Klin. Psychol.* **1996**, *25*, 55–60.
47. *R Core Team R: A Language and Environment for Statistical Computing*; R Foundation for Statistical Computing: Vienna, Austria, 2020.
48. Liu, S.; Zhou, Y.; Palumbo, R.; Wang, J.-L. Dynamical Correlation: A New Method for Quantifying Synchrony with Multivariate Intensive Longitudinal Data. *Psychol. Methods* **2016**, *21*, 291–308. [CrossRef] [PubMed]
49. Dikker, S.; Wan, L.; Davidesco, I.; Kaggen, L.; Oostrik, M.; McClintock, J.; Rowland, J.; Michalareas, G.; Van Bavel, J.J.; Ding, M.; et al. Brain-to-Brain Synchrony Tracks Real-World Dynamic Group Interactions in the Classroom. *Curr. Biol.* **2017**, *27*, 1375–1380. [CrossRef]
50. Mønster, D.; Håkonsson, D.D.; Eskildsen, J.K.; Wallot, S. Physiological Evidence of Interpersonal Dynamics in a Cooperative Production Task. *Physiol. Behav.* **2016**, *156*, 24–34. [CrossRef] [PubMed]
51. Mogan, R.; Fischer, R.; Bulbulia, J.A. To Be in Synchrony or Not? A Meta-Analysis of Synchrony's Effects on Behavior, Perception, Cognition and Affect. *J. Exp. Soc. Psychol.* **2017**, *72*, 13–20. [CrossRef]
52. Haken, H.; Portugali, J. Information and Self-Organization II: Steady State and Phase Transition. *Entropy* **2021**, *23*, 707. [CrossRef]
53. Richardson, K.; Hart, W.; Breeden, C.J.; Tortoriello, G.K.; Kinrade, C. Validating Circumplex Scales of Perceived Impairments and Benefits from Prototypically Problematic Interpersonal Tendencies. *Psychol. Assess.* **2020**, *32*, 415–430. [CrossRef]
54. Newen, A.; De Bruin, L.; Gallagher, S. *The Oxford Handbook of 4E Cognition*; Oxford University Press: Oxford, UK, 2018; ISBN 978-0-19-105435-8.
55. Liu, S.; Rovine, M.J.; Cousino Klein, L.; Almeida, D.M. Synchrony of Diurnal Cortisol Pattern in Couples. *J. Fam. Psychol.* **2013**, *27*, 579–588. [CrossRef]
56. Tschacher, W.; Tschacher, N.; Stukenbrock, A. Eye Synchrony: A Method to Capture Mutual and Joint Attention in Social Eye Movements. *Nonlinear Dyn. Psychol. Life Sci.* **2021**, *25*, 309–333.
57. Koole, S.L.; Atzil-Slonim, D.; Butler, E.; Dikker, S.; Tschacher, W. In sync with your shrink. In *Applications of Social Psychology*; Forgas, J.P., Crano, W.D., Fiedler, K., Eds.; Routledge: New York, NY, USA, 2020; pp. 161–184, ISBN 978-0-367-41832-8.

MDPI
St. Alban-Anlage 66
4052 Basel
Switzerland
Tel. +41 61 683 77 34
Fax +41 61 302 89 18
www.mdpi.com

Entropy Editorial Office
E-mail: entropy@mdpi.com
www.mdpi.com/journal/entropy

www.ingramcontent.com/pod-product-compliance
Lightning Source LLC
LaVergne TN
LVHW070110170726
843515LV00007B/1656

9783036562193